The Government High Line Canal Newspaper Article Archive, *1904 - 1910*

Grand Valley Water Users' Association's History of the Grand Valley Project

The Government High Line Canal
Newspaper Article Archive, 1904 - 1910

—

Grand Valley Water Users' Association's
History of the Grand Valley Project

Compiled by Dr. Hugo Madden.
Introduction edited by Jeffrey M. Widener, Ph.D., University of Oklahoma
and Priscilla Walker, Palisade Historical Society.
Cover Art and Illustration by Gayle Madden.
Photography by Dr. Madden.

With credit to
the Grand Valley Water Users' Association employees
whose foresight has preserved this window into history.

Printed in the U.S.A.

ISBN: 978-0-9786478-4-1

CONTENTS

INTRODUCTION

For more than a century the Grand Valley Project irrigation system has provided a region in mid-western Colorado with a dedicated water source for land and life. This document reveals the planning history of the project from 1904 to 1910 as a first-person account.

In February 2015, Priscilla Walker of the Palisade Historical Society was researching data for a celebration scheduled to commemorate the 100th birthday of the Grand Valley Project's Diversion Dam (the "Roller Dam") dedication. A query to the Grand Valley Water User's Association (GVWUA) unearthed an old scrapbook of Colorado newspaper clippings they had been preserving. The original creator of the scrapbook is unknown. The GVWUA loaned the Palisade Historical Society the artifact. With their permission, volunteer Dr. Hugo Madden worked for three months digitizing, editing and enhancing the scrapbook electronic files to archive this amazing find.

The articles shed light on the barriers, both personal and political, to building the Grand Valley Project, focusing specifically on the Government High Line Canal (Highline Canal). Though the project was envisioned initially in the late 1880s by pioneer settlers, it was not until the Reclamation Act of 1902 that the federal government could legally fund the Grand Valley Project. On September 19, 1912, the federal government released the funding and on June 29, 1915 the first water funneled into the Government High Line Canal from waters diverted from the Colorado River (then called the Grand River) by the Roller Dam. Once completed in 1918, the 55-mile long canal, traversing through three tunnels, plus its 150 miles of piped and open laterals, reclaimed approximately 23,350 acres of previously desert land.

The GVWUA, established in 1905, took over the canal from the Bureau of Reclamation in 1949, and has been managing it since. Water rights in the project area were allocated through "Subscription for Stock," meaning the water rights cannot be sepa-

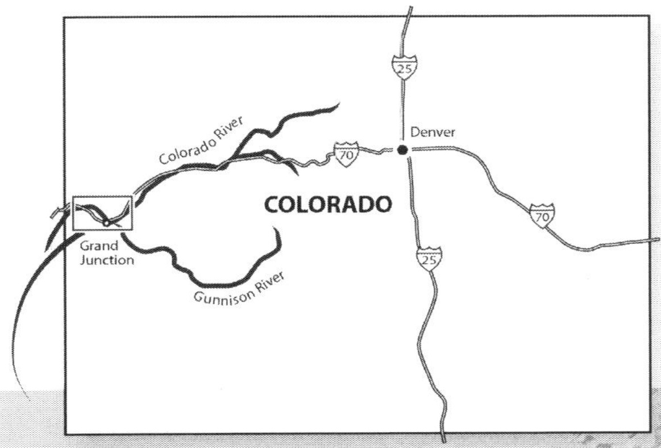

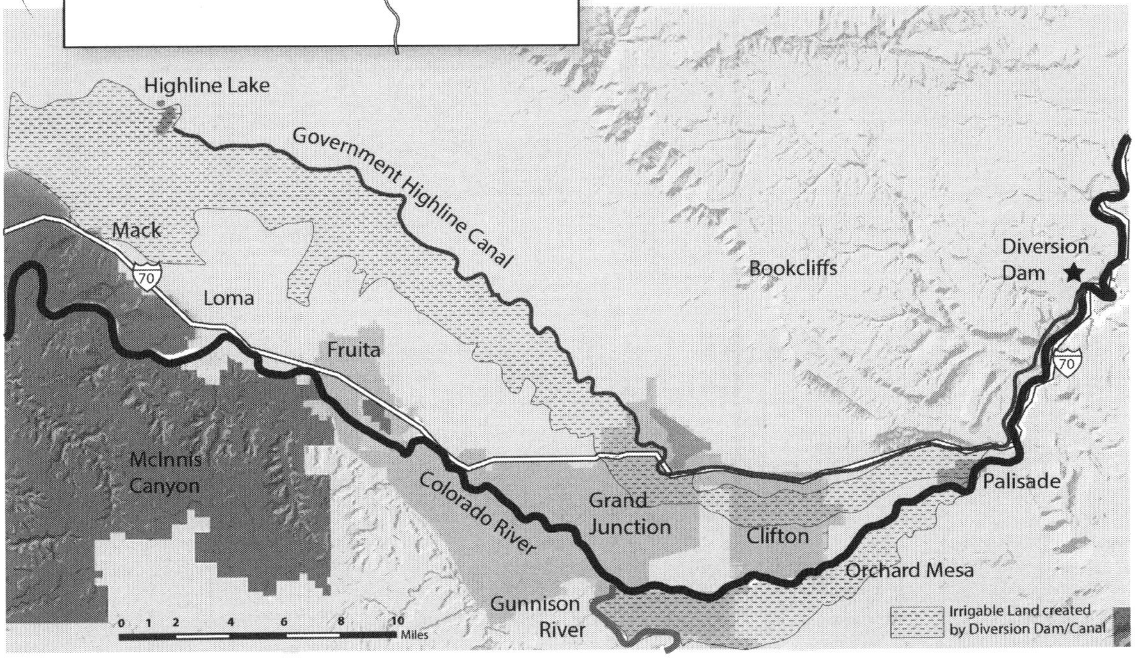

The Government High Line Canal provides water to approximately 23,350 acres.

rated from the land when ownership changes. The project also had to account and plan for the tail water draining off the fields to prevent flooding, seepage, and alkali problems. Therefore, the Grand Valley Drainage District (GVDD), established in 1915, manages around 260-miles of open and piped ditches throughout Mesa County to mitigate possible damage.

We invite you to enjoy the view from this unique window into the genesis of the Grand Valley Project's Government High Line Canal.

NOTE: The clippings are presented as they were organized in the original archive, and are not always chronologically sequential. Please also note, the spelling of names for notable figures in these years was not always consistent. For instance, McConnell was sometimes spelled M'Connell. Spellings were left as they were documented in the original articles.

For further reference:

Grand Valley Project, Denver: Bureau of Reclamation History Program, 1994 by Wm. Joe Simonds, available www.usbr.gov/history/projhist.html,(Accessed October 2015).

The History of Irrigation in Palisade and East Orchard Mesa, Colorado, produced by and available through the Palisade Historical Society, Palisade, CO.

"Every barren butte will be crowned with the burdened orchard bough, every haggard hill will hold the halo of heavy horticultural harvest, and every gashed gulch and gulley will be glorious with overhanging fruitage."

— *George E. Caldwell,* Daily News, *1908*
"Marvelous Growth of the Grand Valley"

Daily News, Sept 15 1910

BOARD HERE SUNDAY

The board of army engineers who have to pass on the merits of the Grand Valley High Line canal will arrive in this city Sunday morning on No. 8, accompanied by F, H. Newell, the director of the reclamation service, according to a telegram received in this city yesterday by D. W. Aupperle from Mr. Newell.

It is not known whether the men will want to work Sunday or, not and a telegram was sent them by Mr. Aupperle asking them about this question. If they want to work Sunday they will be here only during Sunday and Monday and will leave Tuesday morning for Montrose. If they take Sunday off they will be here until Wednesday morning and will be enabled to see a few of the exhibits at the Mesa County fair. However if this falls through with an effort will be made to have them stop off on their way back from Montrose to the west.

The following dispatch will give additional information:

Special to the Daily News.

SALT LAKE CITY, Sept. 14. General W. L. Marshall, chief advisor to President Taft on reclamation matters, accompanied by Major C.W. Kutz and Major Burgess of the reclamation board, arrived in Salt Lake Tuesday on a trip of inspection of the government reclamation projects.

The party is also accompanied by M.A. Cramm, secretary to General Marshall and W. E. Kelley, stenographer of the reclamation board.

Director F. H. Newell of the reclamation board, together with Colonel Biddle, chairman of the board and members, Major Harts and Major W.C. Langfitt, stopped at Ashton's Lake, where a government reservoir site is proposed. They returned to Ashton yesterday morning and will come direct to Salt Lake.

On Friday evening the party will leave for the Strawberry project on an inspection trip.

Edgar F. Gates, special representative of the International Correspondence Schools of Scranton, Pa., will be at the Elberta hotel, Palisade, each evening during this week, from 7 until 12.

Colorado: Partly cloudy Thursday and Friday.

Daily News, Sep 1904

HURRAH FOR THE BIG DITCH

T. C. Henry Admits Private Plan a Failure and Advises Government be Invited to Build it.

CHICAGO, September 7th, 1904.

EDITORS SUN: -- My attention is just called to Mr. McMullen's communication in a recent issue of The Sun. I desire to cordially commend its tone, and, for reasons which follow, the suggestion he makes.

I wish to see the "High Line" built. I have spent more or less time and money off and on for twenty years past in aid of it. I want it built whether I gain any pecuniary return or even receive my outlay, I should like to see the agricultural possibilities of the valley developed and so justify my own work and orify my prophesies. If therefore, the Government be induced to take up the "High Line" project on terms agreeable to the local interests, I shall offer no opposition, but my best wishes rather.

In no event can I handle the Bonds of the District until the question of the constitutionality of the District Law is affirmatively determined by the Supreme Court. Unless the case can be advanced, that cannot be accomplished until at least the Legislature convenes next winter and some somewhat essential amendments are enacted.

I confess ever since the District was so formed as to exclude the Government lands I have been apprehensive that it would be difficult to finance the extension, for that l would involve construction operation and maintenance in advance of settlement and sale of water--a repetition of our Grand Valley Canal experience of the past and a situation which has caused financial disaster to canal investments everywhere. My efforts to market these District Bonds in the east have constantly encumbered the discredit which has arisen from these causes and from our forced rental law rate. Moreover, there has not been quite the zealous local support in favor of the District plan for the "High Line" I had hoped for and expected. That may have been because of a reasonable apprehension that the bonds could not be placed, and too, because it has been thought that better terms might be obtained from the Government. Whatever the cause I have had it to meet in various phases while here engaged in financiering the project in its entirety.

Daily News, Sep 1904

For these reasons and because I do not want to stand in the way of a reasonable hope of construction by the Government I kindly notify all concerned that I will no longer claim the attention of the District Organization.

Thanking you and my many friends in the valley, with my best wishes.

Respectfully,

T. C. HENRY.

Daily News, Sep 1904

S. G. McMullin Advances Strong Argument in Favor of Uncle Sam Building Big Ditch.

GRAND JUNCTION, Aug. 30. '04.
EDS. DAILY Sun:-I have noticed in your paper and The News has at times shown a desire that the High Line ditch be built by the reclamation service of the United States government, but it has shown this in such a way that it would appear that you were afraid of offending somebody by advocating such a measure. I, with others, for some time was doubtful about the' advisability of asking the Government to undertake the building of our High Line ditch for the reason that in its own public works the Government has been notoriously slow.
I cannot refrain, however, from calling the attention of the Chamber of Commerce and the property owners of this county to what is being done in Montrose county. The United States government has now $25,000,000 in cash in this fund and is hunting for meritorious irrigation propositions. This one in Grand valley presented an ideal opportunity in the minds of the Government experts to demonstrate the feasibility of their plan of irrigating a fine body of land through an ample water supply, causing it to be subdivided into farms not exceeding forty acres.
At Montrose the Government has surveyed its ditch on the highest possible level, not seeking in any way to economize. That ditch will cover nearly 180,000 acres of land. The ditch and tunnel will be completed in four years. The Government will run it for thirteen years, and after the third year will charge an annual assessment of $2.50 per acre for ten years, at the end of which time the ditch will be paid for and be turned over to the farmers. Judge Bell thinks that when the time comes they will extend the payments so as make them run twenty-three years, if necessary. The ditch is to be built in in the finest possible manner by recognized experts, and costing no one anything for the first three years will relieve settlers of many of the disadvantages of living under a new ditch.

To anyone who has investigated the situation carefully it must be apparent that there is no other hope of selling the bonds of our irrigation districts at honest figures, and the only possibility to build the ditch is for some construction company to take it at a very high figure and build the ditch at such a price that the bonds will cost them much less than par, and carry the bonds until such time as the district is on its feet, and the safety of the bonds is established, and then sell them.

The building of the ditch is not the only thing, there is also the cost of putting it in successful operation. At Montrose the United States government will do this, here the irrigation district would have to do it and every farmer under the ditch would have to help. We have had an expensive experience with the Grand Valley ditch.
It will not do for us to say that if we do not float the irrigation district bonds we will then go to the Government. I understand the Government officials are very much disappointed at the number of available irrigation schemes and it is quite possible that the opponents of Government irrigation will use the lack of interest which is shown in this plan at some future session of Congress as a means of weakening or repealing the present law.

Daily News, Sep 1904

There is no possible reason why anyone should want to wait longer unless it is promoters who expect to profit out of the building of the ditch by the district. I have talked to a great many and have not met one who does not want to see the Government build this ditch for they could then irrigate Orchard Mesa and develop all our unimproved lands. This is the most important matter before us today, as much so to the merchants who should be interested in seeing from $10,000 to $20,000 a month spent in the valley for two or three years as it is to the land owner under the survey and I write this letter asking you to take the matter up; have the Chamber of Commerce act upon the matter, and then have them call a meeting of the farmers and get such an expression of opinion that the Government authorities will have no doubt as to what the people here want. Our two irrigation districts have not been formed in vain for they can be used as a means of dealing with .the Government in all probability.

It seems to me that we should treat this matter as one of vital importance, and not stand aloof and permit any such misrepresentations of sentiment as have been carried to the Government officials, if all reports are true.

Very truly,

S. G. McMullin.

Daily News, Sep 20, Nov 14-15, 1907

"Grand Junction, November 14, 1907.

"F. H. Newell,
"Reclamation Service,
"Washington, D. C.

"At an enthusiastic meeting tonight of citizens of Mesa county, 700 strong, largely land owners and homesteaders, the offer of Secretary Garfield to order survey and definitely locate canal was unanimously accepted.

"D. W. Aupperle
"E. E. Sands."

D. W. AUPPERLE
To Whom Citizens of Grand Valley Owe Debt of Gratitude.

"Whereas, The Honorable James R. Garfield, Secretary of the Interior, having asked an expression of the people of the Grand Valley, concerning the proposed construction of the High Line Canal by the United States government under the provisions of the reclamation act, it is hereby

"Resolved, by the land holders under the proposed High Line Canal and the property holders in the Grand Valley and the citizens of Mesa County en masse assembled, that it is our earnest and emphatic desire and prayer that the honorable secretary of the interior, James R. Garfield, shall proceed at once along the lines proposed in his letter to Secretary D W. Aupperle, of the Grand Junction Chamber of Commerce, as indicated by his proposition No. 2, "to begin surveys and acquisition of rights of way, preparing detailed plans for the work, with a view of its construction at the earliest possible date, as soon as funds are available"

Seven hundred of the most representative citizens and land holders of the Grand Valley unanimously passed the foregoing resolution at the mass meeting held in the Auditorium last night. By the passage of that resolution and the sending of a telegram to Secretary Garfield by Messrs. Aupperle and Sands the signal has been sounded for the beginning of actual work toward the construction of the High Line Canal by the United States government. The resolution was offered by Professor E. E. Cole.

AUPPERLE CHEERED.

Every available seat in the great Auditorium was filled when Frank A. Lyons, president of the Chamber of Commerce, called the meeting to order. Horace T. De Long was chosen chairman and W. S. Wallace of the Water Users' association, secretary, Chairman DeLong delivered brief introductory remarks painting in glowing colors the future of the Grand Valley when the High Line Canal shall have been completed. He then called on Secretary Aupperle to give a report of his recent mission to Washington and the results thereof. When Mr. Aupperle appeared it was the signal for a tremendous ovation. The audience as one man recognized in him the one who has gathered up all the loose threads of effort and crystallized them into one mighty cable which had pulled the prize within the reach of the Grand Valley. Mr. Aupperle prefaced his report with a brief history of the long years of struggle to secure recognition at the hands of the federal government and of the later and more strenuous efforts to interest the department of the interior in the project which has lain so close to the hearts of the people of this section for so many years.

Mr. Aupperle then presented the following communication over the signature of Secretary Garfield in which he set forth the alternative propositions for the consideration of the people of the Grand Valley.

Daily News, Sep 20, Nov 14-15, 1907

"DEPARTMENT OF THE INTERIOR.
"WASHINGTON.
"November 5, 1907.

"Mr. D. W. Aupperle, Secretary
Grand Junction Chamber of Commerce,
Grand Junction, Colorado.

"Dear Sir:

"Referring to our conversation and previous correspondence regarding the Grand Junction project, I have given this careful consideration, following my brief examination on the ground of the possibilities of the Valley and of the physical conditions, water supply, and other matters affecting the opportunities for: development.

"The extraordinary fertility of the soil, the salubrity of the climate, and the excellent water supply have all imprest me. I know of few localities in the arid region equally attractive for the inauguration of a reclamation project. These conditions have long been appreciated by the Reclamation Service, and the project has been repeatedly recommended as desirable for construction. The reason this has not been undertaken in the past has been the preference exprest by certain of your citizens that construction be undertaken by individuals or corporations, or under the Irrigation District law.

"Conditions are now different from those of four years ago, when the people of the valley practically had the option of immediate construction by the Government or by private parities. At that time the Secretary of the Interior had ample funds to complete the work and could probably have built the works at less expense than now. At the present time the prices of labor and material have advanced, the cost of construction would be much larger; but the most notable difference is that the Reclamation Fund is not in such condition that an allotment can now be made for immediate construction or completion.

"At that time the argument was used that the Reclamation Fund should not be used to anticipate private enterprise. It was asserted by interested parties that private funds were available for building the work, and that the taking up of the project by the Secretary of the Interior would interfere with the plans already made.

"These plans, it is now understood have never advanced beyond the initial stages.

"From your representations it appears that the probability of private enterprise taking up the work is even more remote than in the past. It also appears that the Government has certain rights and privileges which cannot be enjoyed by individuals, or corporations, and that it is the consensus of opinion in the valley that unless the Government undertakes the work it will probably not be done at all or only after great delay.

"If the people of Grand Valley desire that this project be more thoroughly investigated and placed in line for construction through the acquisition of rights of way and beginning of work when funds are available, I am willing to consider it. The preliminary investigations indicate that at least 50,000 acres can be reclaimed at a cost anywhere in the neighborhood of $40 per acre. These figures are given merely for the purpose of discussion and do not in any way bind the Secretary to a definite announcement of cost, which can only be made after the contracts have been let and the necessary data for that purpose have been obtained.

Daily News, Sep 20, Nov 14-15, 1907

THE TWO PROPOSALS.

"I suggest that you lay the matter clearly before the people of Grand Valley and report to me their wishes with reference to the following alternative propositions"

"1. The Secretary of the Interior to withdraw segregations and definitely announce that he does not intend to construct the Grand Junction project.

"2. The Secretary to begin surveys and acquisitions of rights of way, preparing detailed plans for the work with a view to its execution during succeeding years when the money becomes available for this work from the repayments to the Reclamation Fund under the Uncompahgre project, or such other, funds as may become available in future years.

Very truly yours,
"JAMES RUDOLPH GARFIELD,
"Secretary.. •
"F. H. N." •

As a further evidence of the sincerity of the department in placing the matter thus directly before the citizens and land holders of the valley Mr. Aupperle presented the following letter from F. H. Newell, director of the reclamation service, which was received with marked enthusiasm.

"Department of the Interior.
"United States Reclamation Service
"Washington, D. C.
"Office of the Director
"November 5, 1907

"Mr. D. W. Aupperle, Secretary Chamber of Commerce,
"Grand Junction, Colo.
"Dear Sir:

"In accordance with our conversation, and the letter of the Secretary of the Interior of November 5, it is my intention if the people of Grand Junction desire to accept the second proposition of the Secretary, that Mr. E. E. Sands, or some other qualified engineer, be put at work at once to make the detailed surveys and prepare for acquiring the necessary rights of way.

"It is therefore desirable for the people of Grand Junction to take prompt action in the matter and when they have done so I suggest that you wire to me or write at the earliest practical moment in order that our plans may be completed.

'Very truly yours,
"F H. NEWELL
Director"

RESOLUTIONS ADOPTED

At the conclusion of Mr. Aupperle's report, the Water Users' association presented the resolution adopted by that organization endorsing the government's offer to build the canal and also requesting that the Orchard Mesa District be included in the government irrigation project. President Frank A. Lyons of the Grand Junction Chamber of Commerce then read the resolution adopted by that body which are of. similar import as the ones preceding. The resolution adopted by the Fruita Chamber of Commerce in harmony with the foregoing was read by President Merriell.

E. E. Sands, constructing engineer of the United States Reclamation Service, was then introduced and made a most happy impromptu address. He complimented the people of the Grand Valley on their enterprise and perseverance and seconded every claim that has been made with regard to the advantages and desirability of the proposed High Line Canal..

Daily News, Sep 20, Nov 14-15, 1907

"In my report to Secretary Garfield", said Mr. Sands, "I did not say that the land proposed to be reclaimed was the best in the valley nor did I say that it was the best in the state, neither did I claim that it was the best in the United States. **I TOLD THE SECRETARY THAT IT WAS THE BEST LAND IN THE WORLD"**

Loud applause greeted this statement from the government's representative.

Following Mr. Sands' remarks Professor E. E. Cole offered the resolution which appears at the beginning of this article and the same was passed by an enthusiastic and unanimous vote of the entire assemblage. A motion to endorse the Water Users' association's request regarding Orchard Mesa was also carried by a unanimous vote and the meeting which marks the first chapter of the government's undertaking of the High Line Canal project was brought to a close.

AUPPERLE THANKS NEWELL

Following the telegram sent to Director Newell of the reclamation service at Washington advising him of the action of the mass meeting here when the proposition of Secretary Garfield was accepted, Secretary Aupperle of the Chamber of Commerce has sent a written report to headquarters. In this written statement the secretary expresses the thanks of the people of Grand Junction for the courtesies shown him by the federal officials while in Washington as representing the citizens of the valley.

E. E. Sands, the constructing engineer, who also signed the telegram has stated that he is also preparing a written report which will reach Director Newell through~ the regular channels in due time.

Daily News, Sep 20, Nov 14-15, 1907

SEN. SIMON GUGGENHEIM
Who "Took Off His Coat" and Worked for High Line Canal.

I. W. M'CONNELL
Construction Engineer of Reclamation Service, Who Will Have Charge of Work.

W. S. WALLACE
Who Was one of Aupperles's Tireless Co-Workers for Great Project.

Daily News, Nov 21 – 30, Daily Sentinel, Dec 6 1907

Great Prosperity For Grand Valley

(By James B. Young.)
Secretary Colorado State Commercial Association.

"If you are a pessimist and are sorry for it; if because the sun temporarily goes behind a cloud you are inclined to think that it will never shine again, pack your grip, buy a ticket for the Grand valley and effect within a remarkably short space of time a cure for that sour-faced feeling.

"I am not a pessimist and hope that I never will be one" but no one, no matter how grouchy he may be, can make this trip into one of the most prosperous sections of this great and prosperous state without coming back refreshed in mind and body and feeling that good old Colorado is a mighty good place in which to live after all.

"The people in the Grand valley are feeling good these days. The effect of the temporary financial stringency has not been felt 4n that section of the state. The Grand valley is prospering. The fruit crop this fall was not as large as in former years. But what difference did it make? Not a bit. The big price received for the fruit has put as much money in circulation as ever before. The farmers have raised as much, if not more, grain and forage crops than in former years and are getting good prices, thus putting more money into circulation. The yield of beets has been the best ever raised in that section of the state.

"These conditions alone would make any section prosperous. But add to all of this the statement that the United States government has definitely committed itself to the high line ditch proposition and you will readily understand why the people of the Grand valley are feeling so good.

"'The importance of the high line canal proposition cannot be overestimated. There are about 70,000 acres of arid land lying now untilled and practically good for nothing in the vicinity of Grand Junction. At a mass meeting of citizens of the valley held in Grand Junction on Thursday night, E.E. Sands, the government engineer, who will in all probability be in charge of this latest reclamation project, made this statement: "When I made my report to Secretary Garfield on this reclamation project I did not tell him that here are 70,000 acres of good land, or as good land as can be found in any part of the country, lying idle right now; but I did tell him that here were 70,000 acres of the very best land to be found in this wide, wide world, waiting for the government to put water on it.'

"When an expert like Sands makes the unqualified statement that Grand valley land is the best in the world it means something to that valley.

Seventy thousand acres of land will within a few years be planted to fruit; in a few years more they will be in bearing and will be bringing in their golden harvest to the lucky owners.

"The effect of the new high line ditch proposition is already being felt in the values placed upon the land to be watered by the government canal. The man who has say 100 acres of land that is to be watered would have been glad enough a week ago, knowing full well that irrigation, whether by the government or by private capital, was sure to come within a comparatively short time, to sell his property for from $40 to $50 an acre. He now turns up his nose at an offer of anywhere from $100 to $200 an acre. He is sure that it is going higher and when the values of land in that section are considered it is more than likely that the land-holder is right.

"Twelve millions of dollars' worth of property was represented at the mass meeting of Grand valley citizens held Thursday night to ratify the government project.

" 'Fifty thousand population in five years' is the slogan in Grand .Junction now. The metropolis of the western slope is growing; it is going to grow much more rapidly within the next few years. It is a splendid little city now; it is going to be much greater in the near future.

"As Grand Junction prospers so are the other towns in the valley, and so is the slope generally sure to prosper and forge ahead. Palisade, Clifton, F'ruita, Mack, Grand Valley, De Beque and all of the other cities up as far as Glenwood Springs are going to feel the effects of this growth.

"Surely the future is bright for the Grand valley."--Rocky Mountain News.

The $50,000 is Ready

(Special to The Sentinel)

DENVER, DEC 26.—"THE UNITED STATES RECLAMATION SERVICE HAS APPROPRIATED FIFTY THOUSAND DOLLARS FOR THE PRELIMINARY WORK ON THE HIGH LINE CANAL PROJECT AT GRAND JUNCTION, AND THE SURVEY WORK WILL START AT A VERY EARLY DATE."

THUS ANNOUNCED SIMON GUGGENHEIM, UNITED STATES SENATOR FROM COLORADO THIS MORNING IN AN INTERVIEW GIVEN OUT IN THIS CITY. SENATOR GUGGENHEIM IS HERE TO SPEND THE HOLIDAY RECESS.

THIS IS THE FIRST OFFICIAL ANNOUNCEMENT THAT THE APPROPRIATION HAS BEEN ACTUALLY MADE.

THIS EVENING ON THE MONTROSE TRAIN E.E. SANDS OF THE RECLAMATION SERVICE WHO IS TO HAVE CHARGE OF THE WORK ON THE HIGH LINE CANAL, WILL ARRIVE IN GRAND JUNCTION TO COMPLETE ARRANGEMENTS FOR THE STARTING OF THE SURVEY.

Daily News, Montrose Press, Dec 17 1907

THESE MEN WILL BUILD HIGH LINE

E.E. Sands
Who will have charge of High Line Surveyors.

I.W. McConnell
Supervising Engineer of Reclamation Work in Colorado.

Reclamation Engineers Who Will Have Charge of Grand Junction Project are Competent Men

Edward E. Sands, who will be in charge of the engineers' headquarters to be established here for surveying the High Line canal, was, with Assistant Engineer Mathes, the original surveyor of the mammoth ditch. The preliminary surveys were run off by him several years ago and it was then that he had the opportunity to examine the land under the canal and it was this examination which led him to make his enthusiastic report to the secretary of the interior.

It will be remembered that he told Mr. Garfield that the land to be brought under the ditch was not the| best land in the county or state or even in the United States. But Mr. Sands, like a trim "booster," said that it was the best land in the world.

Originally Mr. Sands was with the topographical service in full charge of the permanent location of the south end and west side canals of the Uncompahgre Valley project, known as the Gunnison Tunnel. After the actual construction of that project was begun Sands was made constructing engineer of the reclamation service at that point, which position he now holds.

At that time with a corps of about twenty assistants he went over the location of that project seven or eight times before it was finally and permanently located. Mr. Sands is a graduate of the University of Wisconsin from which institution he came with high honors.

Ira W. McConnell was a mining engineer before entering the government service but upon taking up the reclamation work he was associated with A. L. Fellows as assistant engineer in making the preliminary surveys of the Gunnison tunnel. Mr. Fellows left to take a position as chief engineer of a northern state, while Mr. McConnell was given entire charge over the reclamation work of Colorado, Utah and New Mexico, which position he now holds.

These gentlemen—Sands and McConnell—who will be actively engaged in the actual work of the big high line ditch and under whose direction the work will be done, are both enthusiastic in their praise of the wonderful value of the High Line land. Mr. Sands, is will be remembered added his name to that of Secretary Aupperle on the telegram dispatched to F. H. Newell, director of the reclamation service immediately after Thursday night's mass meeting. That Secretary Garfield considers both as thoroughly competent engineers is evidenced by his correspondence with the local "boosters" in regard to the work.

SANDS IS A GOOD BOOSTER

E.E. Sands of the U. S. R. S. went down to Grand Junction Tuesday to look after the surveying of the proposed new high line canal there. He says the first five or six miles of the canal is through a very rough section, and will probably cost over $650,000 and that the whole area covered will be 60,000 acres. He says the land to be watered is good farming land and will all raise fruit.— Montrose press.

Daily News, Dec 16-17 1907

LAND ENTRIES ARE WITHDRAWN

Congress May Pass Bill at Present Session Making Park National Monument—Another Feather for Chamber of Commerce

---

By a special order of Commissioner Ballinger of the general land office of the United States, Monument Canon, claimed to be the rival of the famous Garden of the Gods, and located just ten miles west of Grand Junction, has been temporarily set aside by the government for a national park or a national monument. It is understood also that a bill will be presented to congress at this session to make the withdrawal of entries permanent and to create and improve the park under government supervision.

This action is another result of the activity of the Grand Junction Chamber of commerce and its efficient secretary. D. W. Aupperle. The Chamber of Commerce took up the matter of making Monument canon a national park over a year ago and have been working unceasingly to that end ever since with the result that success has practically crowned their efforts.

In addition to interesting Congressman Warren A. Haggott and Senator Simon Guggenheim in the project and securing a promise from them that they would do all in their power to have the government take up the matter, the local organization also brought the proposition before Secretary of the Interior James R. Garfield and Commissioner Ballinger. When the two latter gentlemen visited Grand Junction in June to view the land under the proposed High Line canal they were shown photographs of the majestic and inspiring rocks in the canon and became so interested that their assistance was assured. Upon his visit to Washington in the interest of the big ditch Secretary Aupperle also brought up the Monument canon matter again. With the matter still fresh in their minds Secretary Garfield and Mr. Ballinger were then waited upon by Messrs. Haggott and Guggenheim when they arrived to attend the sixtieth congress with the result that an order temporarily withdrawing the canon was issued. These instructions have been received from Bryson P. Blair register of the land office at Montrose who verified the authenticity of The News' information over the telephone last night.

Monument Canon is located between ten and eleven miles directly southwest from Grand Junction and covers an area of about five miles. It is about a mile wide and a little over three miles in length. The canon is of the box formation, walled in on all sides except the entrance. The washing of the water for years has left rocks of gigantic proportions standing in freak fashion throughout the park, some of the pinnacles being 2,000 feet in height. It is proposed to have the government improve the park and open a road to it so that it can be made a scenic spot for tourists to view.

By many, Monument canon is claimed to equal if not excel the Garden of the Gods at Colorado Springs.

Reclamation Men Go to Denver

I. W. McConnell, E. E. Sands and several other reclamation service men left for Denver yesterday where they go as witnesses in the United States court in a federal case against Ormann and Crook, who are charged with violating the eight-hour law, while working on government contracts.

SPLENDID PROGRAM

This is the night of all nights when every earnest and loyal citizen of Grand Junction will make every effort to attend and participate in the fifth annual banquet of the Chamber of Commerce which takes place in Margery hall. Never in the history of that organization nor of the city has a gathering here been so indicative of local progress. With this city's career now fully launched upon the upward grade and every felicitous influence exerted in its behalf it is fitting that the business men of this city as well as their staunch supporters from the urban sections, should gather and not only congratulate themselves upon the progress and present outlook but pledge anew their sincere purpose to leave no stone unturned to place this city where she belongs as the greatest between Pueblo and Salt Lake City.

The program is of especial interest. I.N. Bunting, editor of the Sentinel, will tell of various influences and circumstances which have conspired to make Grand Junction the prosperous little city that she now is. His long residence and active business career in this community make him well fitted for the task.

The subject of irrigation with particular reference to the reclamation of the arid sections will be discussed at some length by I. W. McConnell, of the United States reclamation service, a recognized authority upon such an important subject.

Verner Z. Reed, the multi-millionaire of Colorado Springs who has extensive interests and heavy investments in the Grand Valley will give his impressions of the city as they are received by a non-resident. Colonel W. E. Skinner, president of the Colorado Commercial association, a man as broad mentally as the Centennial state is geographically, will have some telling remarks to make with regard to the industrial and commercial progress of Colorado. Sam. G. Sloane, an eastern capitalist and veteran newspaper man who has

Daily News, Dec 16-17 1907

been a resident of this city but a few months will give some very entertaining views from the standpoint of a tender-foot.

"Grand Junction's Future," is the topic assigned to Edwin A. Sherman, managing editor of The Daily News.

Senator Horace T. DeLong will preside at the head of the table as toastmaster of the occasion.

At intervals musical selections will be rendered by Prof. Boyers' orchestra.

It is especially urged that every member of the Chamber of Commerce be present. Efforts are being made to deliver tickets to each member but it is quite possible that some may not be reached hence it is urged that every member whether he has a ticket or not, attend the banquet.

Daily News, Dec 27, 1907 – Jan 7, 1908

HIGH LINE PARTY SOON TO BE PERFECTED

Surveyors to Work Here Will be Named at a Meeting With Chief

BASE LINE FIRST

That the personnel of the engineering party which is to make the survey for, the High Line canal will be completed at a conference to beheld at Yuma, Arizona, in a short time between A. P. Davis, chief engineer of the reclamation service, and other members of that department, is the: information received by the Chamber of Commerce yesterday.

At the conference in Arizona it is probable that a successor to E. E. Sands at Montrose will be named. The reclamation service, it is said, has under consideration the names of a number of competent and capable engineers who are worthy of promotion and when the various assignments are made one of these men will be advanced to fill the vacancy on the Uncompahgre project caused by the coming of Mr. Sands to Grand Junction.

Headquarters for Mr. Sands were recently leased in the Canon block and as soon as the organization of the engineering party is perfected the actual work on the preliminary survey will begin. While nothing definite has been given out regarding the matter it is understood that most of the surveyors who are to work here will come from Montrose although a few may be impressed into service from other projects throughout the west. The first work of the engineers will be to run the base line after which the topography of the whole valley will be taken for water sheds. The big ditch will then be diagramed by the surveyors after which a levelling party will run the preliminary line. The transit party will follow to establish a preliminary location followed by the plain table surveyors who will determine the exact slope of the land. This work will consume considerable time and at its conclusion he transit surveying party will establish the permanent location followed by the cross sectioning party to get the height of the bank. The next step will then be that of actual construction.

Daily News, Dec 27, 1907 – Jan 7, 1908

SANDS MAKES HOME HERE

Constructing Engineer of Garfield Ditch Arrives With Family to Take Up Residence

PRELIMINARY WORK IS SOON TO BEGIN

E. E. Sands, constructing engineer of the Garfield Ditch Project, accompanied by his family arrived in the city Sunday night and are stopping at the La Court hotel pending the arrival of their household goods. These will come in on No. 315 this evening and as soon as possible Mr. Sands will move into his home at 360 Ouray avenue.

As was announced in The News some time ago Mr. Sands has leased the residence and his family will move in as soon as the house is in shape. He is preparing to begin active work on the preliminary surveys immediately after the Yuma conference which occurs soon.

At that time the personnel of the force for the Garfield project will be named and the successor for Mr. Sands at Montrose will be selected. Other routine work of the service will be attended to but the High Line ditch will be the most important subject to be brought before the conference.

While part of the office furniture for the Reclamation service headquarters in the city has already arrived another consignment from Chicago is now on the way. Mr. Sands made arrangements with the transfer companies yesterday for the handling of it as soon as it arrives. This shipment will contain filing cases and other office fixtures. The surveying instruments have already been assigned for the work as soon as the plans are completed at the Yuma conference. Mr. Sands gives official announcement of the fact that the $50,000 has been appropriated and has outlined his plans for the expenditure and submitted them to I. W. McConnell for approval.

Daily News, Dec 27, 1907 – Jan 7, 1908

This is indicative that the service means business and that there will be no delay in the preliminary work. That there will be no trouble in getting the funds for the actual construction is evidenced by an interview published in The Denver News to the effect that they would be available in 1909 and that the project should be completed in 1910. Mr. Sands states that the work on the Montrose canal is going along smoothly, last month being a record month in the reclamation service for hard rock tunneling. In the month of December 367 feet of the tunnel were removed which is the maximum record for that work. If conditions remain as favorable Mr. Sands is of the opinion that the Gunnison tunnel will be completed by July 1909. A further expenditure of about $20,000 will.be necessary to complete the project.

Daily News, Jan 4 and 9 1908

Furniture Received

Desks and Other Office Equipment For Sands' Force Comes From Provo

----

GOVERNMENT ANXIOUS TO START WORK

Congressman Haggott Says That Federal Department is Enthusiastic Over Garfield Proposition

----

WASHINGTON, D. C, DECEMBER 30, 1907.
I HAD A TALK WITH THE RECLAMATION PEOPLE LAST THURSDAY REGARDING .THE GRAND JUNCTION CANAL PROJECT, AND FIND THEM VERY ENTHUSIASTIC IN THIS MATTER AND VERY ANXIOUS TO BEGIN WORK. THEY SEEM TO THINK THAT THIS 1S THE BEST PROJECT WHICH HAS BEEN SUBMITTED TO THE RECLAMATION SERVICE.

"WARREN A HAGGOTT."

That the government will soon open the headquarters of the Reclamation service in this city was made more apparent than ever yesterday morning when a large shipment of desks, chairs, and other office equipment were received here from Provo, Utah, addressed to E. E. Sands and to be installed in suite 51 of the Canon block.

The conference at Yuma, Ariz., between Chief Engineer A. P. Davis of the reclamation service and other members of the department is to be held soon, it is said, and after this is ended Mr. Sands and his corps of assistants will come at once to Grand Junction to begin the preliminary survey work. A successor for him will be named to take up the work at Montrose on the Gunnison tunnel project.

That the department of the interior is anxious to begin work on the Garfield Canal just as soon as possible is shown in the following communication received yesterday from Congressman Warren A Haggott at Washington:

I had a talk with the reclamation people last Thursday regarding the find them very enthusiastic in this matter and very anxious to begin work. They seem to think that this is the best project which has been submitted to the reclamation service.

"WARREN A. HAGGOTT."

Mr. Haggott, together with Colorado's other representatives in congress, is keeping behind the ditch proposition and in addition to pushing this matter is working with effect for a federal building in Grand Junction. While Durango and Glenwood Springs both have bids in for appropriations for such structures the latest information from Washington indicates that the fight has simmered down to Durango and Grand Junction with the chances in favor of this city to win.

The remarkable increase in population during the past ten years, coupled with the enormous postal receipts and the fact that the building of the Garfield Ditch is recognized by the government as meaning a doubling of Grand Junction's population, are factors which are helping the Haggott appropriation measure, Grand Junction canal project and (remainder missing)

DETAILS OF HIGH LINE

In the January issue of the Denver and Rio Grande Red Book, copies of which were received here yesterday, I. W. McConnell, supervising engineer of the reclamation Service, in writing on the Grand River project, details the work that will be necessary in the following manner:

By I. W. McConnell.

"During the visit of Secretary Garfield to the west last summer he was so impressed with the advantages of the country about Grand Junction that he ordered a report upon the advisability of taking up a complete irrigation of that section under the reclamation act. As a result of that report a sum of money has been set aside, sufficient to cover the preparation of detailed estimates as to the cost of the necessary facilities. "The project proposes the diversion of the waters of the Grand river above the town of Palisade, and the conveyance of the same to the arid lands lying along both

Daily News, Jan 4 and 9 1908

banks of the Grand River, stretching from the mouth of the canon above Palisade to the Utah line. The area to be placed under cultivation will approximate 60,000 acres,

"To accomplish this it will be necessary to construct a series of canals, tunnels, flumes and inverted siphons, conveying water along the foothills of the Book Cliffs and distributing it to the various areas which can be reached.

"The possibilities of the Grand Valley are perhaps unequalled anywhere in the world, and it may be anticipated that the officials of the reclamation service will lose no time and spare no efforts in bringing to completion this project, which will be a lasting credit to themselves and contribute enormously to the wealth of: Colorado."

Daily News, Jan 8, 1908

PLANS ARE ANNOUNCED

Government Makes Known Its Intentions To Clear Away Any Doubt That May Exist

----

ECONOMY IS ABSOLUTELY NECESSARY

That the preliminary survey for the High Line Canal will he started early in March. is the good news received here yesterday from F.H. Newell, chief of the reclamation service, in a letter to secretary D.W. Aupperle of the local Chamber of Commerce.

Mr. Newell states that he has instructed I.W. McConnell, supervising engineer of the west, to complete preparations for beginning the field surveys in March and this means that there will be no delaying the matter.

The conference between Chief Engineer A. P. Davis and the other members of the service is to be held at Yuma, Arizona, about February 12, when the personnel of the working force will be completed and the men will come at once to Grand Junction to take up their labors under Constructing Engineer E. E. Sands. It is said that a Montrose engineer now employed on the Gunnison tunnel and who is one of the oldest men in the civil service has been decided upon as successor to Mr. Sands, but this individual's name will not be made public until the official announcement is forthcoming.

Mr. Newell, in his letter to Mr. Aupperle desires that the people of the Grand Valley be made cognizant of the plan of the department so as to remove any doubt as to the sincerity of the government in the matter and his letter is as follows:

Daily News, Jan 8, 1908

"Mr. D. W. Aupperle, Secretary,
"Chamber of Commerce.
"Grand Junction, Colorado.

"Dear Sir:

"1 have received and carefully noted your letter of December 27. I fully appreciate the situation you describe, and think it would be advisable to announce the plans so as to remove any doubts.

"As funds will not be available for construction for some time, it would be unwise to incur unnecessary expense by undue haste or attempts to make surveys during the short, cold days of winter. Such work can be more economically prosecuted in the spring.

"Mr. McConnell has been instructed to complete his preparations for beginning field surveys as early in March as the work can economically be commenced. It is to the interest of the settlers, as well as the Government, to observe the ordinary precautions of economy.

"'Very truly yours,
"F. H. NEWELL,
"Director."

Daily News, Jan 25 and Feb 17 1908

CHERRINGTON IS NEW PRESIDENT

Chamber of Commerce Names Good Leader - Reelects Aupperle

OTHERS HONORED

NEW OFFICERS.

President—C. E. Cherrlngtnn.
First Vice President--J. H. Rankin
Second Vice President--S. B. Hutchinson.
Treasurer--Frank E. Dean.
Secretary—D.W. Aupperle.

The foregoing officers were elected by the new board of directors of the Chamber of Commerce at its first meeting last night when the organization for the coming year was perfected. The same harmony and enthusiasm that marked the election last Tuesday, was manifest in the proceedings last night and each man was chosen by the unanimous vote of the directors and without a dissenting voice.

The meeting was called to order by Past President Frank A. Lyons, who was immediately placed in nomination for re-election but promptly declined. Mr. Cherrington and Mr. Rankin were then nominated but the latter withdrew and the secretary was instructed to cast the vote of the board for Cherrington who immediately assumed the chair, The other officers were nominated in the order mentioned, no other names being suggested and Secretary Aupperle cast the vote of the board for each. In the case of himself the president was instructed to cast the vote.

The re-election of Mr. Aupperle as Secretary will meet the approval of the citizens of the entire valley as he has, perhaps, done more than any one man in securing government construction of the high line canal. Through his untiring efforts, both as a representative of the Chamber of Commerce and personally, the big ditch was assured and he is better fitted than any other one man in the valley to continue "pushing" this project until it is completed and the fondest dreams of the residents are realized. Familiar with all the details, personally acquainted with the reclamation and other officials in Washington and, in fact, in close touch with the situation in general, he can do much effective work and will leave no stone unturned to advance the interests of the community.

Cherrington a "Live One".

The selection of Mr. Cherrlngton as president of the Chamber of Commerce is a good one and in placing him at the head of this organization that "'does things" the directors showed the same wisdom as that which characterized the election of the members when they named such strong men to direct the affairs of the association. Mr. Cherrington, while a member of the Chamber, has been an indefatigable worker and has always stood for the advancement of Grand Junction and the Grand Valley. He is inoculated with that famous "booster" spirit and is always looking up and ahead, not down and back.

Daily News, Jan 25 and Feb 17 1908

President Cherrington was instructed to report the appointment of his standing committees at the next meeting and the directors voted that in addition to the regular committees, a committee on the government canal and a committee on membership be also named for the year.

The committee appointed to examine the ground offered the city for park purposes by General Palmer made its report, which was heartily and unanimously concurred in by the directors This committee was continued with instructions to circulate a petition among the taxpayers of the city to be submitted to the council Tuesday night urging that body to accept General Palmer's generous offer.

A letter was read from the Southern Idaho Commercial clubs asking that the local Chamber of Commerce indorse the bill now pending in Congress by Senator Hepburn of Idaho providing that an amendment be made to the Interstate Commerce law making the long and short hall clause more specific. It was voted to endorse the measure and Colorado's representatives in congress will be so advised. A number of other communications for membership were read after which J. E. Leaverton and C. W. Teague were accepted.

The agreement to submit part of the Chamber of Commerce office to the fair association for $10 a month was approved. The president was instructed to appoint a committee of three including himself to secure an office for the current year.

Adjourns Two Weeks.

Before adjourning the board moved that the list of the membership as it now stands be given out for publication and the directors then generally discussed plans for the year and outlined some of the work that is to be done. Because of the fact that a number on the lecture course and the Masonic banquet both fall on Friday, December 31, the directors adjourned for two weeks subject to the call of the president in the event that an earlier meeting is required.

All of the new directors were present last night as follows: Loeffler, Price, DeLong, Fletcher, Dean, Rankin, Hutchinson, Buthorn, Cherrington, Lyons and Aupperle.

GUNNISON TUNNEL RECORD

The remarkable feat of the workmen on the river portal of the Gunnison tunnel in smashing the world's record for drilling in hard rock in January as chronicled in yesterday's News, greatly encourages the High Line canal project.

In January the workers drilled through 449 feet in the river portal of the Uncompahgre project, leaving only 6,745 feet yet to construct. This makes it almost certain that the big irrigation system will be completed in at least a year thus making it possible for water to be carried through the bore for the season of 1909. If the January record can be maintained the tunnel will be finished in ten months. If the workmen make two thirds as good progress it will require only fourteen to fifteen months at the outside.

It is not improbable that just as soon as the Gunnison project is done, the force employed at Lujane will be swung down to Grand Junction and put to work on the Garfield ditch. And the sooner the Gunnison tunnel is completed the better the outlook for Grand Junction and its cherished project.

Daily News, Jan 27, 1908

Garfield Opposed to 320-acre Homesteads

Secretary of Interior Announces That His Department Will Not Support Mondell Bill

---

WASHINGTON, Jan. 26. — Secretary Garfield yesterday returned Representative Mondell's bill to enlarge homestead entry acreage to 320 acres with refusal to recommend its passage. In his report the secretary says, in part:

"This department is not prepared to recommend a homestead entry of so extended an area at this time. There are doubtless areas in certain localities within which peculiar conditions would justify 320-acre entries, but in my judgment, there are areas that cannot now be sufficiently identified to warrant such legislation. Rapidly increasing demand for investments and homes on public domain Is fast consuming available lands, nearly 21,000,000 acres having been disposed of during the last year.

"Under these circumstances I am constrained to withhold favorable recommendation for the passage of this bill."

Speaking of Secretary Garfield's report Mr. Mondell said that it is a rather curious position when it is considered that President Roosevelt in his message called attention to the fact that a farm of 160 acres in arid regions is inadequate for self-support, and that Land Commissioner Ballinger, in his annual report, recommended that homesteads should be increased to 640 acres Continuing he said:

MONDELL REPLIES TO SECRETARY GARFIELD.

"'While it is true, as the secretary states in his letter, that a considerable amount of homesteading is being done, it is also true that in almost every instance the homesteader in the semi-arid region is protesting against the small acreage of his homestead, and it is further true that while the area of lands upon which original entries are made annually is quite large, the proportion of this area upon which final proof is made and patent obtained is constantly growing smaller, indicating increasing difficulty of complying with the law, and securing title and increasing dissatisfaction of entrymen, particularly under the homestead law, with the limited area they can obtain an increasing number of cases in which they become discouraged and abandon their attempts to secure a home on the public lands.

"The secretary's suggestion that the class of lands intended to be included in this act will be difficult to identify is not well grounded for it would not be difficult to identify non-mineral lands, non-timbered or timbered non-irrigable lands." Mr. Mondell will modify his bill in slight particulars and press it with a view to passage, regardless of the lack of indorsement by the interior department.

Daily News, Jan 27, 1908

WILL RECOMMEND SECOND ENTRY BILL.

The conference committee which has been considering the second homestead entry bills passed by the house and senate, today reached an agreement and will report the bid as originally introduced by Representative Mondell with a slight amendment.

As reported by the conference committee, the act will provide that any person who has lost, forfeited, or abandoned a homestead entry shall be entitled to the benefits of the homestead act as though the former entry had not been made, provided the former entry was not cancelled for fraud, or relinquished for valuable consideration.

Daily News, Feb 15, 1908

High Line Survey Begins Tomorrow

Outfits Now on Their Way Here

Engineer Sands Announces That Preliminary Work on Big Project Will Be Started at Once

----

CIVIL SERVICE EXAMS FOR EMPLOYEES

ACTUAL WORK ON THE PRELIMINARY SUEYEY FOR THE HIGH LINE CANAL WILL BEGIN IN GRAND JUNCTION TOMORROW WHEN D. L. HENDERSON OF MONTROSE ARRIVES WITH TENTS, INSTRUMENTS, ETC., FOR THE WORK

E. E. SANDS, THE PROJECT ENGINEER, MADE THIS STATEMENT YESTERDAY AND STATED THAT MR. HENDERSON IS NOW ON HIS WAY FROM MONTROSE, DRIVING OVERLAND.

Mr. Henderson is a son of R. C. Henderson of this city and has been concrete inspector for the government on the Gunnison tunnel. He will be one of the right hand men to Mr. Sands and his many friends here will be pleased to learn of his connection with the Grand Valley project as he has worked his way to the front in the reclamation service.

H. C. Essley, who has also been transferred from Montrose to Grand Junction will arrive here soon to take up his work under Mr. Sands

'The reclamation service headquarters in the Canon block have been opened and Mr. Sands is now going ahead with the work as fast as is consistent.

There will be no scarcity of help in the High Line ditch survey according to Mr. Sands, who states that from six to twelve applications are being filed daily. "A new ruling recently made gives the following information: The persons employed in the reclamation service are subject to civil service examination and thus no favors are shown the applicants."

The examination is not rigid, the applicant being required to fill out a blank stating the amount of actual experience, physical condition, age, and general understanding of the work desired.

Daily News, Feb 15, 1908

The marking is based upon these points and the applicant ranking highest stands the first show. The ruling evidently will secure the pick of the applicants for the character of work connected with the project.

Daily News, Feb 18-20, 1908

ENGINEER SANDS DUE HERE TODAY

----

335 Pounds of Mail Received For Reclamation Office

E. E. Sands, constructing engineer in charge of the Grand Junction project, will return today or tomorrow from Yuma, Ariz., where he has been in conference with the other officers of the reclamation service. With him probably will be A. P. Davis, chief engineer, and it is said that as soon as these men arrive active work on the High Line survey will be commenced.

One of the strongest indications of this is the fact that H. C. Essley, an engineer now employed on the Gunnison tunnel, together with several other men there, was yesterday ordered transferred to Grand Junction and it is understood that all are to come here the last of this month or the first of next.

Yesterday 335 pounds of mail was received in the Grand Junction post office for the reclamation service in this city. This mail consisted mainly of blanks, etc., to be used by Mr. Sands and his corps of assistants, who will occupy the suite of rooms already leased in the Canon block.

Mrs. Sands yesterday received a telegram from her husband and stated that he expected to reach here today or tomorrow.

SANDS READY FOR BUSINESS

Continued from page one)

I made numerous inquiries among people engaged in horticultural industries and the same ratio held good. We discussed methods by which the work on reclamation enterprises on arid lands might be furthered and among these was one by which citizens who are employed in the construction of the projects will receive certificates of credit which, when the project is completed will be applied on the citizen's water rights. This is to be tried in Mindoka, Ida. this year and if successful will greatly facilitate the construction and completion of all reclamation projects at an earlier date than could otherwise be expected. If the plan can be used on the High Line ditch there is no reason why we cannot begin construction early in 1909.

"Chief Engineer Davis made the statement at the conference that the largest appropriations would be made to those projects which offered the largest amount of cooperation with the certificate plan. The details of this plan will be announced later. Personally, however, I believe that it will meet with the general approval of the Grand Valley water users. It certainly is an important factor in getting early allotments

"I will leave in the morning for Montrose where I will be gone for a few days. With me on my return will be H.E. Essley, who will act as chief clerk in the offices and S.O. Harper who will be my principal assistant in the field."

Last night the rooms which were once those used by Frank H. Hammond,

Daily News, Feb 18-20, 1908

the missing life insurance agent, had the appearance of neatly furnished offices. On the door was the legend "Department of the Interior, U.S. Reclamation Service." In a few days a fine span of horses will arrive from Provo, Utah and two more teams will follow Mr. Sands from Montrose.

ORCHARD MESA BONDS VOTED

Full of enthusiasm, not only from the general results obtained from fruit growing in the Grand Valley but from his own experience, J. L. Oliver, of Palisade, responded to the toast "Our Fruit Industry." He declared Palisade to be a community where everybody was well-to-do. He predicted that this year's crop would see 1,200 carloads shipped from Palisade and probably 4,000 cars from the entire Grand Valley. He paid a splendid tribute to the quality of citizenship which went to make up the valley and dwelt especially upon the splendid characteristics of those from Iowa. He gave his facts in lucid and convincing manner which won the closest attention of his hearers.

Hen. Howard Morris, of Milwaukee discussing the improvements of the West declared the Grand Valley to be 'the most beautiful in the United States, and he has seen practicality all of them. He emphasized the need of scientific horticulture as being the best means of attaining the highest productiveness. The marked hospitality of the people of Grand Junction, he declared was a surprise as well as most pleasing to the visitors from the East.

Charles E. Cherrington, president of the Chamber of Commerce, delivered a most forceful and convincing address on the opportunities of the present and the hopes of the future. He declared that the harvest was ready, but the laborers few. This valley is looking for investors and the investors are looking for opportunities. The profits to each are sufficiently ample to satisfy the wildest of either. With regard to the position of the Chamber of Commerce, concerning outside capital, he stated that a cordial welcome is extended upon two conditions: First, that all contracts be clean, the investor and the member of the community shall be fulfilled in spirit as well as in the letter. As to the second condition he cited the fact that the organization was pledged to the support of government construction of the High Line canal and would not entertain any possibility of obstacles being thrown in the way of the accomplishment of that project.

'Mr. Cherrington's remarks pertaining to the High Line canal were received with vociferous applause.

"Opportunities Afforded," was the theme with which Senator Horace T. DeLong closed the program. He dwelt eloquently and at length upon the great profits to be derived by intelligent investment and industry throughout the entire valley. He declared that in the very near future the population of the territory immediately adjacent to Grand Junction would number 100,000. The opportunities for irrigation were detailed at length. In discussing the attractions for investment he enumerated climate, which he said was the finest in the world, electric energy, both of sky and earth, and the immense coal deposits of this region. He painted a beautiful picture of the future when this valley from one end to the other should be one vast thriving, contented prosperous community where two and a half acres would be ample for a comfortable living for one family, and yet permit the laying aside of a neat competency. He urged that nowhere on the face of the globe could such tempting opportunities for legitimate investment be found than in the

Grand Valley. In closing he extended a hearty and sincere welcome, not only to the distinguished visitors present, but any others who might come on a similar errand in the future, The success of the oratorical portion of the banquet was due in a large measure to the ready wit and graceful direction of Toastmaster J, S. Carnahan. His introductories were apt and he interpolated remarks which were pertinent and kept the enthusiasm of the meeting at the highest point.

Daily News, Feb 20, 1908

Grand Valley Showing Best of All

SANDS READY FOR BUSINESS

Returns From Arizona Conference With Fine Report and Will Open Headquarters Here Tuesday

----

PEASE IS APPOINTED FOR MONTROSE

When the various project engineers made their reports at the conference including those from Yuma, Salt River and Yakima Valley, not one of them could, in his statement of production, make a showing, per acre, one half so good as that of the Grand Valley. -- E. E. Sands.

E. E. Sands, constructing engineer in charge of the work on the Garfield canal, arrived in Grand Junction Tuesday evening from the conference of project engineers of the Reclamation service in Yuma, Ariz. He at once began work on the new office rooms in the Canon block and spent the larger portion of yesterday supervising the work of putting the offices in shape. All the furniture, desks, chairs, etc., was moved into the rooms, the floor covered and the commodious offices put in ship-shape.

In speaking to a News representative regarding his trip, Mr. Sands said:

"There were thirty engineers from all over the United States present at this meeting, it being presided over by Chief Engineer A. P. Davis. Among the matters which were brought up for discussion was the distribution of the water in the various enterprises and the policies of the service in connection.

"Colorado was represented by C. T. Pease, who succeeds me at the Uncompahgre tunnel; I, W. McConnell, supervising engineer of this district and myself. We were treated royally by the citizens of Yuma and the Southern Pacific favored us with a special train. On the way to Yuma a number of us visited the Truckee-Carson project, There we found the work progressing nicely. While we were in Yuma hardly a week we were shown some of the wonderful enterprises which the service is engaged in building. We saw the Imperial Valley project, commonly known as the Salton Sea.

Daily News, Feb 20, 1908

"Engineer Corey showed us about the work where the men are repairing a big break in the dam. The Colorado river is flowing through a channel a half mile wide and 45 feet deep in which there is 35 feet of water. The Southern Pacific engineers are doing an enormous piece of work at that point. On the 9th we made a trip to the Laguna dam which is 1 1-4 miles long by which the water in the Colorado river is raised for canals on both the California and Arizona sides. The dam is built on several hundred feet of river silt and is therefore substantially constructed. The dam is 250 feet in width at the base and will cost completed a million dollars. Just to show you whether private contracting companies can do this sort of work cheaper than the government, the J. G. White Constructing company, at the time the largest in the United States, took up the project and failed because their bid was too low. If they could have done the work as cheaply as the government is doing it they would have had a substantial profit. "At Phoenix we went in automobiles to Roosevelt, 85 miles above the capital city in the mountains. The Roosevelt dam will hold water enough in storage to cover 1,500,000 acres one foot deep. Added to the normal flow of the Salt river it will hold enough to put 2,000,000 acres in cultivation. This is the largest project that the service has under construction at the present time. Here the finest oranges in the United States are grown and they reach the markets one month earlier than the California product. An interesting industry is that of ostrich raising, over three-fourths of the American ostriches being raised near Phoenix.

"The Granite Reef dam is 40 miles below the Roosevelt enterprise and is 1-4 mile long and 12 feet high. It is for the purpose of diverting the waters of Salt river into a big canal. At Yuma Mr. McConnell and myself were authorized to begin active operations at once on the preliminary surveys of the High Line ditch. This we expect to do.

When the engineers from the various states made their reports, including Yuma, Salt river, Yakima Valley and others not one of them could compare with the record from Grand Valley. in fact the production per acre never came within one half as much as we have shown in numerous instances.

I was two days in Los Angeles and

Daily News, Feb 22, 1908

ORCHARD MESA VOTES IRRIGATION BONDS

MILESTONE IN OUR HISTORY

City Gives Warm Welcome to Visiting Capitalists---Water is Assured For 12,000 Acres of Mesa Land

---

"GREATER VALLEY" KEYNOTE OF BANQUET

Another milestone along the road of the Grand Valley's progress has been erected, the results of which will be felt for years and decades to come. A forward stride has been taken with regard to bringing a larger acreage of arable land under irrigation and outside capital has been given definite manifestation of the welcome which awaits it in this section of the western slope. These two facts working together, will be one of the best advertisements for the Grand Valley which has been placed before the world in many years.

By a vote of 45 to 18 the members of the Orchard Mesa Irrigation District yesterday affirmed the contract with the Grand Mesas Lake Canal and Power company, and voted bonds to the amount approximately of $700,000 for the irrigation of an area comprising nearly 12,000 acres of the choicest land in the entire valley. This means an early and decided increase in the value of the land and the probability of its general settlement and cultivation in the very near future.

The visiting members of the Woodland company of Chicago representing an aggregate wealth running into several millions of dollars were tendered a reception and banquet at the La Court hotel last evening as the guests of the city in general, assisted by the Orchard Mesa Irrigation District and the Chamber of Commerce. No one element or organization could claim the credit for the function inasmuch as the sentiment of welcome, hospitality and encouragement was so general and widespread that nothing short of spontaneous and unanimous expression from the citizens at large could properly comprehend the event.

Daily News, Feb 22, 1908

Eloquence, wit and straightforward business talks were the features of the evening and in the impression left upon the visitors will be found one of the valley's best investments that her citizens have yet made. The gentlemen comprising the Woodland company rank high in commercial circles in the east and where advertising matter of an ordinary nature may be discounted, their statements will carry the utmost weight along hundreds of other avenues of influence and capital. There were over fifty present.

The pretty dining room of the La Court hotel had been attractively arranged by Manager Buthorn and the banquet table was resplendent in cut glass, silverware and floral decorations. The menu was of surpassing excellence and reflected great credit on the purveyor.

Toastmaster Carnahan introduced the program of the evening by a happy reference to the "Wise Men of the East" who had shown the good judgment of coming westward looking for fields for investment. He introduced Mayor J. R. Wentworth who extended a most cordial welcome on behalf of the city, his words ringing true to the ideal of western hospitality.

In the response to the address of' welcome E. G. Eberhart, of Indiana, replied in most appreciative terms. He described the reception accorded the visitors as being most genuine and unexpected. He paid a glowing tribute to the valley, declaring it to be the most awe-inspiring of any stretch of country he had ever seen. The Chamber of Commerce came in for a well merited compliment when the speaker declared that their advertising matter as well as their personnel was full of ginger. He advised them to entertain propositions from no hot air promoters, but only substantial business men of accredited standing'. "No dreamer." he declared, "could possibly foresee the tremendous possibilities to be achieved within the next score of years." His remarks were received with generous applause.

(Continued on page four.)

Daily News, Jul 29, 1908

CATTLE CREEK MEN ACCCEPT TERMS

Elect B.T. Napier President of the Reservoir Association

HOMES FOR MANY

News Special Service.

GLENWOOD SPRINGS, July 28--
The- Cattle Creek Reservoir association was formed yesterday at Cattle Creek with B. T. Napier of this city as president. This means that every farmer in the Cattle Creek district has agreed to the provision of the federal reclamation service and that the proposed building of the- Spring Park reservoir only awaits a favorable report of the government engineers. Engineer E. E. Sands and I. W. McConnell of the reclamation service stationed at Grand Junction, have just completed an inspection of the proposed reservoir site and they are now making out their report to be presented to Secretary Garfield.

There was practically no opposition on the part of the land holders to the government's demands. The relinquishments will be signed when demanded and every effort will be made by the Cattle Creek ranchmen to facilitate the work. In case the reservoir is constructed it will put 25,000 acres of desert land under water and will provide homes for 1,500 settlers.

Daily News, Feb 28, 1908

Survey for High Line is Started

Actual work on the preliminary survey for the High Line canal was started yesterday morning when Project Engineer E. E. Sands and a corps of surveyors began running their lines near Clifton. Mr. Sands stated last night that the work will now continue without cessation and that for about two weeks a small party will be in the field and it is believed that at the expiration of that time a camp will be established in the Canon above Palisade from which point the work will be prosecuted with vigor.

Derry Henderson, with the government outfits, including tents, instruments, etc., arrived overland from Montrose early yesterday morning and went immediately into the field.

Daily News, Feb 28, 1908

ELECTRIC LINE RUNS IN 1909

Preliminary Surveys For Fruit Belt Power Canal are Finished and President Makes Optimistic Statement

----

LONG DREAMED OF ROAD NEAR REALITY

With a large corps of surveyors, Chief Engineer Allen of the Fruit Belt Power Company will today finish the preliminary surveys on the big Plateau Valley canal which is to generate electricity for the interurban system for that section. This ditch is to be six miles in length, 120 feet in width at the bottom and 160 feet at the top and will cost approximately $750,000. This will allow a drop of 62 feet to the mile and Engineer Allen says will furnish enough power for a dozen tramways.

President Richard D. Ewing of the Fruit Belt Power Company was in DeBeque yesterday to get the surveyors' estimate as to cost and also to look over some of the non-irrigated land in the valley for probable purchase. While there he made the statement that the project of his company offered greater possibilities than any he had ever seen before. "We intend to put an irrigation ditch down the Grand river to the Utah line and this will reclaim thousands and thousands of acres of arid lands along the river," he said. "At present the ditch for the power plant will be only 60 feet wide at the top and 80 at the bottom but this will produce power enough for any present use. The actual construction will begin on the ditch April 1. We have appropriated $750,000 for this purpose and the stock has been paid therefor. As soon as the canal is finished we will begin on the power plant. This will be built with ample capacity to do all the work which we have mentioned in our articles of incorporation. This will include the interurban system, telegraph, telephones, lighting, heat and anything else that may come up. No, our project will not interfere

Daily News, Feb 28, 1908

with either the High Line canal or the Orchard Mesa enterprise. We will work independently of them and probably on a larger scale."

Mr. Ewing gave the impression that the railroad should be in running order by April 1909. Already prices are going up in the Plateau Valley. As soon as the actual construction begins property values will be soaring higher still Mr. Ewing refused to state as to whether the electric line will be continued from Collbran to the mouth of the Plateau creek on to Grand Junction or not, but those who live in that section say that it is preposterous to suppose that the line would be dropped at that point when such a region as that from the mouth of the creek to this city would be left untraversed. President Ewing left on No. 6 yesterday for Colorado Springs.

Denver Republican, March21, 1908

RELIABLE INFORMATION MOST NEEDED.

With the taking over by the government of the principal canals in the Uncompahgre valley, for distributive auxiliaries to the Gunnison project, there will have been absorbed nearly three and one-half million dollars of the reclamation fund. Several million dollars and several years more will be consumed before that enterprise is finally launched. It is urged that this large sum, virtually contributed by our own citizens, might better have bean devoted to procuring state-wide information and irrigation data.

The head of a prominent financial house in Chicago recently stated that not one irrigation scheme in ten, brought to his consideration, proved acceptable, because, mainly, of a want of reliable, available information -- particularly lack of stream measurement and order of water appropriation applied. If a fraction of the money which has been expended upon reclamation work were devoted to the careful, systematic and continued gauging of the tributary streams of the state, a basis for safe and intelligent investment of capital and secure home building would, be provided, where now risk, venture and uncertainty are involved.

What is most needed in Colorado, in connection with further irrigation development, we repeat, is reliable official information. Canal construction, can be safely committed to private enterprise and initiative. The government possesses the facilities and is well organized to procure all essential data. It is to be hoped that the reclamation fund may be employed in that way and thus benefit the entire state.

Denver Republican, March 5, 1908

WHERE THE RECLAMATION SERVICE ERRS.

We have ventured some criticisms of the reclamation service in the past because of its attitude toward Colorado interests. We have particularly recited the experience of the San Luis valley in the matter of storage withheld by the arbitrary order of the government which still persists.

Now here comes complaint from the Grand Valley along practically similar lines. The Orchard Mesa irrigation district embracing 12,000 acres, is seeking to build its own system of canals. It has voted $750,000 bonds for that purpose. It asks no aid nor does it propose to invade the rights of others.

The district proposes to acquire certain early filings and franchises all but indispensable. These involve the right of way across a couple of vacant tracts of government land. Although maps and filings as required by the United States statutes have been carefully complied with and months ago forwarded to Washington, the reclamation engineer in charge of the surveys for the High Line canal the government talks of building, now announces that he "will see to it" that the said right of way shall not be granted.

It is also asserted that this employee is personally interfering where private and district enterprise have taken the initiative and where all the lands are already patented. If one-half of what is reported be true, it is full time that a halt be called to this meddlesome arrogance. The laws of Colorado and the right of its citizens to proceed with the development of its resources must be heeded, very young representatives of the reclamation service to the contrary notwithstanding. If the ambition of that department is truly subordinated to the welfare of the public, the fear which has been expressed that the reclamation service is likely to prove curse to Colorado may be averted.

Montrose Press, Feb 28, 1908

THIS CLIPPING IS SAMPLE OF KNOCKING WHICH SOME PAPERS ARE GIVING US

"The Grand Junction people are a little uncertain as to whether Uncle Sam will build that sky-scraper canal for them or not. The Montrose folks on the other hand know that their Gunnison tunnel will be built in course of time but they do not know what the water will cost them. We are guessing this little item at between sixty and seventy dollars the acre. The Cimarron could have been brought over Squaw Hill for $200,000 or an individual expense of' five dollars the acre. It requires good yeast to make good bread but some of the western slopers stand to find raw dough in their baking. The government's policy may be all right but its action is wrong."

The above random knock was a clipping from some paper, but where it was printed we had no means of determining. The clipping was mailed to The Press in a plain envelope, unaccompanied by any explanatory matter whatever. Ordinarily it can be safely assumed that any newspaper that would knock a great enterprise like the reclamation work of 'the government, is too insignificant to merit rebuke, but lest this clipping should have been the utterance of the editor of some newspaper of influence who has been honestly misled as to the subject we reprint the knock for the purpose of branding it as such, and refuting its silly statement about the cost per acre of water from the Gunnison tunnel. Such statements when circulated outside of the immediate vicinity where the facts are a matter of common knowledge would not fail to have a deleterious effect on the movement of settlers to this valley unless they are refuted. The work of the tunnel and its auxiliaries is now so far advanced that almost definite figures as to acre cost of water can be given out, but the cost can be so nearly approximated that the difference between the approximate figure and the actual cost will be so small as to be immaterial. The most careful and conservative men who are in a position to know something about the probable cost of the water place the figure at between $25 and $35 per acre. A mean figure halfway between the minimum and maximum estimates will probably be about right. The highest unofficial estimate has placed the cost at only $40 per acre, which is much lower than the sixty and seventy dollars of our visionary friend of the anonymous editorial.

But aside-from the misrepresentation of the cost of water the article is misleading in its reference to the advisability of the Cimarron water. Every resident of this vicinity knows that the Cimarron could not supply one-sixteenth of the water necessary to make arable the immense tract of land which will be watered from the Gunnison tunnel. This is not a matter of guesswork, but the result of the most accurate and scientific measurements of the water carried by the Cimarron. In fact the Cimarron ditch is now carrying every available drop of water from that river. Our unknown friend should post up a little before saying anything else about the reclamation service. The clipping sounds mightily like it came from the Field and Farm at Denver which carries a seal brown taste in its mouth all the time for anything the government is interested in or in which it takes a part. Its pernicious activity against the government is making it the laughing stock of the country.

Daily News, Feb 29 – March2, 1908

6,000 ACRES MAY SOON BE WATERED

At a meeting of the citizens of Whitewater and vicinity tonight the matter of taking steps to secure water for 6,000 acres of land in that district will be discussed and propositions offered by the Woodland Company of Chicago and the Fruit Belt and Power company considered.

At a meeting Wednesday night H. K. Devereux, of the Fruit Belt Power company made an offer to the farmers on behalf of his institution and previous to the meeting the Woodland company, which will construct the Orchard Mesa ditch, likewise made a proposition. There is a difference of opinion between the farmers as to which is the best proposition and it is for the purpose of reaching some decision in the matter that tonight's meeting will be held.

If water can be secured for the 6,000 acres of Whitewater land now arid it will be another great step forward for the Grand Valley. In order to irrigate this land it will require a big lift to get the water onto it but both of the competing companies say that they can put it through successfully.

EWING IS ENTHUSIASTIC OVER PLATEAU VALLEY ELECTRIC LINE

President of Fruit Belt Power Company, Says Details of Great Enterprise Will Be Announced Soon

----

In an interview with a News reporter last night, Richard D. Ewing, president of the Fruit Belt Power and Irrigation company, said: "While the matter of the electric line we have in view for the Plateau country is a good deal in the future, I want to say that our company expects to do a great deal of work in that section. It offers marvelous possibilities to the investor as well as the promoter and we consider ourselves fortunate in enlisting the good will of the Plateau people.

There are many reasons why some of our plans cannot be made public as yet but as soon as possible we will give you the full details of our project in the valley of the Plateau. Your article in yesterday morning's paper covered the ground pretty well about the ditch. We will get power enough from this canal to generate energy for almost anything we can build. It is an ideal project." Mr. Ewing left for Colorado Springs last evening.

Daily News, Feb 29 – March2, 1908

READY TO START DITCH SURVEY

Louis Henry, a son of T. C. Henry, arrived at noon yesterday to begin drawing plans and arranging the preliminary matters connected with the survey for the Orchard Mesa ditch, which has been watched with much interest of late. Mr. Henry will at once take up the work and expects within a short time to have the men in the field running the first lines. The work will be watched closely by the parties eager to see the commencement of active operations.

Government Acquires State Water Rights

The final steps whereby the United States reclamation service will come into possession of all the existing ditches and water rights in the Uncompahgre valley are being taken by the United States district attorney in this city. Attorney Hartsell is looking over the abstracts preliminary to the final acceptance of the properties. The transaction represents millions.

The ditches and water rights are those out of the Uncompahgre river, the stream which has meagerly watered about 50,000 acres of the valley up to the present time. The Gunnison river will soon flow into the valley through the great six-mile bore under the Mesa Verde and water 100,000 more. The plan is for the government to buy up all the Uncompahgre river rights, consolidate them with the rights out of the Gunnison and then resell the water to the farmers. Denver Times.

March 1-20, 1908

LAND GRAFT IS HINTED BY MANN

Congressman From Illinois Causes Stir in House by Veiled Charge

----

Special to The News.
WASHINGTON, Feb. 29--A veiled charge was made by Mann of Illinois in the house today during the consideration of the army appropriation bill that some government official or employee had yielded to the influence of a land syndicate with a view to selling a government tract of land near Washington, D. C., for a target range. The appropriation of $250,000 for 6,000 acres was provided for, which Mann denounced as beyond its value. He thought there had been undue temptation to somebody. He at once aroused the ire of Young of Michigan, who, springing to the defense of the committee on military affairs indignantly denounced that any member of it had been approached by anyone but the secretary of war or his subordinates. The provision was stricken out on a point of order. The army appropriation bill shortly, afterward passed the total amount carried being $84,757,566.

SANDS WORKING NEAR PALISADE

The engineers with E. E. Sands yesterday were working near Palisade and it is expected that on Monday the first field force will be put to work near Palisade, after which a camp will be established in the canon above Palisade and active work pushed forward with vigor.
It is not known just how many men Mr. Sands intends to put to work on the first field force but applications for positions have been coming in to the local headquarters at a lively rate, and it is believed that every man secured will be of experience as practically all of the applicants have had experience.

M'CONNELL IS IN THE CITY.

I. W. McConnell, supervising engineer of the reclamation service, arrived in Grand Junction from Deerfield, Kan., yesterday and immediately held a conference with Project Engineer E.E. Sands on the High Line canal concerning certain matters in connection with the ditch.
Mr. McConnell will remain here for several days and then go to Montrose where he will look over the work that is progressing on the Gunnison tunnel.
Mr. Sands came in yesterday from the field and announced that on Monday he would make a material increase in his field force so that the work could go ahead on a larger scale. The camp probably will be established in the canon above Palisade.

HIGH LINE LAND IS WITHDRAWN

WASHINGTON: D. C., March 19.--The secretary of the interior, James R. Garfield, announced today that 92,000 acres of land in the Grand Valley of Colorado where the government is building a canal, have been withdrawn from entry. The land covered in the order is as follows: Township 1 south of ranges 1 and 2, east of the Ute principal meridian; townships 10 and 11 south of range 98, west of the 6th principal meridian.

The above announcement means that practically all of the land in the right of way of the government's ditch has been set aside so that the route is thoroughly protected, prevents interference with the government project in any way and is an important step.

WATER USERS TO PUSH THE WORK

WATER USERS WILL ARRANGE TO ALLOW FARMERS TO TAKE CONTRACTS

Want to Push Work on This Project In Every Way Possible.

The board of directors of the Uncompahgre Valley Water Users' association met here Monday and elected officers for the ensuing year. The old officers were re-elected, and are as follows: W. E. Obert, president; L. E. Ross, vice-president; J. J. Tobin treasurer; Ira Monell, secretary.

The directors called a general meeting of the stockholders of the association in this city for April 13, at which time the matter of issuing credit certificates will be taken up.

These credit certificates are designed to be a means toward the end of facilitating the completion of the irrigation project in this valley. The plan has the hearty concurrence of the government officials and unless some unexpected obstacles are encountered, will undoubtedly be adopted and put into execution at the earliest possible moment.

Briefly the plan is to allow farmers who expect to use the water from the system to take contracts for the construction work on the canals, ditches, laterals, etc. Their work will be paid for in certificates of credit issued by the Water Users' association, and these certificates will be accepted in turn by the association in payment of water rights. The certificates will be transferable, and to all intents and purposes will serve as actual money in all transactions touching the purchase of water rights under the tunnel. This plan will enable many prospective water users to purchase water and pay for it in work, who possibly could not rattan the ready money.

Another good feature of this plan is that it will greatly facilitate the work of finishing the system. Lack of funds on the part of the farmers will not operate to delay matters.

Daily News, Mar 11 1908

HIGH LINE CAMP TO OPEN TODAY

Reclamation Engineers to Start Work on Investigating Route for Ditch

----

Project Engineer E. E. Sands of the High Line canal announced yesterday that the camp of the engineers would be established today just across from Cameo and that work will be started immediately to investigate the various routes to be considered by the reclamation service.

Ten men will be placed in the field today when the camp is opened and the investigation will consume about four or five weeks after which the engineering party will move to the lower end of the valley near and just north of Mack. From this point they will survey the government land and then gradually work up the valley until it is mapped in its entirety.

Mr. Sands stated that before the party leaves Cameo they expect to determine definitely the location of the right of way north of Palisade.

The work so far has progressed very satisfactorily. The local headquarters of the reclamation service has been showered with applications for work on the project and a special civil service commission consisting of Messrs. Sands, Essley and Harper has been appointed to pass on the applications and mark them according to the rules laid down by the reclamation service and civil service commission at Washington. This commission has already started the work of grading the papers.

I. W. McConnell, supervising engineer of the west, who has been in consultation with Mr. Sands for several days past, leaves this morning for Montrose where he will look over the work on the Gunnison tunnel. Mr. McConnell states that he estimates that it will take from eight months to a year to complete the Uncompahgre project. .

Daily News, March 15 1908

HIGHLINE FILERS ARE VICTIMS OF SHARKS

Organized Effort is Made to Get Persons to Contest Homestead Entries

---

MAY IMPEDE DEVELOPMENT OF GRAND VALLEY

The Daily News believes that the development of the arid land to be brought under irrigation by the government project is the most important end to be attained and anything which tends to defeat that object, should meet with the most profound contempt of every citizen who has the best interests of the valley at heart and believes in fair play.

Many people have filed on the government land which will come under the canal, making their filings in good faith ready and willing to enter upon the sites, make their homes thereon and fulfill all the governments' requirements as soon as conditions will make the same possible. Such are the real home makers and upbuilders of a community, and they should be encouraged rather than thwarted.

On the other hand it transpires that there is a clique of cheap, curbstone speculators and trouble makers who appear to be encouraging residents and nonresidents to file contests against the homesteaders with the hope of ultimately securing the benefits which should accrue to the original homesteaders.

It is understood that these land traffickers are soliciting people to become contestants for a paltry fee.

The disposition of the land traffickers is to start contests against the persons who made entries three and a half years ago when there was every reason to believe that the government would build the canal. Most of these entrymen have gone ahead and done all they could since then and have stood ready at all times to do everything required by the law as soon as conditions would permit. Many have spent much time and money on the land and a large per cent of them want the land for actual homesteads, but now that there is a prospect that water will soon be available some think that it is a good time to butt in, start contests and get some good claims without much trouble.

This movement is not only unfortunate but ill advised as there will be enough claims to satisfy all, as the unit of acreage probably will be cut to 40 acres and three times as many claims will be thrown open for entry than are now held, and these can

Daily News, March 15 1908

be secured without the expense and trouble of a contest.

It appears on the surface that there |s an organized effort to work up contests and it is said that people are being solicited to start contests by the contest solicitors. The parties who are doing this trafficking are hunting people to file contests against residents of the valley who have been here for some time, who want to stay here and who want to make their homes on the land. These traffickers, it is said, are working up the contests merely for the paltry fee that they secure therefrom, seemingly ready to breed contention in a community for a little money.

This organized effort to thwart the |bona fide entrymen may prove unfortunate to the general situation here and will be of no advantage to the contestants for, if they stir things up much, it possibly will influence the operations of the government in taking up the construction work.

The government was given to understand that everything was harmonious and that there would be no obstacles thrown in the way of federal construction of the project. The contests now being started may be held up in court for several years and when it comes time to contract for the water nobody will be in a position to do so. This condition of affairs might impede the general project.

It is probable that within a very few months the government will be in a position to establish the unit of 40 acres and the parties filing can then hold only that amount of land, and the rest will be open to entry. It is, therefore, clear that there is no justifiable ground for contests against those who have entered in good faith and who are still working in good faith.

As a clear example of the harm that such contests accomplish the case of the Redlands company can be cited. Had there been no opposition to that company there is no doubt but that at the present time there would be at least 40 or 50 houses erected on Red Mesa and actual development work well under way.

It is far better for the people of the Grand Valley to spend their money and energy in developing this beautiful empire than to use it in fighting each other.

It is probable that these land traffickers do not realize that their work may lead to such ill feeling that may impede the development of the valley.

Daily Sentinel, Mar 30-31, 1908

THE MISSION A FAILURE

It is reported this afternoon that the mission of George Smith and other Orchard Mesa Irrigation District men to Denver for the purpose of conferring with government engineers in an effort to get the Orchard Mesa section omitted from the government's plans concerning the building of the High Line canal was a failure. It, is said that representatives of the government have refused to make an effort to have the Orchard Mesa District left out of the High Line canal project.

It is known that the Grand Valley Water Users' Association and other organizations interested in the erection of the High Line canal have opposed the movement to have the government not include Orchard Mesa in its plans.

Daily Sentinel, Mar 30-31, 1908

STATES POSITION

Engineer Sands Makes Plain Government's Relation to the Orchard Mesa Project-Important Statement

THAT THE RIGHT OF WAY ACROSS GOVERNMENT LAND HAS NOT YET BEEN GRANTED FOR THE ORCHARD MESA IRRIGATION DISTRICT CANAL; THAT THE WHOLE MATTER IS AS YET UNDECIDED, AND THAT THE GOVERNMENT HAS NO DESIRE TO INTERFERE WITH THE BUILDING OF THE ORCHARD MESA BY PRIVATE CAPITAL SO LONG AS THE PLANS OF THE GOVERNMENT ARE NOT INTERFERED WITH, ARE IMPORTANT FACTS BROUGHT TO LIGHT IN A MOST INTERESTING AND TIMELY STATEMENT ISSUED TODAY BY E. E. SANDS, ENGINEER OF THE UNITED STATES RECLAMATION SERVICE, WHO IS NOW LOCATED IN GRAND JUNCTION.

The statement will be read with great interest by all directly or indirectly interested. It follows:

Department of the Interior,
United States Reclamation Service,
Grand Junction, Colo.,
March 31, 1908.

To the Grand Valley Water Users' Association and Other Interested Parties:

In view of statements made, verbally and in the press, which have been erroneous, the following statement is made regarding the position of the local officers of the reclamation service in connection with the Orchard Mesa Irrigation District project:

First--The reclamation service has no desire to antagonize irrigation development by private capital provided that the interests of the reclamation service are not endangered.

Second--The original plans of the Orchard Mesa Irrigation District would have conflicted with the plans of the reclamation service in regard to right of way.

Third--The original plans of the Orchard Mesa Irrigation District did not contemplate an economical use of the waters of the Grand river, and had those plans not been changed a shortage of water for the government High Line project might have resulted.

Fourth--Up to the present time a right of way across government land has not been granted for the Orchard Mesa Irrigation District canal.

Fifth--The Orchard Mesa people have altered their plans and believe that they can irrigate their lands without serious interference with the plans of the reclamation service. All the facts in the case, and the proposed plans of the Orchard Mesa Irrigation District, together with their statements regarding their proposed cooperation with the reclamation service, have been sent to Washington for action by the officials of the department. Consequently, at the present time the whole matter is undecided, and any statement to the contrary is made by a person not properly informed.

Sixth--Up to the present time the reclamation service has not decided to abandon any part of the Orchard Mesa, and in any event such farmers as own lands on the lower part of Orchard mesa and desire to do so will be allowed to purchase water from the proposed government canals.

(Signed)

E.E. SANDS,
Engineer U. S, R. S.

ARE TRYING TO HARMONIZE

Private Irrigation Projects and Government May be Able to Operate in Accord---Engineer Sands Outlines Government's Position

----

LOCAL CLASH IS APPARENTLY AVERTED

A long step was taken toward electing a harmonizing of the so-called conflicting interests represented by the Water Users' association, the Orchard Mesa Irrigation District and the United States government.

There has been considerable speculation as to whether there was a possibility of a clash between the government and the private capital which is seeking to install irrigation projects in the Grand Valley.

There have been those who could not see any other outcome than the blocking of the High Line project of the government by the introduction of privately constructed systems. On the other hand there have been those who believed that the government would maintain its rights and yet arrange its plans so that other enterprises might be carried on in accord in almost the same territory.

The Water Users' association held a meeting yesterday afternoon in the office of W. S. Wallace at which United States Engineer Sands of the Reclamation Service was present. The purpose of the meeting was to hear a report regarding the Orchard Mesa situation, and Messrs. Smith and Wallace had Mr. Sands make the report.

According to that report the Orchard Mesa people had agreed to surrender their water rights to the government reserving only enough for their project. It is claimed that the proposition means that of 400 feet of water the Orchard Mesa people will allow 275 feet to be turned back into the Grand river head gate, the remaining 125 feet being sufficient for the irrigation of their territory. By agreeing to a joint right of way, conceding to the government right of way for enlargement and placing the control of the canal in the government, there would be left little opportunity for misunderstanding.

A general discussion of the report was had, as well as of the matter of permitting those who wanted to withdraw their land from any of the projects under consideration. It was decided to grant permission of withdrawal provided the same could be done legally. For the purpose of ascertaining that fact, a committee composed of Messrs. Miller, Delaplain and Wallace was appointed to confer with Attorney Logan, as well as with the Orchard Mesa people and report at a meeting to be held Friday at 2 o'clock in the afternoon.

In the meantime a meeting of the Orchard Mesa Irrigation district was in session at Attorney Logan's office, where practically the same matter was under consideration.

The question of the legality of the suggested permission to withdraw was taken up with Attorney Logan at that time by the special representatives of the Water Users' association and he decided that it would not be possible to accomplish the fact. His decision was based upon the condition confronting the parties. It would be necessary to advertise the proposed withdrawals thirty days and that would bring the settlement of the matter beyond the limit provided for the bond election which will be called for this month. The consensus of opinion seemed to be that those who wanted to have their land cut out should be given the privilege but in the face of the decision of Attorney Logan the matter was dropped.

Engineer Sands gave out the following statement yesterday which materially clears the atmosphere with regard to the position of the government men relating to the irrigation projects which are closely allied to that of the government in purport and territory.

The statement in full is as follows:

Grand Junction, Colo., March 31, '08.

To the Grand Valley Water Users' Association and Other Interested Parties:

In view of the statement made, verbally and in the press, which have been erroneous, the following statement is made regarding the position of the local officers of the Reclamation Service in connection with the Orchard Mesa Irrigation District project.

First. The Reclamation Service has no desire to antagonize irrigation development by private capital; provided that the interests of the Reclamation Service are not endangered,

Second. The original plans of the Orchard Mesa Irrigation District would have conflicted with the plans of the Reclamation Service in regard to right of way.

Third. The original plans of the Orchard Mesa Irrigation District did not contemplate an economical use of the waters of the Grand river, and had those plans not been changed a shortage of water for the government High Line project might have resulted.

Fourth. Up to the present time a right of way across government land has not been granted for the Orchard Mesa Irrigation District canal.

Fifth. The Orchard Mesa people have altered their plans and believe that they can irrigate their lands without serious interference with the plans of the Reclamation Service. All the facts in the case, and the proposed plans of the Orchard Mesa Irrigation District, together with their statements regarding their proposed cooperation with the Reclamation Service, have been sent to Washington for action by officials of the department. Consequently, at the present time the whole matter is undecided, and any statement to the contrary is made by a person not properly informed.

Sixth. Up to the present time the Reclamation Service has not decided to abandon any part of the Orchard Mesa, and in any event such farmers as own lands on the lower part of Orchard Mesa and desire to do so will be allowed to purchase water from the government canals.

(Signed)
E. E. SANDS,
Engineer, U. S. R. S.

Denver Republican, Apr 8, 1908

WHY SHOULD THE RECLAMATION SERVICE STAND IN THE WAY?

We have frequently decried against the invasion upon the State control of our water, threatened by the operations of the Reclamation Service. The refusal of the Department of the Interior to accept reservoir filings on sites occupying Government land on the Rio Grande water shed in Colorado, on the ground that "it would interfere with the plans of the Reclamation Service" to develop the Elephant Butte project, near El Paso, hundreds of miles away, and thus paralyzing further irrigation development in the great San Luis Valley, sustains our warnings. If our representatives in Congress fail to obtain relief through corrective legislation, we hope the enterprising people in that valley will organize to test in the courts their rights under the Federal statutes, to the end that this incubus, blighting their welfare, may, if possible, be removed.

We took occasion recently to comment upon the action of a subordinate representative of the Reclamation Service at Grand Junction, who is strenuously engaged in defeating the independent plans of the Orchard Mesa District, to supply itself with irrigation. Now, it is reported that another subordinate appears on the scene, who, excited by strictures made, "lost his head" and vehemently proclaimed that the waters of the lower Grande River belong to the Government; that the Federal forest reserves are the main source of the water supply; that the Government, through the Reclamation branch, proposes to handle this water as it thinks best, regardless of mere local opinion or private interest.

He announces that no further irrigation development there shall proceed, except it be in line with the views and plans of the Reclamation Department. Therefore, the Government has just withdrawn "from all kinds of entry" certain public lands, across which application for right-of-way for the District's canal had been made. The avowed purpose is to prevent that District from obtaining water, except upon the terms prescribed by this Government agent. In answer to the statement, that there is far more water available than is needed for Orchard Mesa, this spokesman announces that the Reclamation Service would develop water power to be employed, not only to pump water for lands in the Valley above gravity lines, but also to be disposed of for commercial power purposes.

The position assumed is that the water of such a stream, where the Reclamation Service has initiated operations, is to be subordinated to their unrestricted plans is absolutely their property, as the timber on the public domain. The inability to procure rights-of-way for private canal enterprise over Government lands practically gives the Reclamation Service a free hand and Colorado's vaunted ownership and control of its public streams vanishes. Verily, the camel is making headway into our tent.

Denver Republican, unknown date

RECLAMATION WORK IN COLORADO.

The sixth annual report of the Reclamation Service covering the fiscal year 1906-1907 has just been issued, and that part of it which pertains to work in this state leaves little hope that the Uncompahgre valley project, otherwise known as the Gunnison tunnel enterprise, will be completed in time to be of any benefit to farmers in the Uncompahgre valley in the spring of 1909.

On Juno 30, 1907, the work had so far progressed that 19,990 feet of tunnel had been driven, leaving 10,592 feet still to be excavated at that time. The greater part of the work on the distribution system had been completed except the construction of the West canal and the East canal, concerning which the report says that "on account of the prevailing high prices for contract work" further construction of these canals "will be postponed until the season of 1908 or later."

The total expenditure on the project up to June 30, 1907, was $2,384,047.53. Of this amount $525,313.61 has been expended in building the South canal, and $1,536,023.75 had been spent directly in building the Gunnison tunnel. Of the total expenditures, $278,233.84 had been paid for "engineering and administration" and of the last named sum $63,958.10 had been spent on account of the "Washington office." Judging by what has already been expended in comparison with the work done, the completion of the tunnel and of the parts of the distribution system not yet built will add enough to the cost to raise the total expenditure to something like $3,500,000. Very possibly it will be considerably more than that amount.

In regard to the proposed Grand valley project, the report says:

> "No work has been done during the past fiscal year upon the proposed Grand valley project. During the next year it may be practicable to take up some of the preliminary surveys so that at the time funds are available for this project the preparatory work will have been brought to completion. The net expenditures to June 30, 1907 are $8,712.20."

When funds will probably be "available for this project" is not disclosed or hinted at; and we greatly fear that the people of the Grand valley will have ample opportunity to contemplate at leisure the advantages which some of them fancy will be derived from the construction of the proposed canal system by the Reclamation Service rather than by private enterprise at less cost and in shorter time. In the meanwhile, the Reclamation Service continues to block the path of private enterprise and thus checks the development of that part of the state. No doubt, when the Reclamation Service gets around to it, the "Washington office" will secure its full share of the money which the owners of land under the proposed system will have to pay.

Daily Sentinel, Apr 9 & 21, 1908

T. C. Henry and his crowd have bought editorial space in The Denver Republican. That paper is attempting to make the people of this valley believe that the construction of the high line canal by the government reclamation service, is very improbable in the next four or five years. It is attempting to create sentiment favorable to having the canal built by private capital. So palpably purchased is The Republican that its purpose is very poorly concealed. It is a very thin game of T. C. Henry. The editorial matter is being sent to Secretary Garfield to make him believe it is the spontaneous judgment of that paper and, that it is acting for the good of the whole people.

Henry Attempts Hoodoo of High Line

The Denver Republican is indulging in a systematic campaign against the construction of the High Line canal by the government reclamation service. Its editorial articles appearing every few days have the ring of being paid productions. If it was not paid editorial matter, it would not appear with such regularity nor would the columns of that publication be given over to the discussion of a problem in which it can have but very little real interest.

What is being said is inspired by T. C, Henry. Nor is it published to influence anyone in the Grand valley, because the Republican could use an indefinite amount of space without producing any very material effect upon the minds of the people, who are directly interested in the matter.

The purpose of the articles are most sinister and they are given publication, that they may be sent east to Secretary Garfield and the government reclamation officials, to influence them in the construction of the canal. It is the purpose to make them believe that the editorial expression is a reflex of public opinion upon the matter of whether the canal shall be built by the government or private enterprise.

Anyone would be foolish to believe that the Denver Republican would take such extraordinary interest in attempting to befuddle the people upon the High Line canal problem, without there was a purpose back of its endeavor.

And again why should the Republican be interested in the matter at all? It is of no material difference to them financially whether the government builds the High Line canal or it is built by private capital. The only inference then is that it has hired its editorial columns to aid those who have a purpose in defeating the will of the people--which is that the government should build the High Line canal.

T. C. Henry knows that no argument that he could make would change the purpose of the people to have the government undertake the construction of that irrigation enterprise.

He does know, and there is where he is subtle, that if he can convince the government that it is interfering with private enterprise, he can defeat the will of the people here in the Grand valley. He proposes to do this by having these articles appear in a great (?) metropolitan newspaper, which he believes will deceive Secretary Garfield into the belief that the newspaper is reflecting public opinion.

When Mr. Henry was last here he pretended that he did not propose to lay one straw in the way of the work for the High Line canal. He desired an amicable compromise. The Sentinel at the time warned the people that there could be no trust placed in Henry's word. What The Sentinel then said has been proven true, because Mr. Henry and his people have been in Washington doing all in their power to upset the work of the people for the High Line canal project.

If Mr. Henry causes the government to abandon its purpose to take up this great project, which means so much to the whole people, Henry will find things mighty uncomfortable for the prosecution of any project he may undertake in this neck of the woods.

The people don't propose to be bamboozled by Mr. Henry or anyone else for that matter, upon this question which has been so near to their hearts for so long a time. Very happily every move that Mr. Henry has made has been properly checkmated by those here at home, who have the great project under the closest scrutiny and he cannot win in his purpose. He is not now dealing either with ninnies or those who have had any less experience than he has in these matters.

The great surprise is that so reputable a journal as the Republican always has been regarded, would lend itself to so fell a purpose, wherein not less than twenty thousand people are materially interested and looking to the government to do for them, that which has long been hoped for and which is so near fruition.

Daily News, Apr 24, 1908

GREAT ORCHARDS COMPANY FORMED

Woodland Company of Chicago Plans to Develop 20,000 Acres of Land Near Palisade

---

Over $1,000,000 is to be poured into the Grand Valley for the development of the great fruit industry of this section by the Woodland Investment company of Chicago, the representatives of which recently visited Grand Junction and were tendered an elaborate banquet and reception at the La Court hotel.

This morning C. C. Magenheimer of Chicago and E. G. Eberhart of Mishawaka, Ind., accompanied by Robert Morris of Milwaukee, arrived in Grand Junction for the purpose of completing arrangements for the project.

In Denver yesterday the company which is to finance their proposition was incorporated as the Orchard Construction company with a nominal capitalization of $25,000.

According to the eastern capitalists the most of their work is to be done in the vicinity of Palisade and they say that there is unlimited money behind the new company which proposes to construct ditches, reservoirs and power plants to water 20,000 acres of land now above the present ditches.

The company, the men say, expects to spend $700,000 in the work this summer and as much more money will be put in as is necessary to plant orchards on a large scale and to develop them in the most modern manner possible.

As the standing of the Woodland company is known throughout the country and as all the men interested are very wealthy, this project is one of the biggest that has befallen the Grand Valley since the announcement that the government would construct the high line canal.

During the recent, visit of the representatives of the Woodland company to Grand Junction they were accorded a hearty and cordial reception and were so impressed with this city and the entire valley that they decided that there was better investment in irrigation projects here than in any other section of the country. Consequently they have organized a company to develop this section and exploit tile Grand Valley and are willing to spend over a million dollars which certainly shows that they have faith and confidence in this great Garden of Eden.

Mesa County Democrat, Apr 24; Daily Sentinel, Apr 27, 1908

That Henry Ditch

T, C. Henry, some of the Denver papers and the Daily News seem determined to make the people believe that T.C. Henry is greater than the U. S. government and can put his ditch project through whether or no. Mr. Henry knows and the other two can easily find out that the government has positively refused to file Henry's papers allowing him right of way over government land until the right of way for the government high line has been decided upon and the papers filed. Then and not until then can T. C, Henry file his papers and secure his right of way for his proposed ditch.

Working Against the High Line

Mr. C. C. Magenheimer, of the Woodland company of Chicago, which company proposes to construct an irrigation plant on Orchard Mesa or elsewhere, has gone to considerable trouble to make explanations regarding the plans of his company.

His protestations of non-interference with the larger project the High Line canal, are so extended that we are inclined to wonder why the necessity of such exhaustive explanation. If there is no intention of interference in any manner whatever by his company no explanation would be necessary. We might say in passing that Mr. Magenheimer protests too much not to create somewhat of doubt as to his sincerity in the matter.

The truth of the whole matter is that Mr. Magenheimer and his associates are now in contest with the reclamation service of the government over the matter of permission to run their line of ditch over the government land through the canon, which is over the survey for the High Line canal project.

Furthermore, the people concerned in the Orchard Mesa project are everlastingly hold the club of interference by the government with the rights of private citizens, over the heads of the department officers, in order to secure concessions which must more or less interfere with the great scheme which is the high hope of every citizen of Mesa county -- the construction of the High Line canal.

Mr. Magenheimer and his associates know that the club which they are using now, is the one that deprived us of the appropriation from the reclamation service of the government several years ago.

At that time T. C. Henry, A. J. McCune and others were the ones behind the scheme that deprived us of that great boon. They are still connected with the scheme which Mr. Magenheimer and his associates are now working upon.

Mr. Magenheimer has sounded public opinion on the matter and he finds that it is too risky to permit T. C. Henry to be connected with the scheme openly, therefor he has benignly permitted his name to be conspicuous in the management and has suggested that conflicting elements have been removed. Nevertheless Mr. Henry and the men whom Mesa county have abundant reason to condemn bitterly for their interference heretofor when the High Line canal project was so near fruition, are fighting in the background. If they have success with the government in securing rights in the canon, Henry and Mr.

Mesa County Democrat, Apr 24; Daily Sentinel, Apr 27, 1908

Magenheimer and his associates may forever bar the construction of the High Line canal by the government.

If Mr. Magenheimer, T C. Henry, A. J. McCune and their local associates were simply proposing to construct an irrigation project that would reclaim land on Orchard Mesa and would secure a supply of water from sources that would not affect in the slightest the proposed source of supply and right of way for the High Line canal project, they would not be interfered with for one minute. There would be no necessity for extended explanations whatever and everything would be peaceful. But they are not doing anything of the kind. If they can beat the High Line canal project they would do so in a holy minute. And we believe we know what we are talking about, when we make that assertion. It is not their purpose to be satisfied with watering 12,000 acres of land on Orchard Mesa.

We want Mr. Magenheimer and his associates to spend every dollar of money for the betterment of the irrigation measures of this valley that they can lay their hands on, we want them to do it though without interfering with the High Line canal project.

Mr. Magenheimer and his associates know that the men who are working for the High Line canal project know every move that they have made in Washington - we know it, else we would hardly make the statements we do.

If we had not had Secretary Garfield on the ground we would not stand much of a show against Mr. Magenheimer and his associates, which includes Mr. Henry in the number, in their campaign to influence the government that they must be permitted to have a right of way and sufficient water for their project from the Grand river.

However, we believe that Secretary Garfield understands one thing of paramount importance and the most convincing of all arguments to him in favor of our project for the construction of the High Line canal and that is that the people desire but one thing first, the construction of the High Line canal by the government and no representations by others can convince him otherwise that the whole people desire to see the High Line canal constructed not by private capital, but through government aid.

Nor are the Woodland people inactive. Mr. Magenheimer has been to Washington, has been in close conference with the people and came near convincing Secretary Garfield and the members of the reclamation service into believing that we had agreed to some things, which we had not in manner agreed upon

If Mr. Magenheimer and the Woodland people are here to construct and further an irrigation enterprise that will not interfere with the High Line canal project they are a thousand times welcome, if they are here in the interest of an undertaking that means the abandonment of that enterprise in favor of their own, we wish them confusion in their enterprise and they will find out that they can never bring their scheme to fruition, if their purpose is to kill ours.

The mere fact that the Denver Republican has warned our people that it may be many years before the government will undertake to build the High Line canal and that if we know when we are well off it were well for us to enter into a contract with private corporations to build that enterprise for us, is sufficient for us to understand that there is a deep and well defined plot to defeat the people's will in this matter if possible.

Fortunately we understand from whence the editorial came, being inspired by T. C. Henry and written by Bob Gauss, an editorial writer on the Republican, we could readily unravel the reason for the Republican's intense interest in a matter, which ordinarily would be a matter of so little consequence to it.

Daily News, Apr 23 1908

HENRY FILES DITCH PLAT

Makes Map for Canal to Run From Point Above Canon in Plateau Valley to Palisade

----

PLACES COST OF DITCH AT $1,000,000

County Clerk R. E. Starr yesterday received for filing a map for the Book Cliffs Canal to cost approximately $1,000,000 and which is a project of T. C. Henry.

According to the map the new canal starts above the tunnel east of Palisade in the Plateau country and runs down to Palisade. The headgate is to be located on the left or east bank of the Grand river and the land to be watered is given as 100,000 acres, for which 1,800 cubic feet of water are claimed, a portion of which is to be used for the purpose of generating power for pumping, manufacturing heating and other purposes.

The depth of the canal is given as 12 feet carrying ten feet of water. The width is 60 feet at the top and 40 feet at the bottom while the grade is shown as .17 per 1,000 feet. The total length of the ditch is 54,298 feet and with a carrying capacity of 1,800 cubic feet per second.

According to the sworn statement of Henry, the survey was made January 31, 1908 by his son George Henry. The map was filed and approved by State Engineer T. W. Jaycox on March 30, 1908.

Special to The News.

DENVER, April 22.--By the dismissal of the suit of T. C. Henry against H. A. McIntyre et al. for a one-third interest in the La Junta Land company, by Judge Bliss of the district court yesterday, one of the most amazing legal tangles in the history of Denver courts was finally disposed of,

Henry sued McIntyre, Rice Means, as administrator of the estate of George E. Ross-Lewis, the La Junta Land company and the Monmouth Investment company. At the time that Ross-Lewin, formerly with the First National bank, and McIntyre were partners. Henry interested them in a land proposition in the

Daily News, Apr 23 1908

western part of the state. Ross-Lewin died and there were several years of legal controversy in the courts. McIntyre finally bought out the claims of Ross-Lewin and assumed the Henry litigation. The case came up yesterday and McIntyre was given a decision.

Montrose Press, May 8 and Daily Sentinel, May 9, 1908

CANNOT GET THE TUNNEL MONEY

TRAVELERS' INSURANCE COMPANY WILL NOT AGREE TO DEFERRED PAYMENT.

Company Had Promised to Use Canal Payment in Making Big Loan.

Project Engineer Pease received word from Consulting Engineer McConnell a few days ago that the Travelers' Insurance company, owner of the Montrose and Delta Canal, had declined to accept the proposition of the Uncompahgre Valley Water Users' association that the payment for this canal be postponed for eight or ten months on the understanding that the water users should pay the interest until such time as the next year's appropriation for the Gunnison tunnel project becomes available.

The company assigns as its reason for refusal the fact that it has agreed to loan the entire amount on or before August first at five percent for a long time and that prevents their accommodating the people of this valley at this time, however much they would like to.

A meeting of the board of directors of the Water Users' association was called to hear the report of Mr. Pease, assembling Monday at the Court house. A full discussion of the subject was brought out by the members present and plans were under consideration for raising the extra amount needed to push the work vigorously this year. One was to call a meeting of the stockholders of the association and urge an amendment to the articles of incorporation-which would be approved by the secretary of the interior - authorizing the issuance of certificates of indebtedness that would relieve the present financial stringency.

Whether the association would agree to this, however, the board is in doubt, still the question was referred to Millard Fairlamb, one of the directors of the board, to prepare an outline of the steps necessary to take in order to bring this plan before a meeting.

With the approval now of the abstracts for the Montrose and Delta canal and Loutesenhizer canal, now in the hands of the interior department, the sum of $140,000 will have to be taken from this year's allotment of $460,000, leaving a balance to carry the work thru the year of only $320,000

It is said that from several other projects small sums have been borrowed—that will not be used during the year by them—aggregating $75,000, and this sum will be transferred to our project in order to expedite operations. This will be mighty good news and do much toward furnishing the means necessary to increase the force on the tunnel. Already one additional man has been placed on the air drills on each shift in both headings.

Montrose Press, May 8 and Daily Sentinel, May 9, 1908

GARFIELD MAY COME

James Rudolph Garfield, secretary of the interior, may visit Grand Junction again this summer.

It is announced from Washington that the distinguished official will visit a number of the irrigation projects in the west and northwest during the summer and it is believed that the Grand Valley High Line canal project, the Gunnison tunnel and other projects in Colorado will be among those projects to be visited by Secretary Garfield.

Already the Montrose people are busy with plans to get Secretary Garfield to inspect the Gunnison tunnel, and the Grand Junction Chamber of Commerce will make every effort to bring him to this city for a longer stay than he made last summer, when he was in the city only about an hour while passing across the state.

Daily Sentinel, May 9, 1908

THE PLOT IS FOILED

Efforts Made to Knock Out Filing on Water for the High Line Canal —Who Was Back of it?---A Shady Movement

IT LOOKS VERY MUCH, INDEED AS THOUGH THE STATE OFFICES HAVE BEEN BREACHED IN AN EFFORT TO THWART THE WILL OF THE PEOPLE ON THE CONSTRUCTION OF THE HIGH LINE CANAL BY SOMEONE WHO SEEMS TO HAVE INFLUENCE SUFFICIENT TO HOLD BACK IMPORTANT FILINGS SENT FROM THIS COUNTY IN THE INTEREST OF THOSE WHO ARE WORKING FOR THE HIGH LINE CANAL PROJECT.

Just take note of the following and draw some conclusions for yourself:

On the 23rd of April, the Water Users' Association of this county sent filing maps of the water to be used for the High Line canal, the water to be taken from the Grand river, to the state engineer. According to a return postal received for the registered package, the maps were received by the state engineer on April 25. The filing was on 1,300 cubic feet of water per second.

On the evening of the 25th of April there was sent from Palisade by the Orchard Mesa Construction Company maps for filing on water for power for that company. That package of maps sent by the Orchard Mesa Company was received by the state engineer two days after the maps sent for filing by the Water Users' Association, or on the 27th of April.

Notwithstanding telegrams were sent to the state engineer, also letters from the Water Users' Association, asking why he did not re-

Daily Sentinel, May 9, 1908

turn the maps so that the filing could be completed by filing them with County Clerk Starr, no answer was received until a letter came to W. S. Wallace from Governor Buchtel, to whom a telegram had been directed stating that the maps had not been returned because the state engineer questioned the authority of the Water Users' Association to file. That was, of course, all poppycock, as the state engineer has no authority to decide the qualifications of anyone who desires to file on water in the streams of this state.

Three days ago the maps for the filing of the Orchard Construction Company that had been received by the state engineer two days after the maps sent by the Water Users' Association had been received, were returned to this city to have the filing completed; but as no fees accompanied the maps County Clerk Starr did not make the filing, as he had no right to do.

Someone, whether the state engineer or other parties concerned, feeling that it was perfectly safe to return the maps of the Water Users' Association, forwarded them from Denver last night and they came to hand today at noon.

W. S. Wallace, acting for the Water Users' Association, took them at once to County Clerk Starr, paid the fee, and the scheme that was hatched by someone in Denver to defeat the filing of the Water Users' Association and give the advantage to the Orchard Construction Company were foiled in their pretty little game, because of a very important oversight which they had made of not remitting the money for the filing of the maps. Therefore, the maps of the Water Users' Association ante~date those of the Orchard Company, and their water is assured.

Now, who was back of this game? It looks very much as if it was an attempt to beat the government through the Water Users' Association of this city, as the maps had to be placed on file here in the county to complete the filing.

Furthermore, the state engineer, who took it upon himself to attempt to decide the qualifications of the Water Users' Association of this city, violated the law in sending back for filing the Orchard Construction Company's maps. One of the provisions of the law is that in filing maps for water with the state and also with the county, those maps must show the ownership of every tract of land crossed. The map of the Orchard Construction Company did not show this, and if any map should have been held up by the state engineer because of fault, it was those maps. The maps of the Water Users' Association showed the ownership of every tract of land and were without fault in the least.

There is a big nigger in the wood pile this time. Someone very familiar with the workings of the state engineer's office got in his fine work, but it was too fine.

Does anyone feel that Mr. Henry and his party had not something to do with this matter and that it is all being done to frustrate the work of the High Line canal project?

Daily Sentinel, May 9, 1908

THE PLAT FILED

Water Users' Association Put Formal Description of High Line Canal on Record – Another Plat Also Filed

The map and specifications of the Grand Valley High Line canal were filed today by the Grand Valley Water Users' Association with County Clerk Starr.

The map, plat, etc., have been approved by State engineer Jaycox.

The map is accompanied by a statement from Engineer Sands to the effect that the survey has been made.

The specifications state that the headgate is to be located on the left bank of the Grand river at a point in the upper end of the valley. Sixty five thousand acres of land are to be watered. The canal is to be 10 feet deep, 72 feet wide on top and 32 feet on the bottom, and will have a fall of .014 foot per 1,000 feet.

The length of the ditch is to be 34,655.5 feet, The carrying capacity will be 1,300 cubic feet of water per second.

The estimated cost is $850,000.

The map of the canal is accurate in every detail.

The water filing has also been recorded.

Also this afternoon the Orchard Construction Company filed a plat of the Narrows Power canal, with headgate on the Grand river near Palisade.

15,000 acres on Orchard mesa are to be watered by this canal, and the estimated cost is $210,000.

The canal will be 9,285 feet in length and the survey work has been done.

C .A. Gibbs of Delta is in town today.

Today T. H. Boyd purchased from William Campbell lot 8 in block 124, Grand Junction, for $1,000.

The formal transfer of the H.G. Chissey orchard near Palisade to F.E. Swisher of Kansas (6 acres for $24,000, announced some week ago) was made today.

Daily News, May 9, 1908

GREAT DITCH IS TO BE BUILT

Orchard Company Sends Plat For $210,000 Canal

PART OF BIG SCHEME

A plat for the Narrows Power Canal, a project of the recently incorporated Orchard Construction company which is financed by the Woodland company of Chicago, was sent to County Clerk Starr yesterday to be placed on record.

The proposed canal is to irrigate 15,000 acres of land in the vicinity of Palisade and is to cost approximately $210,000. The headgate will be located on the left bank of the Grand river in the canon above Palisade where a big dam is to be built. The water will then be carried down to a point below Palisade. A power house is to be constructed and a plant established and power is to be utilized for pumping, lighting, heating, manufacturing, etc. The depth of the canal will be eight feet, the width 172 feet on top and 150 feet on the bottom. 'The grade will be 0.13 per 1,000 feet and the total length of the ditch 9.285 feet. The carrying capacity is fixed at 3,425 cubic feet per second.

According to the plat the survey for the project was made April 21. C.C. Magenheimer signs the plat, as president of the Orchard Construction company.

Daily Sentinel, May 12, 1908

WON'T PERMIT HIM TO INTERFERE NOW

Frank H. Newell, director of the Reclamation Service, has taken steps to prevent any possible interference with the High Line canal of this valley, by the building of a project which J. H. Allison, well known here, has under contemplation.

The following dispatch sent out from Denver last night will be read with much local interest as the great project so near to the hearts of the people of this city and valley, is concerned:

Denver, May ll. There promises to be somewhat of a clash between the government and local ditch builders in eastern and western Colorado over the position that has always been maintained by the government that it controlled the waters of the streams upon which it is constructing irrigation schemes. This time the clash will come between a project headed by J. H. Allison, who has a scheme to water several thousand acres of land near Pueblo. and the government high line ditch proposition near Grand Junction.

About a year and a half ago, Allison filed on the headwaters of Eagle river on the western slope. Eagle river being tributary to the Grand river, his intention is to divert 300 cubic feet of water per second from the Eagle river to the Arkansas river on the eastern slope. He will construct a ditch across Tennessee pass, driving a tunnel through the mountain a portion of the way.

He will then bring the water down to a point a little below Pueblo where he has a reservoir site from which he will irrigate a large tract of land.

He made his filings many months before the government started in the high line ditch on the western slope.

In building his ditch from the Eagle river to the Arkansas he had to cross government land. He made the correct application to the government for the right of way and it was referred to F. H. Newell of the reclamation service.

Newell has written for the records of the flow of the Eagle river and wants to know if this diversion will interfere with the reclamation project in Grand Junction, This indicates that the government intends to investigate and if this would interfere with the project on the western slope to refuse him the right of way across government land.

Allison is already to start his project. The government cannot build its Grand Junction project until after it has money from the Gunnison tunnel project sufficient to counteract the former. In the meantime Allison will have to wait until the government ascertains whether or not his project will interfere with it.

Daily Sentinel, May 12, 1908

Caught in the Act

That was a very close shave that the Water Users' Association had a few days ago on their water filing. If it had not been for the little oversight of neglecting to send the clerk's fees, in all probability Henry and his crowd would have given that slight signal of contempt to the Water Users' association, by placing their fingers at the end of their noses saying to them "What are you going to do about it, with our filing on the water?" Some one should have gone to Denver to find out just why the state engineer was so reluctant to send the filing of the Water Users' association back to this city. so that the filing could be completed, but took such great care that the filing of the Orchard Construction company should get back in advance. We confidently believe that if Governor Buchtel could be informed of the true reason for the withholding of that filing that he would make it extremely warm for the state engineer, or whoever was responsible for the manner in which the Water Users' association filing was juggled.

The Orchard Construction company did not attempt to make any filing on the waters of the Grand river for power purposes, until they found out that the Water Users' association had made one, when theirs was rushed off to Denver as fast as possible, They thought by sharp tactics that they could complete theirs before the one made by the Water Users' association.

Had their filing been completed before that of the Water Users' association, they could have used that bugaboo to the government, of interference with individual or corporate rights,

Under the present circumstance they had a run for their money and even with a pull with the state engineer's office of some kind, by which they succeeded in having the filing of the association held up on the pretense that the Water Users' association were not responsible filing agents, they were beaten and should take their medicine and not "faunch" round as they have been doing.

No blame can attach to Governor Buchtel for being made a tool of by Henry and some complaisant official who attempted to befuddle him because he relied on the integrity of his appointive officers when he sent the answer to Mr. Wallace of the reason for the delay.

It ought not be hard for those who have believed that Mr. Henry and his associates have been pretty harshly treated, to understand matters now, if they just examine the little game that was attempted to foil the High Line canal project.

The Sentinel is first and last for the High Line canal, if Mr. Henry and the Orchard Construction company were honest in their purpose of non-interference with the High Line canal project, why should they adopt the means they have to foil every move of the citizens of Mesa county who are working to forward that project.

Then it is all poppycock for the Orchard Construction people to say that Henry has no connection with their schemes or that there is any attempt on the part of those concerned to foil the work for the High Line canal. Every move that has been made for that sole purpose and it don't fool The Sentinel, if it does fool others.

Daily Sentinel, May 12 & 13, 1908

FIGHT OVER HIGH LINE

State and Government Officials Make Serious Charges

DENVER, May 12,--The projective government High Line Canal in the Grand Valley has caused a serious quarrel between the state and government officials and charges are made that the government is giving private users the worst of the bargain. The private users say the government out of a selfish motive has made filings for the high line canal although it knows that nothing can be done for years. The government says the state engineer has held up government filings so as to benefit private enterprises. A government agent a few days ago wired Governor Buchtel asking why the state was postponing action on the filings of the government for so long a time. Assistant State engineer Beach said yesterday that the delay was caused by the inability of the attorney general to look over the filings at once because he was too busy.

State Engineer Jaycox has received a letter from Ed Allison of Grand Junction in which he says he arranged to build a dam at the headwaters of the Eagle river to divert 300 cubic feet of water over Tennessee pass into the Arkansas river. He says he applied for a right of way over government land but this was refused by Director Newell of the reclamation service. Allison declares that the government cannot use the water for years. Assistant State engineer Beach said today: "The government can do nothing in the Grand Junction project until the Gunnison tunnel has been completed."

The high line people are highly indignant over the trend of affairs and declare it is merely a scheme to block government construction of their long cherished project.

Daily Sentinel, May 12 & 13, 1908

GOVERNMENT IN ACCUSATION OF STATE ENGINEER

ALLEGES IN FILINGS THAT INDIVIDUALS ARE BETTER TREATED

GRAND VALLEY CANAL SAID TO HAVE BEEN HELD UP AND COMPETITOR PASSED.

Private Water Users in turn accuse Former With Like Dealings.

Filings for the government High Line canal in the Grand valley have started charges that the state engineer's office has been "double-crossing" tile government and counter charges that the government is "double-crossing" certain private water users on the western slope, It is claimed by the latter that simply to prevent them from diverting water across government land, the government from a selfish motive has made filings on the canal, although it is aware that it can do no work for several years. The government agents accuse the state engineer's office of supporting the water users by holding up the filings made on the project as long as possible, and acting immediately on another and rival scheme proposed by a private company opposing the government.

It is alleged that the government's filings were made about three weeks ago. The early part of last week the private company is said to have entered filings on its proposition, and two or three days afterward the state engineer acted on it The government's agents, hearing of this, are said to have wired Gov. Buchtel asking him why the state was discriminating against the government and postponing decision on its filings for so long.

According to Deputy State Engineer Beach, the deferring of action on the government's filing was due to the fact that it was necessary first to have the attorney general look over them and that official was too busy to do so, State Engineer Jaycox has received a number of complaints, he says, from private water users who are resenting the government's attitude, greedily keeping all the water to itself.

F.H. Allison, one of the most prominent water users in the valley, claims in a letter to Mr. Jaycox to have built a dam on the headwaters of the Eagle river for the purpose of diverting almost 350 cubic feet of water over the Tennessee pass to the Arkansas river. He applied for right of way over the government land and his application was turned down by Director Newell of the reclamation service. Not for years, Allison claims, will the government make use of the water. According to Mr. Beach. the government can do nothing on the High Line canal until the Gunnison project is completed and it can get money from water users in that section to appropriate toward its plans in the Grand valley.

Denver News, May 12, 1908

PRIVATE DITCH WILL INTERFERE WITH U. S. WORK?

F.H. Allison of Las Animas Faces Wait of Years in Completing Project,

----

The old question of the respective rights of the government reclamation service and a private irrigation enterprise in Colorado is involved in a matter laid before State Engineer Jaycox yesterday.

One and one half years ago F. H. Allison of Las Animas filed on the headwaters of the Eagle river, which is tributary to the Grand river. The filing contemplated the boring o£ a tunnel through Tennessee pass to being water down the Arkansas valley to a few miles from Pueblo.

The survey of the proposed ditch is in part over government land. Allison made application to the government for a right of way, and the application was referred to F. H. Newell, director of the reclamation service. Newell indicated that the building of the Allison ditch would interfere with the government irrigation project near Grand Junction, the High Line canal.

According to the terms under which the government will prosecute the Grand Junction enterprise, it cannot be begun until the Gunnison tunnel is completed and enough land under it has been sold to warrant beginning work on the Grand river. This, it is estimated, will be from six to ten years.

Daily Sentinel, May 5 1908

Some people have an idea that the government, through its reclamation service, is very desirous of building the Grand valley canal project, over a half dozen other projects equally as important. Mr. Henry is doing overtime, in Denver in attempting to make the state officials believe that the government is abnormally anxious to build the High Line canal. There is no advantage to the government in any way or to any official of the reclamation service in the building of the High Line canal.

From the very frantic appeals of the Denver Republican, that paper is to have a big slice of something, if they succeed in inducing the government to withdraw from its purpose of constructing the High Line canal. The attitude of the state officials is also to the effect that people from the outside are to be protected to the exclusion of our home people.

If the High Line canal is built by the government the land pays the cost of construction without interest. If a private corporation should undertake the building of the canal the land must pay a substantial profit together with interest on the money from the day the work begins.

Is there any reason why the people of this valley have not as much right and more, to file on the waters of the Grand river and protect their own water supply, as some one individual in Chicago or Denver or some other place?

The filings on the waters of the Grand river are not made by the the government for its use, but by the citizens of the valley for the benefit of the lands in the valley.

Daily News, May 17 1908

MILLION DOLLARS TO BE SPENT HERE

Government Grants Right of Way to Orchard Mesa Project

SURVEYORS OUT

WHAT IT MEANS.

Spending of $1,000,000 in Grand Junction.

Circulating of money in local business channels.

Employment of hundreds of men,

Building of new homes.

Buying and selling of land.

Settlement of contention between private projectors and high line canal.

A general boost for the Grand Valley.

That the alleged quarrel between the government and state officials over the filing of water right maps for private and federal projects has been brought to an end, was indicated yesterday when George Smith received a telegram from C. C. Magenheimer announcing that Secretary of the Interior James R. Garfield had signed a contract granting a right of way to the Orchard Mesa company for its project.

It was the filing of the Orchard Mesa company's plat with the state engineer and the fact that it was returned to the county clerk here before one of the Grand Valley War Users' association, that caused contention. A charge was made that the state officials had given the map of the private company preference over that for the high line canal.

It was understood yesterday, however, that the government does not desire in any way to interfere with the private project and Secretary Garfield having satisfied himself that the Orchard Mesa enterprise will in no way conflict with the high line canal, readily granted the right of way. The government, it is said, stood ready to include the Orchard Mesa district in its project if necessary but as long as the private corporation will positively build it the latter will be allowed to go ahead with its work unmolested and the government will continue its high line canal along the route already laid out.

The granting of the right of way is conceded to mean, much for Grand Junction and vicinity as it assures the construction of the Orchard Mesa project which will irrigate thousands of acres of land. Surveyors are already in the field and the route is now dotted with the flags of the engineering parties. M. C. Hinderlider is in charge of the party of 18 men.

Orchardists who have suffered losses as a result of the freezes will have an opportunity to secure employment under the new company. The money to be spent on the project will be put in circulation in local business channels and the entire proposition will be of great benefit to this city.

It will mean the bringing in of new settlers who will take up land under the irrigated sections. It means the building of new homes and, in fact, the building up of that entire district which is tributary to Grand Junction and the development of which is of as much importance to this city as the district itself.

PAGE PHOTO: *Daily News, Great Things are Assured*

Daily News, May 22 1908

Great Things Are Assured

Street Car Line and Early Construction of High Line Canal are Promised at Big Smoker

MEETING SHOWS CITY'S ADVANCEMENT

The Meeting Summarized

Announcement that work on an electric street car line in Grand Junction will start in October.

Announcement that high line canal construction work will probably be taken up immediately after the engineering work has been completed.

Announcement that Secretary of the Interior, James A. Garfield will visit Grand Junction in August and personally go over the project.

Resolutions adopted denouncing The Denver Republican for publishing articles detrimental to the high line canal and reflecting on the reclamation officials

"Progress." That was the keynote of the big smoker of the Chamber of Commerce held in Margery hall last night at which over 200 of the members of that valuable organization were present and entered into the discussions in an enthusiastic manner.

Perhaps no session of the kind has ever been held in Grand Junction at which so many concrete plans for the advancement and betterment of Grand Junction have been announced than last night's gathering.

I. N. Bunting presided as Chairman and made a few remarks after which J.H. Rankin submitted the report of the city and county committee for the year showing what had been accomplished and what was still under way. A discussion followed during which paved streets and good roads were dealt with. The ladies quartet composed of Mesdames Greer, O'Hare and Hubbard and Miss Florence Brooks then rendered a beautiful selection which was heartily received.

A bass solo by E. A. Sherman delighted the audience as did also readings by Leon J. Chapman. Following this Secretary Aupperle submitted his report on the high line canal.

Mr. Aupperle said:

"There is no necessity for bringing up any ancient history concerning the high line project. A year ago this month conditions became such as to make it possible for us to go after the ditch in a practical way. The heads of the reclama-

Daily News, May 22 1908

Garfield Coming

"I believe the plans of the reclamation service should be developed with care and forethought. The question of government interference is always a matter for wise thought, and the interests of |he people must not be sacrificed for the prospective profit of some promoter. I will pass through Grand Junction in August and hope to be able to go over the ground more thoroughly with the people of the Grand Valley. I have a great personal interest in the work there and will do all in my power to complete it."----Secretary of the Interior James R. Garfield.

Sands Determined.

"Whenever any proposition comes up that I think will interfere in any way with the high line canal I am going to report against it to Washington; I am going to notify the Water Users' association and all the rotten newspapers in Denver and all the political patronage in the state of Colorado won't stop me."

E. E. Sands, project engineer.

tion service were found to be more interested in our project than we had imagined. Secretary Garfield was here in person and got a birdseye view. He was impressed with the feasibility and desirability of the canal. After his departure we laid plans to take the matter up direct at Washington. On the morning that I planned to go I received a letter from Secretary Garfield that was discouraging, to say the least. However, we had our hearts set on this matter and I laid the letter away and left for the capital. The reclamation officials were ready to receive me and we reached an agreement as to the proposition they would make to the people of the Grand Valley." At this juncture Mr. Aupperle read the proposition which was submitted to the masses meeting in the auditorium here about a year ago. He laid particular stress on the last clause in the proposal which said that construction work was to start when money became available from the repayments on the Uncompahgre project or other funds.

"This clause has misled some and has been taken advantage of by others said Aupperle. Director Newell prepared the proposition and Garfield signed it. In discussing it with Mr. Newell he said: "We will build your project as soon as we can. I hope this will be soon but I am not going to put the department in such a position that your people could say that we have encouraged them by putting engineers in the field and should go ahead with construction work. Mr. Newell said that if the government's proposition was accepted or rejected it must be done on the face of it."

Newell Optimistic.

Mr. Aupperle then declared the reference to the repayment on the Uncompahgre project was simply a safeguard for the position the officials had taken. "Director Newell." said Mr. Aupperle, "never had any idea that the high line canal would be built from any funds received from the Uncompahgre project but he had to put that clause in the proposition to protect himself. You

Daily News, May 22 1908

D.W. Aupperle

E.E. Sands

must notice the last words 'and other funds.' After I got this proposition in writing I called his attention to the fact that the department had not committed itself as to when construction work would start.

I will use Newell's own words, which were: "Personally I am optimistic and believe we can make arrangements to follow the engineering work with construction but will make no promise."

"Secretary Garfield, after signing the document, said: "We are not in a position to commit ourselves. There are so many people who are clamoring for work all over the country that we dare not make any definite promise or we should have them all on our neck""

These remarks were received with great applause and when this had subsided Mr. Aupperle continued: "There has been much said about the position of the government in this matter. Some have claimed that it would be 5 or 10 years before work would be started because there would be no money coming in from the Gunnison tunnel before that time. I want to say right here that I think construction work will be taken up in a very much shorter time than five years."

"When in Washington Mr. Newell told me of about 25 projects then under way which by the end of this year will be on a self-supporting basis and I feel sure that Mr. Garfield and Mr. Newell will find ways and means to take our project in the very near future."

Cooperative Plan.

"We also are offering to help them. A plan which is under consideration, but which has not yet been fully worked out is a cooperative scheme. This plan is considered feasible and where the people are anxious for construction or extension and no funds are available, they are permitted to do work or give money to the water users' associations to have the work done. In return for this, certificates are issued redeemable in payment for water. The Grand Junction Chamber of

Daily News, May 22 1908

Commerce has said that if the plan is worked out we are ready to do our part and give all the assistance in our power. I am sure that the landholders under the canal are not only willing, but anxious and able to do so."

"Another matter in connection with this is the contention that has arisen between the government project and private enterprises. Some criticism has been made of the Chamber of Commerce and the Water Users' association for their action in the matter."

"The position the Chamber of Commerce is simply that the government must not be interfered with in the construction of the high line canal. After three years of' persistent and hard work and every legitimate representation having made that the way was open we got the government to take up the project. We must protect it. When the capitalists from the east were here a few months ago and the Chamber of Commerce was asked to assist in entertaining them President Charles E. Cherrington, in his address, stated the chamber's position and it has been maintained ever since. It was that we believed that the two projects could be worked out without conflicting with each other and we hoped that both could be completed at an early date."

"There has been some unpleasant contention over this but we take the stand that we will protect the government ditch if at the expense of the Orchard Mesa ditch, but this has not been necessary. We are not opposed to the private project and we told Secretary Garfield so in a letter. We told him that we need men and money and invite them to come here but we can't stand for them interfering with plans that are greater than their own.

Cancellation Clause.

"I understand the contract has been signed by Secretary Garfield granting the Orchard Mesa ditch a right of way to carry its water along the same route as the high line canal. If so there will be no conflict between the two projects. I am sure that that contract contains a clause that any attempt to interfere in any way with the government canal will cancel the license."

Mr. Aupperle's remarks, giving further and more encouraging information were generously applauded and E. E. Sands, project engineer was called upon and said:

"I believe that the Orchard Mesa and government fight incident is now ancient history. About the cooperative scheme I wish to say that this was talked over at the conference of project engineers held at Yuma, Ariz., with Chief Engineer Davis. And, by the way, whatever Davis says, goes. He advises Newell and you can bank on his word. I told Mr. Davis that the people of the Grand Valley would take hold of the cooperative proposition quicker than those in any other section of the United States. I also told him that the people here were more anxious for their ditch than the people anywhere else on any project in the country. "The scheme originally was to have work done and not receive cash but the water users' associations now are in a position to employ labor when money is paid them. In many cases where the settlers are on the ground they use their teams for the work and receive credit certificates redeemable in the payment for their water.

Daily News, May 22 1908

The World – When the old man gets started in that game I can go hang for all he cares until he has finished

“In reply, Mr. Davis said: “Those projects showing the greatest interest and desire to help themselves are going to get the biggest and quickest allotments in Washington"

"The subsidized sheets of Denver," continued Mr. Sands, "are getting information from clerks in the state engineer's office and spinning yarns attacking the reclamation officials. We have never gone by such clerks and we are not go-

Daily News, May 22 1908

ing to now. The knockers on the outside are not the people who are to benefit by this high line canal. I am not going to conduct a campaign against anybody who is opposing us, either."

Square Dealing.

"Our dealings with the Orchard Mesa Irrigation District have been mostly with George Smith and I want to say that I found him always square. He has not talked behind my back and when he had anything to say he came to me and said it. Magenheimer and others interested in that enterprise are the same way but there is somebody under the surface, apparently not identified with the Orchard Mesa people in any great degree, who is inspiring dirty articles in Denver papers. My first stand was that if the Orchard Mesa people could get the money and construct their project without interfering with the high line canal, it was heartily approved."

"When Mr. McConnell, myself and others went to Denver for a conference with the Orchard Mesa promoters Mr. Magenheimer made a proposition that he would divide his water with us as we had the right of way and he did not. When this was agreed we dropped our fight and have not fought the Orchard Mesa project since."

"I am satisfied that the contract granting them a right of way and signed by Secretary Garfield, states that the water is to be divided."

"Whenever any proposition, however, comes up that I think will conflict with the high line canal I am going to report against it to Washington. I am going to notify the Water Users' Association and all the rotten newspapers in Denver and political patronage of the state won't stop me.

"As long as we have the backing of the people of this community we are going to stay here and we are going to do things. All we ask is that you people support us in our efforts to prevent the diversion of any of the headwaters of the Grand river away from the western slope."

Mr. Sands was given a tremendous ovation when he sat down and Mr. Aupperle immediately arose and read another letter received from Secretary Garfield in which the latter said" "I believe the plans of the reclamation service should be developed with care and forethought. The question of government interference is always a matter for wise thought and the interests of the people must not be sacrificed for the prospective profit of some promoter. I will pass through Grand Junction in August and hope to be able to go over the ground more thoroughly with your people. I have a great personal interest in the work here and will do all in my power to complete it."

When Mr. Aupperle had concluded I.M. Slocomb read a set of resolutions, which he moved be adopted.

Republican Rapped

The resolutions as read by Mr. Slocomb were adopted by a unanimous vote of those present after an addition which was suggested by Attorney J. S. Carnahan, on the call for the question,

Mr. Carnahan said that he believed that the question of building the high line ditch should be left with the people of the Grand Valley and not for the Denver newspapers to determine. "We are capable of attending to our own affairs and do not need the dictation of the papers of that city," he said.

Copies of the resolutions were ordered mailed to each of the Denver newspapers. The resolutions are as follows:

Daily News, May 22 1908

"WHEREAS, The people of the Grand Valley for a period of three years or more, earnestly and persistently petitioned the Reclamation Service of the United States to accept the Grand Valley irrigation project and

"WHEREAS, The Secretary of the Interior, Hon James R. Garfield, and Director of Reclamation Service, F. H. Newell, at the request of the petitioners accepted the Grand Valley project and are at this time having the survey made by government engineers preparatory to taking up the construction work at the earliest possible date, and

"WHEREAS, Certain publications have made unjust and misleading statements about the Reclamation Service,

"THEREFORE, BE IT RESOLVED, by this meeting of the general membership of the Grand. Junction Chamber of Commerce held Thursday evening, May 21, 1908, that we deprecate these unjust and misleading statements made by these publications concerning the officials of the Reclamation Service and be it further,

"RESOLVED, by the Chamber of Commerce of Grand Junction, that it is the sense of this general meeting assembled that it is the right of our own valley and people, and. that this right is not delegated to, nor possessed by the Denver Republican or any outside newspaper hostile to our plan of government reclamation; and that the people of the Grand Valley appreciate the actions taken and the work done by the Reclamation project and pledge again their hearty support at all times and in every way possible, that assistance may be rendered in furthering the interests of this project, and

"BE IT FURTHER RESOLVED, that it affords us great pleasure to learn of Secretary Garfield's contemplated visit to Grand Junction in August and that we request that he give us as much of his valuable time as may be, in order that he may get acquainted with the conditions and our people.

"AND BE IT FURTHER RESOLVED, that a copy of these resolutions be furnished project engineer Sands to be forwarded to Secretary Garfield and Director, Newell"

Street Car Line.

Chairman Bunting called upon Mr. J. Noble who has been in the city for several weeks in the interest of the proposed electric street car line which will extend through the valley from Palisade to Fruita. Mr. Noble said that he was not a prophet but he would prophesy that an electric car line was practically assured for the Grand Valley and Grand Junction. Upon his arrival in the city to secure the signatures of property holders along the proposed route, for which a franchise is to be asked, he at first consulted the city council and having gained their consent framed the petition which has been so generously signed by the various property owners. He was pleased to state that the required number of signatures demanded by the statute had been obtained, and that at the first meeting of the city council in June a franchise will be applied for. The proposed car line will be adjunct to the power plant of the Fruit Belt company which is contemplating the erection of a large plant near DeBeque. Upon the arrival of the parties interested in the Fruit Belt company Mr. Noble said that the proposition of building an electric line had not entered their mind. Upon viewing the situation, however, they were convinced that a splendid opportunity presented itself, and they at once took the matter up for consideration with the result that a line is assured.

Daily News, May 22 1908

The franchise which will be asked for is to provide for the beginning of construction work by October, next and with limitations for the completion of certain portions of the line.

Mr. Noble was vociferously applauded.

Attorney Benjamin Griffith, formerly of Montrose, made a few brief remarks upon the impressions of the newcomer in the valley. Halsey Rhoades who enjoys the distinction of being the oldest newspaperman in the state of Colorado having for 42 years been identified with newspaper work in this state, was called upon for a few remarks and responded in a pleasant way.

The meeting then adjourned to the banquet room where a splendid lunch was served by John Armstrong.

Daily Sentinel, May 22 1908

ELECTRIC ROAD SEEMS CERTAINTY

Important Announcements at Chamber of Commerce Smoker---Resolutions Denouncing Denver Republican Passed-High Line Canal Discussed

----

There were about 125 representative citizens present last night, at the smoker held under the auspices of the Chamber of Commerce at Margery hall. The smoker was a very pleasant and a very enthusiastic affair and its influences for the good of the city and valley will long be felt.

Mr. I. N. Bunting was the chairman of the smoker and opened the meeting with a brief statement as to its purposes. He introduced the various speakers and musical numbers on the program in an appropriate manner.

J. H. Rankin was the first speaker. He made a report for the city and county committee of the Chamber of Commerce showing what had been accomplished during the past year. The report was well received and some discussion followed. Paved streets and good roads were the subjects of the greater portion of the discussion.

The ladies' quartet, composed of Mesdames O'Hare, Welles, Greer and Miss Brooks, delighted the crowd with two pretty numbers. E. A. Sherman gave a pleasing bass solo and Leon J. Chapman entertained with a refreshing reading.

Secretary D. W. Aupperle of the Chamber of Commerce then addressed the smoker at considerable length on the high line canal project, the situation as it stands today, the outlook and the work being accomplished. Mr. Aupperle gave considerable new and hitherto "inside" information concerning the project, and it was made pretty plain that there is a strong probability that the actual construction work on the canal will begin immediately after the engineering work has been completed.

Secretary Aupperle announced that he had been informed by Secretary James R. Garfield of the interior department that he would visit Grand Junction in August and make a personal and thorough inspection of the project. This announcement, as well as all other statements made by Secretary Aupperle were most enthusiastically received.

Project Engineer E. E. Sands was then introduced, and he had something of interest to say regarding the High, Line canal, and he made it plain that he is for the construction of the High Line canal above any other local project, and if any proposition comes up that may conflict with the High Line project he is determined to report against this opposing proposition to the authorities at Washington. He does not look for any controversy injurious to either side, to come up between the High Line and the Orchard Mesa projects.

Goes After Republican.

Ed M. Slocomb then offered a set of resolutions condemning the Denver Republican in no uncertain terms for knocking the High Line canal project in a direct or indirect manner. These resolutions, which follow, were adopted by a unanimous vote:

"Whereas, The people of the Grand valley for a period of three years or more, earnestly and persistently petitioned the reclamation service of the

Daily Sentinel, May 22 1908

United States to accept the Grand valley irrigation project; and,

Whereas, The secretary of the interior, Hon. James R. Garfield, and the director of the reclamation service, F. H. Newell, at the request of the petitioners, accepted the Grand valley project and are at this time having the survey made by government engineers preparatory to taking up the construction work at the earliest possible date; and,

“Whereas, certain publications have made unjust and misleading statements about the reclamation service; therefore, be it

"Resolved, By this meeting of the general membership of the Grand Junction Chamber of Commerce, held Thursday evening, May 21, 1908, that we deprecate these unjust and misleading statements made by these publications concerning the officials of the reclamation service; and, be it further

"Resolved, By the Chamber of Grand Junction, that it is the sense of this general meeting assembled that it is the right of our own valley and people, and that this right is not delegated to, nor possessed by the Denver Republican, or any outside newspaper hostile to our plan of government reclamation; and that the people of the Grand valley appreciate the actions taken and the work done by the reclamation project and pledge again their hearty support at all times and in every way possible, that assistance may be rendered in furthering the interests of this project; and, be it further

“Resolved, That it affords us great pleasure to learn of Secretary Garfield's contemplated visit to Grand Junction in August and that we request that he give us as much of his valuable time as may be, in order that he may get acquainted with the conditions and our people; and, be it further

"Resolved, That a copy of these resolutions be furnished Project Engineer Sands to be forwarded to Secretary Garfield and Director Newell."

Electric Line Coming

Mr. Charles Noble, one of the officials of the Fruit Belt Power Company, and who has charge of .the local offices of the company and who has been making investigations and arrangements looking to the building of an electric railway in the city and valley, was then introduced by Chairman Bunting.

No remarks made during the evening created more enthusiasm or were of greater importance than the remarks Mr. Noble had to make. It was a plain, conservative statement to the effect that the outlook for the early construction of an electric line in this valley is very favorable indeed. Many were surprised to know that so many advanced steps looking toward the construction of the electric line had been taken by Mr. Noble and the strongly organized and splendidly financed company which he represents.

The city council will be asked for a franchise at an early date, and under the circumstances it will doubtless be granted. Hundreds of property holders have willingly signed right of way privileges, and many other preliminary arrangements have been completed. There is no question but what the Fruit Belt. Power Company, whose plans concerning a great power plant at DeBeque were outlined by The Sentinel some weeks ago, means business.

Mr. Noble believes that construction work on the road will begin within a period of six or eight months.

Mr. Noble was tendered a regular ovation by the citizens present.

Daily Sentinel, May 22 1908

A very entertaining and interesting brief address by Attorney Ben Griffith, and an equally as delightful informal talk by Halsey Rhoades of Denver, the veteran newspaper man closed the program.

A delicious luncheon was then served under the direction of John W. Armstrong.

Such gatherings as that of last night mean much for the progress of any community.

Daily News, Denver Post, May 22, Daily Sentinel, June 2(?) 1908

CHAMBER OF COMMERCE

AT JUNCTION DENOUNCES THE DENVER REPUBLICAN

[SPECIAL TO THE NEWS.]

GRAND JUNC'TION. Colo., May 21- Resolutions denouncing the Denver Republican for publishing articles detrimental to the government high line canal were passed at the meeting of the chamber of commerce held here tonight. The resolutions point out that the newspaper has been publishing matters detrimental to the officers of the reclamation service and the canal, and condemn the paper for its attitude. It was announced at the meeting by the Fruit Belt Power company that work on the interurban street car line would begin here in October.

GRAND JUNCTION BODY DENOUNCES REPUBLICAN

Grand Junction, Colo., May 22. The Grand Junction Chamber of Commerce has adopted resolutions denouncing the Denver Republican for publishing articles detrimental to the government highline canal. The resolutions point out that the newspaper has been publishing matter detrimental to the officers of the reclamation service and the canal, and condemn the paper for its attitude. It was announced at the meeting by the Fruit Belt Power company that work on the interurban street car line would begin here in October.

CONFIDENCE IN GRAND VALLEY

Reclamation Officers Say Only Best Feeling Prevails

"It is hoped that the construction and operation of the Grand Valley project can be carried out with the continuance of the mutual confidence and good feeling between the department and the people of Grand Junction."--A. P. Davis, director of the United States Reclamation Service.

Two letters, one from Secretary of the Interior James Rudolph Garfield and one from A. P. Davis, acting director of the reclamation service, were read at the meeting of the Chamber of Commerce last evening by Secretary D. W. Aupperle which demonstrated the friendliness and support of these public officials in the Grand Valley project. The letters came as replies to the resolutions which were passed at the recent smoker of the Chamber of Commerce in which the work of the service and the Department of the Interior was praised and the continuance of their support solicited.

The letters are as follows:

"Washington D.C. June 1, 1908.

"President Chamber of Commerce,

"Grand Junction, Colo.

"Sir:

"'I am in receipt of a letter of May 25 from Mr. E. E. Sands, engineer, transmitting preamble and resolutions passed by the Chamber of Commerce on May 21 1908. One copy of these resolutions has been forwarded to the Secretary of the Interior as requested.

"I wish to express my hearty appreciation of this action of the Chamber of Commerce, as there are few who will take the trouble to express themselves that they approve our actions while our enemies are, of course, less modest with their criticisms.

"It is hoped that the construction and operation of the Grand Valley Project can be carried out with the continuance of the mutual confidence and good feeling between the Department and the people of Grand Junction.

"Very respectfully,

"A. P. DAVIS,
"Acting Director."

"Washington, June 1, 1908.
"The President Chamber of Commerce,
"Grand Junction, Colo.
"'Sir,

"I am in receipt of a resolution transmitted through the Reclamation Service from the Chamber of Commerce regarding the recent criticisms of that service and of this department that have been made in the Colorado press.

"I am much pleased to receive these assurances of the confidence and cooperation of the people of Grand Valley, and earnestly trust that the relations at present existing between your people and the department may continue throughout the construction and operation of the government project.

"Please convey to the Chamber of Commerce my thanks for the expression of their confidence and support.

"Very respectfully,
"JAMES RUDOLPH GARFIELD,
"Secretary."

Denver Post, June 18 1908, Daily News, June 19 & 26 1908

U. S. MAY WATER 12,000 MORE ACRES

Glenwood Springs, Colo., June 18—I.W. McConnell, formerly of the United States reclamation service, and chief engineer of the Gunnison tunnel project, accompanied by E. E. Sands, who is now constructing engineer of the High Line canal of Grand Junction, in company with B. T. Napier, left this morning for what is known as the Spring park and Cattle creek district, where they went to examine the feasibility of construction of a reservoir to impound the flood water of the latter stream. Should they find it possible, such a reservoir would irrigate 12,000 acres of rich, fertile soil, suitable for raising alfalfa, sugar beets and potatoes, but too high for fruit. The gentlemen are going merely to make a preliminary survey, but upon their report will depend whether or not any further action in the matter is taken by the government.

SANDS INSPECTS GLENWOOD PROJECT

GLENWOOD SPRINGS, June 18—For the purpose of determining the feasibility of an irrigation project to reclaim from 5,000 to 12,000 acres of land on what is known as Spring Park near Glenwood, I. W. McConnell, supervising engineer of the reclamation service and E. E. Sands, project engineer in charge of the Grand Junction high line canal, arrived here today.

The engineers have received instructions from Washington to make an investigation of the project and submit a report to headquarters. They left immediately for Spring Park accompanied by B. T. Napier. It is proposed to construct a huge reservoir on the site of the ranches of the McLean and Hetz ranches, the water to be secured from the flood supply of Cattle Creek.

Denver Post, June 18 1908, Daily News, June 19 & 26 1908

NEWELL TO BE HERE SUNDAY

Reclamation Service Heads to Visit Grand Junction

WILL SEE WORK

Directors of the Reclamation Service, Frank H. Newell and Chief Engineer of the Reclamation Service A. P. Davis will arrive in Grand Junction on Sunday evening and make a stay in this vicinity for several days, They come here in the interests of the government reclamation service and will spend the greater part of their time in this city in conferring with Grand Junction men in regard to the proposed work.

Both Newell and Davis will come to Grand Junction directly from Montrose where they will spend part of Friday and all of Saturday looking into the present condition of the Gunnison tunnel.

The two government officials are now in Denver attending the meeting of the national convention of American Civil Engineers and following its sessions they will visit the two Colorado points. From Grand Junction Director Newell will go to San Francisco where he will meet Secretary of the Interior Garfield, who is now in Honolulu but who will return within the next two weeks. The two officials will then return to Colorado, after having visited Oregon, Idaho, Yellowstone Park and Wyoming and at that time Garfield will make a personal examination of the Grand river project.

On Monday there will be a called meeting of the directors of the Water Users' association to meet Director Newell and Engineer Davis. At this meeting the various phases of the project will be thoroughly discussed.

Daily News, date unknown

HIGH LINE DITCH CONTRACT A PLEASURE

"No official act of Secretary of the Interior James R. Garfield has given him such pleasure, in my knowledge, as the signing of the contract for the taking up of the Grand Junction High Line canal Project"

This was the statement made at the public reception in the Odd Fellows hall last night by F. H. Newell, director of the reclamation service who with Chief Engineer A. P. Davis were the guests of honor.

The seating capacity of the hall was taken up by the crowd that turned out to meet and talk with the government officials who showed that they were well pleased.

District Judge Sprigg Shackleford presided. The first speaker was Senator Horace T. DeLong who made a few introductory remarks and paid some compliments to Messrs. Newell and Davis.

Mr. Newell who spoke next discussed the work of the service in general and took up the certificate plan proposition somewhat in detail telling of its success, where undertaken on other projects, He declared that the demands are now in excess of the reclamation funds but that if the work on the high line canal could be taken care of next year funds undoubtedly would be available to push it through in 1910.

He said that the present funds are needed to complete big projects now under way, which when finished will be revenue producers

W. S. Wallace, of the Water Users' association spoke next telling of the plans of those in charge of the work here and stating that sufficient headway had been made to assure that those interested would assist in making early construction possible by adopting the certificate plan

Chief Engineer A. P. Davis spoke of his early knowledge of the valley as both he and Mr. Newell have visited here before. I. W. McConnell, D.W. Aupperle and R. P. Davie followed. Mr. Davie declared that his company, the Western Sugar and Land company, was vitally interested in the project and would lend all assistance in its power.

Following the reception President Cherrington and Secretary Aupperle of the Chamber of Commerce held a conference with the officials who leave this morning for Provo Utah and other northwestern projects on their way to California.

Much good is expected to come from their visit here.

Daily News, date unknown

CITIZENS ARE TO MEET VISITORS

In order to give the people of Grand Junction and the Grand valley an opportunity to meet the distinguished reclamation officials Director F. H. Newell and Chief Engineer A. P. Davis, who arrived in the city yesterday evening from Montrose, a meeting will be held this evening at the Odd Fellows hall, under the auspices of the Grand Junction Chamber of Commerce. Secretary Aupperle, however, wishes it distinctly understood that the meeting is open to all, whether members of the Chamber of Commerce or not.

Messrs. Newell and Davis have expressed a desire to meet the people interested in the construction of the High Line canal, and the meeting tonight will give the people an opportunity to get in touch with these gentlemen, whose good will means so much for the valley.

Messrs. Newell and Davis are on their way from Washington to San Francisco, where, on or about July 22, they will meet James R. Garfield, secretary of the interior, on his return from his trip to Honolulu.

With Secretary Garfield, Messrs. Newell and Davis will then make a tour of the west, visiting all the great reclamation projects which the government now has under way.

From San Francisco they will go to Portland, and thence to Idaho, where they will visit several great projects--Boise, Nampa, Minidoka, Twin Falls and Shoshone. Three days will then be spent in the Yellowstone national park, after which the party will visit Cody, Wyo., and Edgemont and Belle Fourche, S. D. The party will then come directly to Montrose and Grand Junction, three days being spent in these two cities.

It is expected that they will be here from the 14th to the 16th of August; but it may be a few days later or earlier. Announcement as to the date will be made later.

All the places at which the secretary is to stop are making extensive preparations for his entertainment, and Grand Junction will not be at all behind in this matter. Already tentative preparations are under way for a banquet and big public reception.

Messrs. Newell. Davis and McConnell, accompanied by Mr. Sands, spent the day looking over the Grand Valley High Line proposition. They left early in the morning for Palisade in an automobile furnished by the Chamber of Commerce, and did not return until after 4 p.m. The party investigated carefully the preliminary surveys made in the canon under the direction of Mr. Sands on both sides of the river.

At tonight's reception there will be no formal program. It is possible that there will be informal addresses by one

Daily News, date unknown

or two local citizens, and perhaps one or two of the reclamation officials may be prevailed upon to say a word or two, although they are not in the habit of publishing their views abroad. The chief purpose of the reception, however, is to give the people of the city an opportunity to meet the officials. It is hoped that there will be a large attendance.

It behooves Grand Junction and the entire Grand valley to show Mr. Newell and his assistants that in their opinion the High Line project is the most important proposition now before the people of this section.

Messrs. Newell and Davis will leave tonight on No. 5 for Provo, where they will spend a day or so in inspecting the Strawberry reclamation project. From Provo they will proceed directly to San Francisco.

Daily Sentinel, August 2 1908

TUNNEL SIX MILES LONG IN ARID LAND

People Southwest of Denver Are Watching Work of Reclaiming Territory.

WATER KING IN THE WEST

Over 150,000 Acres of Land Will Be. Opened on Completion of Tunnel.

BY T.W. BRAHANY

As the crow flies it is little more than 100 miles to the southwest from Denver to the heart of the Uncompahgre valley. Railroad tracks in Colorado must find their way along the mountain ledges, and so it is nearly 200 miles by rail from Denver to that point in the valley where the Gunnison and Uncompahgre rivers flow.

The people of this valley are watching with expectant eyes a small hole at the base of one of the giant peaks of the Rockies which rises at this point more than 2,000 feet. Some day there will gush from this tiny opening a torrent of water which will spread over the valley and convert into blooming fields what is now about as barren, desolate and hopeless looking a spot as can be found in the great western desert.

Way back inside the hole in the mountain a gang of men are performing a modern miracle. They are digging a tunnel six miles in length through the toughest kind of granite. Soon they will break through the mountain and into the Black canyon of the Gunnison. Then they will compel the waters of the Gunnison to depart from the path they have followed since the beginning of time and pursue the new and artificial route under the mountain and into the valley of the Uncompahgre.

The United States government is responsible for the scheme to play a practical joke on nature by switching one of the biggest rivers in the state of Colorado into new and strange channels. The work is being done by the reclamation service and is one of the largest and most difficult of its numerous projects for providing water for the vast areas out here which require only moisture to make them available as homes.

"Water is king in the west, just as cotton is king in the south and money is king in Wall street. We will have plenty of water in this valley soon, and where you see nothing now but sage brush, there will one day be found rich farms, valuable orchards, ripe wheat fields and a vast city."

The prophesy was made to the correspondents by a Colorado homesteader, tanned and parched by a sun which shines here with unrivaled fierceness throughout a long summer and which is hidden less than thirty days a year. It was difficult to believe that the prediction could ever come true. Stretching away from the foot of the mountain was a dry desert. The ground was of forbidding adobe, baked hard as brick and curled into small cakes. The inevitable sagebrush, its harsh and scraggly leaves coated with dust, broke the monotony of the gray expanse. Not a blade of grass was to be seen nor a single tree. And to add to the dismal prospect, the sun beat

Daily Sentinel, August 2 1908

down with cruel intensity, seeming to threaten with death by scorching any green thing that might have the temerity to push its head above the baked earth.

"Look there," said the Colorado man, stretching his hand toward the west. That's the way this country will look before long." He was pointing to a broad band of green which, some twelve or fifteen miles away, ran through the desert like an elongated oasis. It was divided into square patches of green of varying hues, a most pleasing relief to the eyes that had been wearied by the miles and miles of adobe and sagebrush. The emerald ribbon covered the alfalfa and grain farms and orchards of Montrose, all made possible by irrigation.

With the completion of the tunnel bringing the waters of the Gunnison river into the valley of the Uncompahgre, there will be opened to settlers 150,000 acres of land. A channel has been prepared for the imported waters which will be turned into the bed formed by the Uncompahgre river. The Uncompahgre is now unequal to the demands made upon it for irrigating purposes. The increased volume it will receive from the Gunnison will furnish ample water for the entire valley.

In spite of the forbidding aspect of the acres and acres of barren land in this neighborhood it contains just the elements needed to produce large agricultural crops, provided moisture is brought to it. The government chemists discovered some years ago that the dead looking adobe was rich with the elements necessary to grow grains, alfalfa and fruit, and that the sage which thrives on it is always a sign of good earth.

"Wherever you find sagebrush growing luxuriantly," say the Colorado arid land conquerors, "there you will find rich soil, providing water can be had. The sage is like the Kansas populist, it cannot thrive under prosperity. Water kills the sage, just as it vitalizes these arid lands."

Uncle Sam has concocted a scheme by which settlers may have the full benefit of the government irrigation service on the easiest possible terms. The total cost of the Gunnison river project--about $4,000,000--will be divided by the whole number of acres to be watered. The farmers will get the water at once and will then be taxed pro rata for the service, the total charge to be paid in ten annual installments. It is expected that the water cost will be about $30 or $35 an acre.

Many Have Lost Lives.

A faint idea of the danger attending this tunnel digging business may be gained from the statement that many adventurous men have lost their lives attempting to push through the Gunnison river canyon. Only two men have passed through the great chasm and lived to tell the tale of their experiences. The mountain stream tears through the narrow pass like a race horse. The slippery rocks lift themselves toward the skies to heights ranging from 1,000 to 2,500 feet.

The reclamation service enthusiasts lowered themselves into the chasm by means of ropes, surveyed the treacherous cliffs and decide where the tunnel should be bored. Then they built a wagon road over the 2,000 foot mountain to Cedar creek, some twelve miles distant on the Uncompahgre valley side. Over this trail every pound of material needed in establishing a large excavating plant was hauled by horses, In places the road clings to the narrowest kind of ledges and follows sharp angles and breaks along the mountain that tax the nerves of the strongest. How the engineers succeeded in getting the heavy boiler motors down this frightful trail is a mystery to those who have felt their hair rise in terror while riding down it in a safe and comfortable wagon.

A trip into the tunnel is an experience to be remembered. Five of the six miles of the connecting tube have been dug. A trolley line runs from the opening on the valley side to the point where the half naked and thoroughly drenched men are laboring with pick and shovel. As you enter the hole your first sensation is one of relief to get away from the burning sun and dust laden air. The shaft drips water at almost every step, the seepings coming from springs, small creeks and subterranean lakes which have been disturbed by the borers. A mile or so inside the water runs a foot deep.

After you have left the entrance far behind you and only the electric lights furnish illumination you begin to think what a stuffy, uncomfortable death you would have if there should be a cavein between your car and the outside. You try to get rid of this cheerful thought by centering your attention on the underground workers. They are attired in a waterproof coat, long rubber boots and little else.

"We will take out the last ton of rock some time next year," said Engineer McConnell. "Before the water is turned on we will cement the entire tunnel, thus providing a perfectly smooth bore from end to end. We will take from the Gunnison about 1,300 cubic feet of water per second, which diverted through a system of canals and laterals will furnish moisture for 150,000 acres."

Denver Post, Aug 2 1908

SECRETARY OF INTERIOR HERE TO SEE THE SAN LUIS FARMERS GET WATER THAT IS DUE THEM

Commercial Bodies of Denver and Colorado Have Chance to Reclaim Thousands of Acres to the State.

Action Needed at Once If Centennial State is to Get Its Rights, Instead of Having Them Lost to Help the Claims of Mexico and Others

TOMORROW Secretary Garfield will be in Colorado on the most important mission that ever brought a cabinet officer to this state. He is going to personally investigate the conditions existing in the San Luis valley, involving the right of the people of Southern Colorado to impound the waters of the Rio Grande river for the purpose of creating immense reservoirs.

If he decides in favor of Colorado thousands upon thousands of acres of arid land will be turned into flourishing fields of grains; the population of the San Luis valley will be greatly increased, and millions will be added to the wealth of the state. The rights of Colorado are clear in the case and if the question is properly presented to Secretary Garfield a great victory will be won. It is the duty of the State Commercial association, the Denver Chamber of Commerce, the governor and Colorado's representatives at Washington to see that the state's case is not neglected.

Colorado's future is involved and it should be somebody's business to see that it is well taken care of.

NO RED TAPE NOW

Cabinet Officer in the West to Do the Right Thing by Agriculturists.

Secretary of the Interior James R. Garfield came all the way from Washington to Colorado to see that the thousands of farmers of the San Luis valley were treated fairly by the federal government in the matter of distribution of water for irrigation purposes from the Rio Grande river.

This progressive government official, recognized as one of the strongest in Roosevelt's cabinet, did not rely on the reports of subordinates heads of departments but came personally to see that justice was done.

Denver Post, Aug 2 1908

His visit to San Luis valley means that millions have been gained for the state. Secretary Garfield has not told an outsider of his intentions, but The Post is reliably informed that as soon as he returns to Washington he will issue a proclamation in which he will go over the complaints of the San Luis valley settlers and the treaty with Mexico thoroughly. In this proclamation he will announce the joyful tidings that he has found a way not only to placate the farmers of the valley, but to satisfy the demands of the international treaty.

When the industrious farmers of this rich valley presented their case to the secretary some time ago he went over the entire ground carefully and the original order limiting the supply of water was so amended that the husbandmen were able to get all of the water they needed, temporarily at least.

But Garfield is imbued with the idea that they are worthy of more than they have been getting and after his personal inspection they are liable to get it, and all the water they need for all time will be placed at their disposal.

It is his idea that Colorado settlers are not the ones to suffer, as the border line of this state does not touch Mexico, and if there is to be any diminution in the supply of water the farmers of other states will have to stand for it and not Colorado.

SUPPLY IS SUFFICIENT.

The supply at the headwaters of the Ro Grande is sufficient to supply all of the Colorado farmers without affecting the amount that is supposed to go to Mexico, according to the terms of treaty

If there is any dispute in the future as to the diminution of the supply by other states, Garfield can be depended on to settle it to the satisfaction of all concerned.

He is doing things the like of which no other cabinet official ever attempted, and if he stays in public life it is the opinion of government officials of Denver and the West that he will be a strong factor in the next presidential race.

Garfield has been secretary of the interior less than a year and a half, yet in this short time he has made two personal trips of inspection to all of the big projects of the West, and has gone over Indian affairs, the great irrigation and reclamation Schemes and has visited all of the national parks, including the Yellowstone and the national park in Montezuma county, Colorado.

The Gunnison tunnel project, one of the biggest ever attempted in the country, is a subject of special study by the secretary on his Western visit.

Garfield does not take the word of his subordinates for important recommendations, as is instanced by his trip to the Hawaiian Islands.

Denver Post, Aug 2 1908

GARFIELD IS HERE TO GIVE COLORADO FARMER FAIR PLAY

SECRETARY JAMES R. GARFIELD, **Roosevelt's Cabinet Officer, Who Has Come to Colorado to Personally Insure the Farmers of the San Luis Valley That They Will Get All the Water They Need**

Attorney Frank C. Goudy says that there are at least 300,000 acres of land possible of reclamation by means of proper storage reservoirs. The number of acres now under cultivation in San Luis valley is not much in excess of 300,000, so that the population and wealth of the valley will be doubled if Secretary Garfield's ruling is in favor of Colorado. Some of the proposed reservoirs which will be built if the people of Colorado are permitted to impound flood waters of the Rio Grande and its tributaries are as follows:

Toltec reservoir, capacity 40,000 acre feet; Elk creek, 20,000 feet; the Magote, 30,000 feet; the Alamosa, 20,000 feet; the La Jara, 15,000 feet; Farmers' Union, 23,000 feet; South Fork, 5,000 feet; Prairie 10,000 feet, and the Rio Grande, 45,000 feet.

The secretary of the interior has heretofore held up filings for reservoir sites which were filed subsequent to March 1, 1903. He gave as the reason for not placing his O. K. on them that a treaty which the United States government had made with the republic of Mexico preempted him from taking any other action. The United States government has promised Mexico that sufficient water to irrigate 40,000 to 50,000 acres of land in Mexico will be permitted to flow down the Rio Grande river.

The government proposed to construct a large reservoir on the Rio Grande river in New Mexico for the purpose of supplying the shortage of water to Mexico, and at the same time impound water for the reclamation of a large acreage of land in Texas.

The residents of San Luis valley object strenuously to this. They say that private capital should not be denied the right to build reservoirs, just because the government wishes to build reservoirs on the same stream. They also allege that an act of congress passed in 1891 gives them the right to reservoir sites, and that the secretary of the interior has no discretion in the matter. All he can do is to approve or disapprove the maps and charts filed with him. As the head of a bureau he cannot legislate. That remains for congress. He has, the San Luis valley people allege, failed to obey the law in not approving maps for numerous reservoir sites in the valley.

Denver Post, Aug 2 1908

GARFIELD TO REACH PUEBLO TOMORROW

Pueblo, Colo., Aug. 1.--Secretary of the Interior James Rudolph Garfield will arrive in Pueblo Monday morning in a special car en route to Denver after inspecting the government irrigation projects in Mesa and Montrose counties. The Pueblo Business Men's association has invited Mr. Garfield to stop over here and examine local reclamation enterprises.

Daily Sentinel, Jul 30 1908

"THE CANAL WILL BE BUILT"

MOST significant and important utterance ever made for the upbuilding and future growth of the Grand Valley, was that uttered by Secretary James R. Garfield at the breakfast given in his honor this morning---Positive assurance that the canal will be built has now been made by Secretary Garfield ---The value of the assurance is worth untold thousands to the land owners of this valley and to every other citizen residing here---More than pleased with the outlook---An Epoch making day in the history of the valley

As No. 4 of the Rio Grande came to a standstill this morning at the union depot a sun-tanned cowboy-looking young man stepped from Assistant General Passenger Agent Wadleigh's private car, Ballyclare, and the honor of taking his hand was anxiously sought by a large representation from the Chamber of Commerce. This man was none other than James R. Garfield, secretary of the department of the interior and the man at the head of the United States Reclamation service.

Secretary Garfield and party, consisting of F. H. Newell and Mr. Garfield's stenographer, Mr. Brown, were escorted by the representatives of the chamber to the LaCourt hotel, where a breakfast was served in his honor, fifty-nine covers being laid. Prominent among those present who assisted to entertain the secretary was Verner Z. Reed.

Of course a breakfast of this character does not consist merely of going through the process of stowing away-food, the better part of it always being the mental "feast." Secretary Garfield was called upon for a speech, in which line he is an adept, and. among the many other details thereof he said, and with emphasis: '"We are going to build this ditch." These words brought forth a roar of applause that might have been continued for a longer period than even the silver-tongued Bryan was cheered in the Denver convention but for the fact that a report got out that the journey would be continued to Montrose by 9:40. That was enough. The audience didn't care to hear of anything else but their ditch; they were satisfied.

Later on a reporter for The Sentinel interviewed the Secretary at the depot while awaiting the word "All aboard," but which did not come on account of the continued delay in the rebuilding of a bridge at Dominguez, which was washed out last night by the storm. Learning that the line would not be open before noon, Mr. Wadleigh ordered a crew and attached the Ballyclare to the engine and they took a little spin up to Palisade, accompanied by Project Engineer E. E. Sands.

Mr. Garfield said that the board of engineers of the reclamation service would have the report of Mr. Sands in its hands by the last of August, and at that time the details would be given out; that so far everything had been reported favorably upon by all officials who had anything to do with the project, and there was no doubt that it would be speedily passed upon when it reached the board.

Mr. Garfield left Washington the day congress adjourned and has been in Honolulu and the west ever since.

He said:

"Business conditions were very bad in the east when I left, but I understand that many large manufacturing and wholesale houses back there now have sent out a full line of their wares and a full force of drummers, there being a demand the country over for as much merchandise of all kinds as before the flurry of last fall.

"In the west business conditions are running along in the even tenor of their way and seem to be very good--better, in fact, than 1 had expected to find. Throughout Wyoming, where I have spent considerable time inspecting various projects, the farmers are in excellent shape financially and the crop prospects are better than for years, much rain having fallen where but little has been known heretofore, and they are paying off their mortgages and making extensive improvements on their ranches.

"In Honolulu I found the large plantations were beginning to be cut up into smaller ones, giving the small farmer a better chance to get a foothold. Much irrigating has been done there for the rains all fall during the rainy season, and, on account of the steep slope of the watershed, the moisture quickly runs off. The watershed rises abruptly at the coast to an altitude of 14,000 feet above the sea"

Regarding the candidacy of ex-Secretary Taft for the presidential chair, he said:

"Taft will be elected without a doubt, as nearly as I am able to determine from the opinions I have heard in the west during my trip. Not having been east since I left there several weeks ago, I cannot say how the east feels about it, but I know the west is favorable for Taft.

"I think that there is much work for congress at its next session. There should be no backing down as regards the regulation of the giant corporations. The government must have an affirmative control over all railroads, Standard Oil, harvester and other corporations that are of a quasi-public nature. There should be even more publicity given to the conduct of their respective businesses than there is. I have no fear of a big corporation just because it is big, if it is properly controlled by the government so that it will be properly conducted."

Speaking of the decision of Judge Peter S. Grosscup of the supreme court reversing the famous Landis decision, he said:

"I have not read the opinion yet, so would not wish to enter into discussion of it.

"No, I was not necessarily surprised at the announcement. Lawyers are never surprised at anything in the way of court decisions."

When asked what would likely be done regarding the Indian school here, which it had been reported some time ago was to be discontinued, he remarked:

"Indian Commissioner Luke has recommended to congress that non-reservation schools for the Indian are expensive and more or less impractical, and that they should be abandoned. The Indians who receive a good education frequently go back to the reservation and take to their blankets when educated in the ways of the white men far from their homes, and it is apparent that some other plan should be adopted. Just what that will be no one can guess. Congress took no action on the matter at the last session."

Daily Sentinel, Jul 30 1908

THE GARFIELD BREAKFAST.

The breakfast tendered in honor of Secretary Garfield this morning by the Chamber of Commerce and citizens, was an entire success.

At the table sat more than 50 of the city's representative citizens, including a number of prominent citizens of the valley.

Senator DeLong was introduced by Secretary Aupperle of the Chamber of Commerce, and he made a very excellent speech introducing Secretary Garfield, whose utterances upon the High Line project were, of course, the most important of all the things that were said by the various speakers.

In his speech Secretary Garfield reviewed the various projects now being undertaken by the government, the competition for money to undertake reclamation work and what were to be considered the meritorious projects in the eyes of the government.

"THE CANAL WILL BE BUILT:" This was the keynote of the address by the secretary. There were no ifs nor ands about this emphatic declaration, save that he said that, if the engineer's reports bear out the contentions of the people of this valley, there would be no question upon the building of the project.

The unqualified statement of Secretary Garfield is the most important utterance that has been made in this valley, and means more to the people than volumes of words.

He said it would be impossible to say when money would be available or when work could begin on the project, but that it would be ultimately built was his satisfying assurance to those gathered around the board.

But he said that before the government shall ever stick a pick in the ground there must not be any entanglements of any kind whatever. The right of way and all obstructive methods must be satisfied before the government would undertake to strike a shovelful of work on the project.

Furthermore, Mr. Garfield said that he had a peculiar personal interest in the project and that in his judgment it was one of the most feasible and best in all the west.

Verner Z. Reed, who sat beside Mr. Garfield, was the next speaker. He said that in the words of Mr. Garfield, the canal would be built; that it was the most important saying that could be made for this whole valley, and was of inestimable value.

Engineer F. H. Newell spoke very shortly. He said that he realized that the people in western Colorado were favorable to the government undertaking the projects of canal building; but he deplored the fact that some newspapers had misrepresented matters and that these clippings were sent to Washington and that senators and representatives of other sections of the country, who were working to have the government undertake reclamation projects used this adverse comments for their own private ends.

Following the breakfast the party hurried to the union station, only to be told that a bridge had burned out at Dominguez, and then it was decided that the party, at the request of Secretary Newell, should visit the famed orchard lands of the Palisade district.

Those who accompanied the party in Mr. Wadleigh's private car were Mayor Wentworth, Edwin Price, Secretary Aupperle, M. O. Delaplain, J. H. Schiesswohl, Senator DeLong, John F. Moore, F. H. Newell, E. E. Sands and O.

Daily Sentinel, Jul 30 1908

G. Coen, and several others whose names were not obtainable.

Arriving at Palisade, the party was driven through the orchards, the scene presented being a revelation to Mr. Garfield, to whom was explained the value of the land before orchards were planted and its present day value. It was explained to him that the product of one acre of land in that fruit region was equal to the selling value of the land in any other reclamation project now being built by the government.

The party returned to this city at about 12 : 10 p.m., thoroughly pleased with their trip.

Certainly it has been a most important day in Grand valley history.

Montrose Press, July 31 1908, Denver Republican Sep 18 1908

GARFIELD ENTERTAINED AT BANQUET LAST NIGHT

Special to The Press.

GRAND JUNCTION, July 30.--Secretary of the Interior James A. Garfield was entertained at a banquet last night at the La Court hotel, by the local commercial association. 60 covers were laid.

Secretary Garfield assured the people that the high line canal would certainly be built in the near future. The reclamation service has the reports of Engineer E. E. Sands, showing the feasibility of the project, and as soon a few preliminary details carl be arranged dirt will begin to fly.

The secretary left here at 1 o'clock this afternoon on a special train for Montrose.

THE GUNNISON TUNNEL PROJECT.

When the Gunnison tunnel was projected estimates of the work placed the cost of that work and the canals to be constructed in connection with it in order to supply 100,000 acres with an adequate quantity of water at $25 per acre. As the work progressed it was seen that this estimate was too low, and that in order to cover the cost of the enterprise it would be necessary to increase the assessment per acre by at least $10, and it is probable that it will have to be greater than that amount.

The farmers in the Uncompahgre valley are now considering the matter of agreeing to this increased assessment; and it seems that they have no choice,, unless they are willing that the project shall be abandoned, which is out of the question.

There is, doubtless, some disappointment respecting the cost of the work; but the land owners may reflect that even though the final assessment should be twice the original estimate per acre, it still would be profitable to pay it in view of the enormous increase in the value of the land which will result from the application of an adequate supply of water. Land under irrigation in a valley like the Uncompahgre is worth a great deal more than $50 per acre, but without water it is worth practically nothing.

The whole "Uncompahgre" project includes not alone the tunnel, but also three extensive canal systems which when constructed will supply water for considerably more than 100,000 acres. It is expected that the tunnel

Montrose Press, July 31 1908, Denver Republican Sep 18 1908

will be completed sometime next summer, if there are no further delays, but a longer time may be required to complete the entire canal system. This will mean a great deal to that part of the state, and every consideratlon suggests that the people themselves should throw no obstruction in the way.

In considering this and other projects in the hands of the reclamation service, it should be borne in mind that the income from the sale of public lands in the arid and semi-arid states is not sufficient to supply funds as rapidly as the speedy completion of the projects now in hand calls for. It may be said that the reclamation service handles an enormous amount of money; but it also is doing an enormous amount of work, and the claims of one project are about as strong as another.

The necessary result of this is delay in completing the projects already in hand and insistence upon repayment by land holders already benefited. Hence the Uncompahgre project has been delayed; and unless congress comes to the rescue, the farmers holding land under it will have to pay their proportionate share of its cost, even though it may exceed the original estimate.

Daily Sentinel, Jul 31 1908

Best Boost the Valley Ever Had

The High Line canal project had perhaps the largest boost given it yesterday in the history of the efforts to secure the building of that great irrigation waterway.

Engineer Newell well stated that he believed the people of the western slope were favorably inclined toward the government, and that it was their desire that the government undertake the construction of the High Line canal project. Nor was his strictures at all out of place against those newspapers which for personal reasons opposed the work being undertaken by the government. We have reason to know that the articles which appeared in the Denver Republican directed against the government undertaking the construction of the High Line canal project, were inspired by T. C. Henry for a purpose.

And we believe furthermore that the governing head of the Republican did not understand the significance of those articles, but that they did not have his sanction and were written by a person very familiar with the situation in the Grand valley and who was to profit in some manner by attempting to knock the proposition.

It is an ill wind that does not blow somebody good and the burning of the bridge on the narrow gauge yesterday gave the people of this city an opportunity to present the claims of the High Line canal proposition in such manner as they could not have hoped to do in the very short time that Secretary Garfield had planned in his itinerary to be given to this city.

An actual presentation of the matter to him by means of personal inspection was worth a thousand times more than to endeavor to interest him through descriptions and by speeches, giving in detail the opportunities presented to the government for reclamation work. In his trip to Palisade he could be shown the territory to be covered by the canal, the source of supply and other matters which could better be presented through the eye than by any other means.

There is one thing most admirable about Secretary Garfield. He does not hold out false hopes. He does not say when money will be available or when work can begin. He prefers to make that indefinite, rather than to raise false hopes in the mind of anyone. He says the canal will be built. He fully assures us of that matter. But he shows us also the many hindrances which beset the way of the government in forwarding their many undertakings.

But we feel assured that the High Line canal project has the ear of Secretary Garfield. We believe that, aside from his official relations to such a project, he has a peculiar interest, not born simply in the performance of his duties as the head of the reclamation service. We place our high hopes for the future of the canal in that personal interest in the project, which Secretary Garfield said he entertained while speaking of the matter yesterday at the breakfast at the La Court.

Those who have labored and struggled for the consummation of this project year in and year out may sometimes feel somewhat discouraged; but so far they have not been able to abate one iota of the interest that these men and women of the valley have taken in this matter; and, furthermore, we do not believe they will ever be able to do so, until the work is under construction and they see and feel that their hopes are being realized.

Daily News, Aug 10 1908

CAN USE IT ALL

It is not at all strange that Pueblo should look with longing eyes upon the priceless waters of the Grand River, and desire to turn them into her own dwindling Arkansas, but it is not at all likely that the wish will ever be realized. The Western Slope knows as well as the Eastern the value of the Grand, and, when the present great irrigation projects are consummated, there will be a use on the Western Slope for all the water that the Grand and its tributaries will be able to furnish. There are some things that the Eastern Slope will find itself unable to take from the Western, and the Western's abundant water supply is one of them.

Daily News Aug 11, Pueblo Star Journal Aug 25 1908

LAND WITHDRAWALS

RECENT ORDER NOW EXPLAINED

Land Official Explains Recent Ruling of Department Regarding Withdrawals Under Government Reclamation Projects

----

There has been very much discussion as to the scope of the order recently promulgated by the department of the interior withdrawing lands under the government irrigation projects at Gunnison and Grand Junction from entry. The News has had many inquiries as to the meaning of the order, and today secured from E. E. Sands, in charge of the reclamation service in the Grand Valley, the following explanatory statement:

"In 1903 all public lands in the Grand Valley which were to be irrigated by the high line canal were withdrawn from all forms of entry except what is known as the second form. This means that these lands could be entered only under the homestead and reclamation acts. The reclamation act authorizes the secretary of the interior to divide the public lands to be irrigated into farm units varying in size from 10 to 160 acres. During the last six years a large part of the public lands under this project have been entered. Some of these entries have been bona-fide and the laws have been complied with while others have been made for speculative purposes.

The validity of an entry is a matter to be determined by the land officials and the reclamation service takes no part in any contest initiated either by individuals or by the land office. However, the work of deciding the size of the farm units is done by the reclamation service. This work is about to be undertaken. Until this work is accomplished all public lands under the project not previously entered are withdrawn from all forms of entry, under orders from the secretary of the interior bearing date of July 28, 1908.

Daily News Aug 11, Pueblo Star Journal Aug 25 1908

This order in no wise affects previous entries, but simply prevents new entries being made until the lands are again restored to entry under the second form. At the present time it is not known how soon the work of making the farm unit plats will be completed. When this work is finished those who have made filings of 160 acres will receive notice from the land office to appear at said office and select a farm unit out of the 160 acre filing. The remainder of the 160 acre filing will once more become vacant public lands and may be entered by others as soon as the lands are restored to entry.

"The methods to be adopted by the land office in opening the remaining lands to entry have not yet been determined."

TO START WORK ON ALLISON PROJECT

W. H. Allison, who is the leading spirit in the proposed Allison irrigation tunnel, arrived in Pueblo today and announced that the work of constructing the tunnel would be commenced. The project is to be financed by local capital.

According to the plans, which were reported favorably by the committee of Pueblo business men who investigated the project, the tunnel will be 4,000 feet long and will have dimensions of four by seven feet. It is to penetrate the mountains and tap Eagle, Ten Mile and seven other smaller streams and turn the water into the Arkansas river, from whence it will be converted through irrigation ditches to fertile lands in the Arkansas valley on this side of the range.

Mr. Allison says the tunnel can be driven without great expense, as most of it will be in soft soil, with very little rock or blasting ground to penetrate.

There will be obtained for irrigation purposes 500 to 550 feet of water through the months of June, July and August, which will be sufficient to irrigate a large area of land on this side of the range. Mr. Allison announces that he will go ahead with the project and that he is assured of sufficient capital to carry out the plans.

Daily News Sep 29, Daily Sentinel, Aug 28 1908

ASKED TO FIGHT FEDERAL SERVICE

That there is an organized plot to secure the passage of resolutions through the next National Irrigation Congress which meets today at Albuquerque. N. M. to condemn the workings of the Reclamation Service and to attempt to limit its field of Operation, was confirmed last night by C. C. Magenheimer of the Woodland Construction company, who is now in Grand Junction closing up a negotiation with the Orchard Mesa Irrigation district for the watering of those lands.

"I was approached by a number of capitalists in Chicago recently," stated Magenheimer last night, "and asked to go into a movement with them to secure the passage of resolutions through the National Irrigation Congress that would be detrimental to the service. I promptly refused to do anything of the kind. In my dealings with the Reclamation Service I have always found those gentlemen most ready and kind to forward the interests of any private enterprise that was conducted on an honest and efficient basis and I am heartily in sympathy with their work. My relations with the service have always been of the most pleasant. Since that time I have heard that the men opposing the service were completing their plans to fight the service before the congress but I have had nothing to do with them. I think that the Reclamation Service has accomplished wonders in the brief six years of their existence. They have 29 projects moving at present with the High Line canal here next on their program. I am pleased to hear of the success of the Water Users' association in raising funds by which they may join in with the government in the early stages of the work."

It is said that among those planning to fight the service at Albuquerque is T. C. Henry, a Denver capitalist, who has long been interested in the private development of the High Line canal.

Daily News Sep 29, Daily Sentinel, Aug 28 1908

Let the Work go Ahead

So long as the government interposed the least objection to private capital engaging to promote irrigation enterprises in this valley, because of some possible interference with the future plans of the reclamation service, we joined in such objection. As the government does not now or will not in the future make any objection to the plan to irrigate the lands of the Orchard Mesa, through irrigation enterprises erected through private capital, we can see no force in attempting to put a bar in the way of the consummation of an enterprise, which the government has tacitly said will not interfere with their plans for the construction of the High Line canal.

The plans for the irrigation of the Orchard Mesa lands have been submitted to government engineers and there being no conflict, the promoters of' the project stand ready to do their part, according to their own assertions. The matter of money to undertake this vast enterprise which will bring under cultivation 12,000 acres of the choicest land in all the west, has been solved by the offer of the Woodland Investment company to supply same and they go even further and will bid for the work of construction.

The plans have been submitted to the most expert engineers for the land owners and also Captain Wray, for the Woodland company, who assert that they are most complete and assure us that Mesa county will have one of the finest irrigation projects, when completed, in all the west.

With all the: apparently insurmountable obstacles that: confronted the projectors of this irrigation project out of the way. we can see no reason whatever why we should not all join to further the consummation of this project, as well as the High Line canal scheme. Both are for the upbuilding of the valley and if the conditions of conflict have been removed, it is puerile to hold enmity because of former differences.

President Magenheimer says that if his company, the Woodland Investment company, received the contract to perform this work, he will set 500 men to work to prosecute same to the limit during all the time this winter that it is possible to work.

This sounds good to us. Delayed projects don't find us in the best possible mood, we have arrived at that condition, where we want to see something done to make this valley reach up to our anticipations. We see in this, one of the projects that means for the upbuilding of both the city and the valley and here's to it, may it be a splendid success.

Daily News Sep 29, Daily Sentinel, Aug 28 1908

WOODLAND CONSTRUCTION COMPANY'S BID ACCEPTED BY DISTRICT DIRECTORS

C.C. Magenheimer For Firm Refuses to Give Out Figures But it is Said to be Over $750,000

NEW WORK WILL BE PUSHED AT ONCE

Shortly before eleven o'clock last evening negotiations were practically closed between the Woodland Construction company and the Orchard Mesa Irrigation district whereby the Woodland Construction Company contracts to take over the big Orchard Mesa Irrigation project and complete it. Although Mr. Magenheimer refused to give out the exact figures of his bid it is known that it is over $750,000. The final details of the matter will be closed today.

C. C. Magenheimer, general manager of the Woodland company of Chicago arrived in Grand Junction yesterday for the purpose of representing the interests of its subsidiary company, the Orchard Construction company in the matter of the proposed Orchard Mesa Irrigation Project which has long been occupying the attention of the public. It will be remembered that the district adopted plans and specifications for its system of irrigation at a meeting of its board of directors held August 17 last, subsequently the board called for bids on the work and the time for putting in bids expired at three o'clock yesterday afternoon.

Mr. Magenheimer was seen at the La Court hotel, last evening and made the following statement: "It is true that in behalf of the Orchard Construction company, I put in a bid this afternoon for constructing the Orchard Mesa irrigation system. Ever since last spring when it was found that the plan originally proposed would not accomplish the object aimed at, I have kept in close touch with the project and have employed expert engineers meantime to go into the matter carefully. The district found it necessary to increase the scope of its enterprise very materially and moreover it was essential to secure some governmental action without which it is now known the entire project could not have been successfully prosecuted. In brief, I have now succeeded in putting the Orchard Construc-

Daily News Sep 29, Daily Sentinel, Aug 28 1908

tion company into a position, legally and financially, which will enable it to give ample water supply to the orchard district at a minimum of cost and-time."

Mr. Magenheimer was asked to name the figure embraced in his bid, but declined to enter into any discussion of that phase of the matter. He stated that the board was still in session and that various questions still required to be dealt with and that the proprieties of the case appeared to require of him that full and public announcement of the details should be left to the authorities of the district and that he was not willing to forestall publicly such statement as it might seem proper for them to make.

Daily News, Sep 30 and Oct 1 1908

IRRIGATION DISTRICT DIRECTORS ISSUE FULL STATEMENT TO ELECTORS

Magenheimer Declares That He Hopes to Complete Construction in Record Time

----

"With their interests amply protected by the terms of our contract and with the almost certain rise in value of their lands everywhere from 300 per cent to 1,000 per cent, I cannot see what valid objection the property owners of the Orchard Mesa Irrigation District can have from expressing their approval of the work of the board of directors of the Irrigation District and in approving of their contract with us in the election next month."

This was the statement of C. C. Magenheimer of the Woodland Construction company, whose subsidiary organization, the Orchard Construction company entered into a contract with the Irrigation District to build a complete irrigation system on the Orchard Mesa by April 1, 1910.

"If the contract is approved" stated Magenheimer last night, "we intend to make the irrigation system a model one. It is our desire to be able to use the Orchard Mesa system as our best advertisement both in the quality of the work and the time consumed in its construction. I leave tomorrow for Chicago, but will return next week with my brother, G. J. Magenheimer and together we will remain here until after the election in order to answer questions and to take care of our interests. Immediately after the contract is approved of by the people we will start to work. The construction company will have to pour money into this valley until the job is done. The completion of the ditch will mark a big forward step in the history of the progress of the valley."

The following statement was issued yesterday afternoon by the directors of the District explaining fully the details of the transaction whereby the Orchard Construction company secured the contract to do the work:

"TO ELECTORS OF ORCHARD MESA IRRIGATION DISTRICT AND OTHERS WHOM: IT MAY CONCERN:

The undersigned, constituting the Board of Directors of the Orchard Mesa Irrigation District, deem it proper at this time to lay before the parties entitled to the same a brief statement of the results thus far accomplished towards putting the lands of the District under the ditch.

Said District was organized in the year 1903. Its officers and directors in the early part of the year 1907 were told by parties upon whom it appeared that reliance could properly be placed that water could be supplied to the lands in question by means of a canal constructed in the canon of the Grand river, delivered into the heads of two subsidiary canals traversing Orchard Mesa, one of which would water about five thousand acres by gravity, and the other of which would require only seventy-two second feet of water to be pumped a distance of only one hundred and twelve feet and watering about seven thousand acres of land. As you are aware various negotiations were had during the early part of the present year with The Grand Mesas Land, Canal and Power company, which finally resulted in the adoption of plans and specifications for the proposed irrigation system on March 26 last, and in a call for bids.

The parties with whom such negotiations were had owned a sufficient water right, and certain of them claimed that all necessary steps had been taken by them to insure proper right-of-way for the proposed canal over government lands. While the call for bids was being published, the contrary became manifest, and the financial backers of the parties interested in The Grand Mesas company, from which a bid was expected, having been unable to secure the right-of-way and having also met with certain obstacles in perfecting their internal arrangements for providing the moneys necessary to bid whatever was received, and the entire project was thus left stranded.

ACCURATE SURVEY.

In the meantime some rumors had been put in circulation that the proposed gravity canal above mentioned was so located that it would water only from fifteen hundred to two thousand acres instead of five thousand acres as originally claimed; and that even if it could be constructed at all as an engineering proposition, it would involve a far greater expenditure than had been estimated, and when constructed, would be substantially impossible to maintain for the reason that for a long distance it would require to be excavated on the precipitous southerly bank of the Grand river which in that locality rises almost perpendicularly from the river bed. These rumors, if true, affected the action of the District as well as that of any intending bidder, and the fact that the Government Engineers, as we then learned, had investigated and abandoned the proposed gravity canal in that location, made it particularly necessary to proceed with caution. Accordingly on April 25 the undersigned arranged for an accurate survey of the entire situation by engineers skilled in the problems to be met, to be accompanied by detailed plans and specifications.. On August 17 last, we were provided with full plans and specifications, and all necessary drawings, maps, profiles, etc., and after full consideration, formally adopted the same as the basis for further proceedings in the premises; and by appropriate board action, called for bids for construction, which by the terms of the call were to be opened at three o'clock. p.m., of Monday, September 28.

At that time, the board received only one bid and proposal, viz: That made by The Orchard Construction company, which named the price of Six Hundred and Seventy Thousand ($670,000) Dollars for the completed work of construction. At meetings held on that day, the undersigned, in behalf of the district, en-

Daily News, Sep 30 and Oct 1 1908

tered into a contract with The Orchard Construction company for the construction of the proposed irrigation system for the amount named, and also for a contract for the purchase of a 450 feet second water right from Grand river, having the proper point of diversion, and for the entire right-of-way over all public lands traversed by the proposed canal of the system for the sum of One Hundred and Thirty Thousand ($130.000) Dollars. As provided by law, these contracts are now before the electors for appropriate action.

The Orchard Construction company will receive under the two contracts referred to the sum of	$800,000.00
In addition to these principal items the District must, under the law, provide other sums for the project (right-of-way, etc.) which are estimated to require during the coming year about	55,000.00

Making a total of	855,000.00
In order to realize the sums above mentioned and sustain the discount of 5 per cent on bonds sold at 95 per cent the minimum required by law, which amounts to	45,000.00
The District must issue bonds for	900,000.00
The annual interest on this issue is	54,000.00
The estimated annual cost for maintenance is	12,000.00

Making an estimated total cost of	66,000.00

Or an annual charge of less than $6.00 per acre of the land to be watered when the lands properly to be watered shall be embraced within the District.

RUSH TO COMPLETION

The contractor undertakes to complete the entire system by April 1, 1910 and failure so to complete same (unless prevented by causes specified in the contract and specifications) is visited with a penalty of $100 per day. In making the surveys, the Board required that the contours of the entire District should be laid out so as to assure to every owner of land under the proposed ditches the most accurate information obtainable as to the grades and levels of his land with reference to the distributing canals on the Mesa. This will enable each land owner to lay out his laterals scientifically, and with a view to obtaining the best results most economically.

We think it to be fortunate for the District that the same company which owns the water right and right-of-way over government lands has also made a bid for the construction of the system covered by the plans and specifications adopted by the Board, as it materially facilitates the speedy completion of the system. The government lands traversed by the proposed canals in the canon of the Grand river and elsewhere were withdrawn by the Secretary of the Interior from the operation of the land laws relating to rights-of-way for irrigation ditches, and this action having been taken at the instance of the Reclamation Service, the acquisition of the right-of-way over the lands in question involved direct application to the authorities at Washington. The grant of rights-of-way finally obtained assures to the enterprise now under construction, the friendly cooperation of the federal authorities. The proper authorities of the government have stated that the Reclamation Service will not undertake any project which can be cared for by irrigation districts formed under local laws. This attitude establishes beyond doubt the proposition that the owners of land included within the Orchard Mesa Irrigation District, as well as lands which ought to be included, need have no expectation of getting water from the High Line Canal when that project is put through.

GOVERNMENT IN ACCORD

The contracts made by the undersigned, acting as directors of the District were made only after giving the entire subject the most careful consideration, and because of the paramount necessity of acting at this time. It is a matter of common knowledge that various projects are under way for the diversion of water of the Grand river for use on the eastern slope for the purpose of supplementing the waters of the Arkansas river, already found to be seriously insufficient to supply the demand. If one or more of such projects should succeed before the Orchard Mesa District is successful in carrying through its project, it is quite possible that we shall lose the opportunity of obtaining a sufficient supply of water at a reasonable cost. In fact, it is not at all impossible that we might lose our last chance of obtaining any adequate supply of water at any price whatever. It is a matter of common notoriety that the chief opposition to the present project has arisen from the apprehension of those interested in irrigating lands which it is proposed to put under the Government High Line Canal, that the prosecution of the Orchard Mesa project would prevent the government from supplying them with an adequate amount of water.

The grant of the right-of-way over the public lands was carefully framed by the Secretary of the Interior in view of the impending scarcity of water, and contains express provisions which recognize the necessity of conserving the waters in such manner that in times of low water the Orchard Mesa project may, under existing conditions, have to make concessions for the benefit of lands on the north side of the Grand river under the proposed government High Line canal. We wish to impress these facts upon our constituents for the purpose of calling their attention to the urgent necessity of doing now all things which must be done in order that they may be assured of water supply.

Daily News, Sep 30 and Oct 1 1908

WATER BASIS OF VALUE

The consequences of a different course are almost certain to be serious. At the present time the raw lands on Orchard Mesa have a substantial speculative value, which we all know is due to the expectation that they will soon be put under the ditch. If this expectation be not realized, it is certain that this value will greatly diminish, and it is also certain that if unfortunately the District should finally be unable to obtain water, the raw lands therein would cease to be of any value whatever. On the other hand, it may confidently be expected that if the electors authorize and approve the contracts which have now been made by the undersigned, acting in their behalf, the raw lands alone will increase in value within the coming year two or three hundred percent; and upon the completion of the enterprise there is no reason to doubt that the lands of the Orchard Mesa will rank with any in the Grand Valley. In view of the increase in value which has uniformly resulted from putting lands under the ditch, it appears to us that it would be extremely short sighted if those interested in the creation of large additional values in their property should not now seize the opportunity which after long waiting is presented to them to secure the benefits of water at a price which is small in comparison with the increase in value which will follow.

BIG DEALS ON.

We personally know that large investments in Orchard Mesa lands are now under contemplation at present prices conditioned upon the consummation of the present project. It is highly significant that investors should base their purchases upon the condition that the proposed bonds should first be issued. On the other hand, it can be safely asserted that no intending purchasers will look further into the matter on the basis of prevailing prices for Orchard Mesa lands if the electors at the approaching election shall refuse to authorize and approve the contracts for construction of the irrigation system and the acquisition of the water right and rights-of-way over government lands, and the issue of bonds proposed.

DISTRICT PROTECTED.

It will be seen that the contract for construction is carefully framed to protect the interests of the District, and effectually insure the completion of the entire system. This is accomplished first by the provision which defers payment of upwards of One Hundred Thousand ($100,000) Dollars to the contractor until the completion of the entire system in accordance with the plans and specifications. In addition to this safeguard, the contractor is required to furnish the District a bond not for Sixty-Seven Thousand ($67.000.00) Dollars as is required by law, but in the sum of One Hundred Thousand ($100.000.00) Dollars. No circumstances can arise which will find the interests of the District unprotected.

We are aware that some objections to the Orchard Mesa Irrigation project have existed, mainly growing out of hopes and anticipations that in some way, and at some time, it would be found possible to obtain water from the proposed government High Line Canal. We desire to record ourselves, both personally and officially, as most heartily in favor of the prosecution and completion of that enterprise. We became convinced, however, long since that the owners of lands on Orchard Mesa would do well for themselves not to depend upon the construction of that canal. We were made aware of the great demands upon the government in other directions and of certain serious complications relating to the construction of its proposed ditch immediately west of the canon of the Grand river, and of the further fact that very considerable bodies of land lying above the proposed High Line Canal cannot possibly be watered except by means of private pumping plants deriving their water supply from the proposed High Line Canal. These considerations coupled with the acknowledged scarcity of water long since convinced us, both as officials of the District and as individuals, that it was incumbent upon the owners of the Orchard Mesa lands to act for themselves if they expected to reap any benefits from their properties within any reasonable time, if at all. Further investigation and study into the conditions have merely strengthened these convictions.

In conclusion, we strongly urge upon our neighbors and all who are interested in the development of the Grand Valley, and especially in the creation of values in the lands comprising the Orchard Mesa Irrigation District to seize the present opportunity, and by their influence and their votes at the approaching election to bring to a complete fruition the project for irrigating the Orchard Mesa which has after so long a time, been worked out on right lines and submitted to them for approval.

FRED W. BARBER.
N. V. FOSTER.
GEORGE SMITH.
Directors of Orchard Mesa Irrigation District

Daily News, Sep 30 and Oct 1 1908

HIGH LINE DIRT WILL SOON FLY

Money Necessary to Begin Work is Now Subscribed

WATER USERS MEET TODAY

With a considerable amount over the sum needed and with still others desiring to subscribe, the canvass for subscriptions to the fund to he applied to construction work on the High Line canal closed yesterday afternoon and the Water Users' association will meet at 9 o'clock this morning to pass upon the matter. The exact amount cannot be stated until the board has canvassed the returns, but it is learned from reliable source that the amount secured exceeds $100,000.

It is probable that a representative of the association will be despatched to Washington at an early date to lay the matter before the Reclamation Department and it is possible that work upon the enterprise will begin at a very early date. The land owners have fulfilled their part of the agreement and it only remains for the government to appropriate its share for the actual construction to begin. The agreement existing between the Water Users' association and the Reclamation Service is, that upon the advancement by the association of a certain amount of money - which money is to be applied in part payment of their water right - the government will appropriate a like amount and work will begin at once.

This proposition was made to the department by the association and - verbally, at least - accepted by the heads. Those in a position to know state that there will be no difficulty whatever in completing arrangements with the Reclamation Department and dirt from he canal is expected to fly even before January 1.

The work of securing the subscriptions has been almost entirely in the hands of Secretary Aupperle and the short time in which the amount was raised stands out a monument in his honor.

The amount raised by the citizens of Fruita was surprisingly large - in fact more than the association had expected from that end of the valley.

Daily News Oct 9 1908

WORK ON HIGH LINE MAY BE INAUGURATED BEFORE CHRISTMAS DAY

Secretary Aupperle Will Leave For Washington Monday--Subscriptions Steadily Increasing

---

WATER USERS ASSOCIATION REPORTS

Active work on the High Line canal may begin in the month of December. This is the present prospect according to the men who are closest to High Line canal affairs. The success of the Grand Valley Water Users' association in securing a subscription of $117,000 with a bright prospect of bringing the total up to $125,000 and the favorable attitude of Secretary Garfield of the Department of the Interior towards the project makes the closest observers of the situation believe that dirt on the long looked for ditch will begin to move before Christmas day.

Secretary Aupperle of the Water Users' association will leave Grand Junction next Monday to go to Washington for a conference with Secretary Garfield and at that time the contract with the government, providing for the early opening of the work will be signed. Last evening Mr. Aupperle received a telegram from the department of the interior stating that Secretary Garfield was out of the office and would not be back at the Capitol until October 12. This caused Secretary Aupperle to change his plan of leaving Grand Junction today and he will not leave until next Monday The secretary is steadily receiving notices from slow subscribers that they will come in for their pro rata of the subscription. Several telegrams were received yesterday by Mr. Aupperle from out of town realty owners asking that they be assessed for their share. Among these requests came one from Robert C. Williams, an employe in the Reclamation Service in Washington who wired his subscription of $1,600. Williams has long been in the department and while in favor of the project his attitude indicates that the feeling in the department is one of great friendliness for the High Line canal.

Daily News Oct 9 1908

One subscription of $5,000 has been received by Secretary Aupperle from Clifton landowners whose land now lies above the ditch. This subscription has been accepted conditionally, but is not included in the total announced at yesterday's meeting of the Water Users' association or the new total given out last evening by Mr. Aupperle. One of the principal reasons for believing that the work on the High Line will be started in December is based on the fact that about $30,000 of the subscriptions received comes in the form of labor. Labor is accepted by the department of the interior on par with cash. These men who have offered their labor on the canal in lieu of money are anxious to get to work this winter before their work on the ranches begins next spring. The Department of the Interior has always shown a disposition to meet such a situation promptly and these ranchers will probably be out with shovels in their hands within the next ninety days.

At the meeting of the Water Users' association yesterday Secretary Aupperle reported that $116,999 had been subscribed. Of this about $38,860 came from Fruita, where the results of the canvass made by the Fruita committee, was especially good. Secretary Aupperle was then instructed to notify all of the subscribers that the amount necessary to make their subscriptions binding had been received by the association. It was then decided that Secretary Aupperle should go to Washington at once and close up the final details with Secretary Garfield.

The work of securing the funds has been an exceptionally well handled campaign. Secretary Aupperle. W. S. Wallace and H. H. Beach worked the Grand Junction and outside districts.while Rev. G. W. Belsey with the aid of an active committee canvassed the Fruita district. The Fruita committee worked in pairs. The team of J. N. Whittemore and D. W. Lynch and A.J. Lee were very successful in their canvass about that town.

In order to catch the out of town men circular letters were sent notifying them of the plans of the Water Users' association and requesting their cooperation. Towards the end of the campaign more letters were sent out and the telegraphing was resorted to in certain cases.

Daily News Oct 9 1908

SANDS UTTERS SHARP DENIAL

"If there is a single correct statement in tonight's evening paper concerning my alleged discovery in regard to the amount of water necessary for use in a Grand valley orchard I have failed to find it," declared Engineer Sands last evening.

The paper declared that Sands had made a discovery that it would only be necessary to use one-sixth of an inch of water in irrigating scientifically water orchards. "Such a declaration is foolish," declared Sands last evening. "I have conducted no series tests and the statement going out will injure the valley and delude probable investors who will grow angry when they learn the truth. I believe that it is necessary to use from one-third to one-half an inch of water per acre to properly water a full grown orchard.

The orchard mentioned in the evening paper is a young one and does not use the usual amount of water. The test I mentioned took fifteen minutes of' time and was made four months ago. It could not in any sense be considered as a standard in any way."

Denver Republican Oct4 1908

RECLAMATION SERVICE MEN ON THE CARPET

GROSS INCOMPETENCY IS THE CLAIM OF ED. F. BROWNE

SPECIFIC CHARGES IN CASE OF THE TRUCKEE-CARSON PRO JECT IN NEVADA.

Officials and Engineers Place Government in Bad Light--Colorado Project Criticized.

ASPEN, Colo., Oct. 3.--(To The Denver Republican):--The "mistake" made in the figures regarding the construction of the Gunnison water tunnel and the irrigation project in the Uncompahgre valley is but a continuation of a series of "blunders" made by the reclamation service, which should put a stop to all of this Socialistic attempt on the part of our government to do any business or compete with private endeavor in business matters.

If a private corporation had undertaken that venture, the citizens who have been deceived into locating land and guaranteeing future payments for water, would either get their water at the contract price or seize on the tunnel and complete it themselves by legal process.

However, the government cannot be sued for breach of contract and the poor locator of land under that ditch must stand it and be robbed, or accept a loss of all of his investments.

The "boobys" who get the position of "irrigation expert engineers" in the reclamation service of the United States, really are the limit, even if their names do happen to appear in the magazines.

It appears that this government expert missed his guess by a million on a $2,500,000 job. A private corporation would "can" such an engineer on the instant and settle with its customers in some fair way--or go into bankruptcy Not so with the reclamation service. The bureau sends out a fewer circulars at the people's expense eulogizing the blunderer, and then sends him off to some other field to mix things up and retard private citizens in their endeavor to open up new country.

"Rummaging Around the Land offices"

Since this venture of the government was started on the Gunnison over 500,000 acres of land have been put under irrigation by private citizens in the state of Colorado.

The greater part of the time of those "higher up" in the irrigation department is used in rummaging around the land offices and planning with the interior department some way to prevent private citizens from getting permits or securing the necessary land and right of way to make improvements of this character under what is known as the Carey act, the only American way of doing such things under the law.

The reclamation bureau within the past few years has grown particularly obnoxious and have prevented the sale of bonds and improvement of more land than all of the government ventures put together.

Down near Grand Junction these men succeeded in holding up an enterprise intended to irrigate more land on the east bank of the Grand river than would be reclaimed by the Gunnison tunnel, and did so because they said the general government might in the future want to do the same thing that the private corporation stood ready and willing to do immediately.

When Mr. Moffat started to grade his road through Gore canon, these same men held him up for six months, because they represented at Washington that they wanted to build a dam at that point and conserve water, not for Colorado. but to regulate the flow of the Colorado river in Arizona, Utah and California. These men never could have seen the Colorado river, or they must have known that if their puny "reserve" at Gore canon should all be released in a day, it would not affect that river one inch at Needles, Parker or Yuma; and they don't want the water anyway out there.

However, the government employes delayed the construction of the road until the courts told them to keep quiet.

Opportunities in Colorado Under Private Control

We have great possibilities in Colorado for irrigation and power schemes. The waters of our streams belong to the state, and this position has been upheld by a hundred decisions of the supreme court; and for a citizen of Colorado to either invite the general government to take part in developing these natural resources, or even allow a usurpation on the part of these bureaus at Washington, it proves him to be a socialist or a fool.

These bureaus have seized all of our timber and coal, they are planning to seize our water power, and some foolish men in the state say it is all right.

I say it can only end in imperialism or socialism and marks the commencement of the end of free government. It is throttling the progress of our state, and by the silly use of "permits" and "leases" it is creating a form of serfdom more obnoxious than in Russia, because these men were once free, while Russia always harbored serfs.

Truckee-Carson Project Called the Worst of All

The foregoing gives a few instances of blundering government engineers in Colorado, but it chanced that I stopped a short time in Hazen, Nev., this summer, and of all the ridiculous fiascos in irrigation ever attempted, the so-called Truckee-Carson irrigation project of the United States reclamation service takes the cake. This venture partakes of three forms of rascality. First, venal, petty graft on the part of some of the promoters; second, stupid blundering on the part of the government engineers (?), or worse; third, the outrageous "buncoing" of the few farmers roped into the silly venture.

After looking over all. the ground carefully and investigating this work, I frankly report that the government has lost every cent it ever has put into that venture; will never receive a dollar's interest on its cost, and that the engineers who recommended it and Mr. Newlands one of the senators from Nevada, should be prosecuted for taking part in throwing away about $5,000,000 of the people's money.

However, Newlands stands well with the irrigation forces at Washington and, although a Democrat, is President Roosevelt's adviser in such matters. If this Gunnison engineer (and, mind you, I understand fully that no civil engineer of

Denver Republican Oct4 1908

great ability could be secured at the wages paid by the government) made a little blunder in his estimate, out in Nevada they blundered every possible way.

The Many Mistakes of the Truckee Project

The so-called "engineers" estimated that 300,000 acres of land in the Carson sink and Lower Carson river could be "retained" by constructing a great canal which would carry the water from the Truckee river over into the sink. The government appropriated about $4,000,000 for the work, and it was intended that a charge of $13 per acre for the water, payable in ten installments, would return the money to the government. Settlers were invited to use their rights as citizens to locate homesteads under the lines of the ditch and agree to this arrangement. To provide for a permanent supply, lakes or reservoirs were to be established and the flood waters drained therein and conserved for use when needed.

Mr. Newlands takes pride in being the father of this project and it appears that he also took first profit from the venture because he sold 6,000 acres of desert land, which I estimate as being not worth over 75 cents per acre, to the government for $30,000, for one of these reservoir sites. It will never be used. because the engineers now recognize that the scheme is a failure.

This irrigation scheme has now been completed two years, and the results are simply startling.

Commencing the work by contract, as all government work should be done, the "engineers" inaugurated a system of persecution and regulation that "busted" every contractor on the job.

Then, you should have seen the administration buildings constructed fine teams purchased and reckless throwing away of the people's money. Foolish magazine writers pictured the stupendous feats of engineers, and the venture only cost a million or so more than was estimated but at last the water was run over the point desired.

The only credit I can give those engineers is that they really had sense enough not to try to run that water uphill. From my own observation I see that it runs down a gradual descent, I also see that it has not cost over twice as much as it should, if a private company had constructed it; which is doing well for a government venture.

Dismal Results From Such a Costly Venture

Now, I will give results. The water having been carried to the point desired two years ago, I find that about 2,000 acres are being irrigated with a show of possible profit this year. The engineers have revised their figures, and instead of 300,000 acres being available, they now estimate but 54,000, and as they apparently have made a liberal reduction, I, being a more liberal man, will estimate that not over 10,000 acres are available for successful irrigation. Judge Hallett in the early days in a homely way expressed the value of the 290,000 acres now in the "discard" when he said that a certain tract of land would be valuable if it could be arranged so that two showers of rain and one of manure could be given to it each week for twenty years.

The sloping mesas of volcanic ash would not raise a crop of weeds if washed each year with the waters of Lake Superior, but the irrigation engineers of the government failed to notice

Denver Republican Oct4 1908

such a little thing as that, and invited a lot of poor farmers to disaster. In a foolish attempt to "save their face" in Washington they now have raised the price of water to $30 per acre. They have cut down the size of location from 160 acres to 80 and 40, according to locality but all this is useless and it will pay the United States to sell this venture for $5,000, if they can get no more for it, and put $4,995,000 over into a profit and loss account, headed "Loss by the government attempting to practice Socialistic theories."

Last year it cost $75,000 for administration and repairs, and in June 1908 52 families were using the water

Fourteen farmers and their families are going to abandon their farms and seek work this season. Some of these poor dupes of the government have lost from $1,000 to $3,000 and two years' time, and depart broken hearted men, with no recourse, and they also have lost their rights as citizens to locate homesteads.

Innocent Farmers Suffer in Many Other Ways

It appears that when these men chanced to" make a location on land unsuitable for farms, the officers of the government would not allow them to change to other land, some of which was tillable, but by the refusal they take away the right of location which every American citizen is supposed to have.

The glowing literature sent out by these fellows has attracted many innocent farmers who knew little or nothing of desert land, and by poor choice of location now finds that he has not only lost his money and his time, but his right to locate a homestead. If a private company had perpetrated such an outrage it would be obliged to fix that man with tillable land or pay damages, but as it is the government, officers try to make him out a fool and are willing to let him starve,

There are little spots along this ditch where crops could be raised if stones were attached to the seed and roots, so that the sweeping winds would fail to blow the plant away, and a protection put over the farm so that the ashes and sand would not cover over in a single night the results of two months' work but this Truckee-Carson business is about the most contemptible outrage ever perpetrated on the people of the United States. There is not a single redeeming feature to it.

This government was not formed to conduct business, and every attempt, naturally, will end in scandal and ultimate loss.

We have progressed and have passed all other nations by granting business freedom and encouraging personal incentive and inventive genius but we now face a decline through meddling of government and regulative interference with business affairs.

ED F. BROWNE.

Daily Sentinel, Oct 15, Palisade Tribune, Oct 24 1908

NEARLY $300,000 READY FOR THE HIGH LINE

As was generally expected, owing to the promise already made by the reclamation service that the government would appropriate a sum equal to a sum subscribed by Grand valley land owners to start, and push the work on the High Line canal, Secretary Garfield has approved the contract desired by the Grand Valley Water Users' Association and actual work on the great project, so near to the hearts of the citizens of this valley and which means such a wonderful development for this section, will go forward at once, as it was expected could be the case after land holders in the valley had subscribed nearly $120,000 to get the work under way.

About ten days ago Secretary Aupperle of the Grand Junction Chamber of Commerce, who is also secretary of the Grand Valley Water Users' Association, left for Washington to officially inform Secretary Garfield of the interior department of the raising of a fund of approximately $120,000 by the land owners of the valley for the purpose of meeting the government's promise to appropriate an amount equal to the sum subscribed by local people and to get work under way at once on the great canal. The fund subscribed by the local people is to be paid back to them in the future in water rights, etc. About $25,000 of the local subscriptions are to be paid in labor by many of the subscribers. Those who made the cash subscriptions are ready to meet them as necessity demands.

The following message was received last night by E. E. Sands, project engineer, from Secretary Aupperle, telling of his success in Washington:

Washington, D. C., October 15.

E. Sands, Engineer Reclamation Service, Grand Junction, Colo.

Secretary Garfield today approved of the contract covering everything asked for, subject to the approval of the board of project engineers. He has asked Chief Engineer Davis to appoint the board at once. Approves under same conditions, allotment will be $125,000. $25,000 available at once. Have Wallace call board meeting for afternoon of October 23.

D. W. AUPPERLE.

With the big fund subscribed by local people, the $125,000 allotted by the government and the $50,000 previously set aside by the government for the permanent survey work on the canal, close onto $300,000 will be available for expenditure on the work on the big project this winter and during the coming spring.

Daily Sentinel, Oct 15, Palisade Tribune, Oct 24 1908

It is very probable that Chief Engineer Davis will announce the personnel of the board of project engineers in a very short time. These men will be selected from among the consulting engineers of the reclamation service.

The surveys for the big canal are about completed, and everything is practically in readiness here for the taking up of the actual construction work. The rights of way are arranged. As announced by The Sentinel some days ago, no trouble will be experienced in getting the canal through the railroad's right of way in the canon above Palisade. It is probable that the construction work will start within the next eight weeks.

The High Line canal, for the construction of which the citizens of Grand Junction and the Grand valley have long struggled, will bring between 40,000 and 50,000 acres of the finest land in the world, now arid, under irrigation, and will provide homes for thousands of families. Its construction means an unparalleled growth for this valley and this city.

It will take about two years to complete the canal, which will start from a point on the Grand river above Palisade and wind its way through the valley to a point near the little town of Mack, embracing a vastly rich but undeveloped territory.

While it was practically certain that Secretary Garfield would approve the contract presented to him by the Water Users' Association, the formal announcement that such approval has been made and that an early starting of work on the canal is now an absolute certainty makes everybody feel good and is a bugle call for the greatest march of progress and growth that this richly favored valley and city have ever experienced.

EVERYBODY INVITED

E. E. Sands, government civil engineer, will speak at Bower's hall, Monday evening, October 26, at 7:30 o'clock. Mr. Sands will bring charts, diagrams, etc., and will inform the people concerning the details of the new High Line government ditch.

Mr. Sands comes under the auspices of the Palisade Chamber of Commerce which organization cordially invites everyone interested in the project to be present. After the subject is presented Mr. Sands will be glad to answer any questions or consider any suggestions. The purpose of this meeting is the promotion of good fellowship and a mutual understanding of the facts in the case.

PALISADE CHAMBER OF
COMMERCE, Palisade, Colorado

Daily News, Oct 20 1908

SURE TO FOLLOW

CONDITIONS ARE SUCH THAT CAPITAL WILL BE EAGER TO COME INTO GRAND VALLEY

Garfield Definitely Declares That Government Will Not Interfere With Private Magenheimer Project.

SENDS LETTER OUTLINING THEIR STAND

"If the Orchard Mesa Irrigation project contract is approved by the electors of the district next Monday the construction of a street car line from Grand Junction to the Orchard Mesa and Palisade within a year is an absolute certainty. This does not mean that we are in any way interested in the Devereaux franchise for we are not but with the construction of the Orchard Mesa ditch conditions will have so shaped themselves in this valley that the construction of the line is as sure to follow as the rise of the sun follows night."

This was the message that C. C. Magenheimer of the Orchard Construction company brought back to Grand Junction from Chicago when he arrived here yesterday and which he communicated to the Daily News. With C. C. Magenheimer came his brother and partner in the enterprise G. J. Magenheimer, who will remain in Grand Junction in case the election is favorable to the company, and will supervise the work of construction.

"And you can say for me," declared George Magenheimer, in discussing this phase of the deal, "that I will pound that job as I never pounded on a contract in my life. I'd like to see it well and properly accomplished within the next six months. We are here to make a record on the Orchard Mesa project."

"The street car line to the Orchard Mesa is sure to come," continued C. C." Magenheimer, "because of the fact that all of the elements that go to bring about the construction of a line will be present when the ditch is completed. When a corporation goes into a street car deal, and we have many of them, they take certain things into consideration. The essentials are the number of people living along the line and the amount of traffic they will produce. With a rich and populous section like the Orchard Mesa will be when it is irrigated within close touch of Grand Junction and with Palisade and the rich Plateau country as a goal it will be easy to secure capital to advance such a project. Mr. Devereaux and his associates will certainly have no trouble in financing their line when the Orchard Mesa builds up as it surely will."

For the next week both of the Magenheimers will be located at the La Court hotel in order to be on hand to answer any questions and explain any particular points concerning their proposed ditch construction work on the Orchard Mesa.

Absolute declaration from Secretary of the Interior Garfield, that the government did not intend in any way to sell any water to ranchmen living under the line of the proposed Orchard Mesa ditch was also developed yesterday upon the arrival of the Magenheimers. They brought with them a letter from Secretary Garfield to Attorney Howard Morris of the irrigation district which distinctly stated that the government had not now any desire to interfere in any way with the carrying out of the present plans of the district directors and the Magenheimers. The letters that passed between the irrigation district directors and Secretary Garfield are appended. They outline the policy of the department in relation to the Orchard Mesa project very clearly.

Washington, D. C., Oct. 8, 1908.

Hon. James Rudolph Garfield, Secretary of the Interior, Washington, D. C.

Sir:— The undersigned are the directors of the Orchard Mesa Irrigation district, a corporation under the laws of Colorado, organized for the purpose of providing said district with an irrigation system. We beg to refer you to the contract dated May 16, A. D. 1908, made by you with the Orchard Construction company, by which said company obtained, for the benefit of the Orchard Mesa Irrigation district, a right-of-way upon certain conditions over the government lands in the canyon of the Grand river and on Orchard Mesa.

On September 28, 1908, the undersigned, in behalf of the district, entered into contracts with the Orchard Construction company which, if ratified by the electors of the district, will result in providing the district with its completed irrigation system, adapted to water not less than twelve thousand (12.000) acres of land.

Under date of March 31, 1908, the local engineer of the Reclamation service at Grand Junction, officially announced that, regardless of our district, owners of lands on lower Orchard Mesa would, if they desired, be furnished with water from the Government High Line canal, expected to be constructed on the north side of Grand river. He used the following language in that connection:

"Sixth - Up to the present time, the Reclamation service has not decided to abandon any part of the Orchard Mesa and in any event such farmers as own land on the lower part of Orchard Mesa and desire to do so will be allowed to purchase water from the proposed government canal."

Some owners of land on Orchard Mesa appear to be relying upon this assurance. Maps of our proposed distributing canals have been filed with the Reclamation service, and show that our

Daily News, Oct 20 1908

canals will water all of the lands of the lower or westerly part of the Orchard Mesa which are or ought to be included in our district and doubtless will be so included when the policy of the government is thoroughly understood. We understand that policy to be that the government will foster and encourage, local capital in providing irrigation facilities by keeping out of territory which corporations, formed under local laws, can adequately provide with irrigation facilities. We are informed that some of the qualified electors in our district will oppose the ratification of the pending contracts by the electors solely upon the ground that the government stands ready to furnish them with water from its proposed High Line canal. We believe they would favor ratification if the official announcement, made by the local engineer in March last, should now be so modified as to make it clear that, if our project is carried out the government will leave them to rely upon the Orchard Mesa Irrigation district for their irrigation facilities.

We are advised that not less than fifteen thousand (15,000) acres of public lands susceptible of irrigation lie above the proposed Government High Line canal, and that such lands cannot get water from any other source. Of course such lands should be cared for. Since the Orchard Mesa district is provided with water sufficient to irrigate the entire Orchard Mesa, we think that all of the owners of private lands on the Mesa should know that they must rely upon our district for their irrigation facilities.

We therefore respectfully suggest an official statement in the premises of such tenor as you deem proper and efficient to prevent our local enterprise from being prejudiced in any manner by the misunderstanding, as we believe it to be, occasioned by the statement of the local engineer of the Reclamation service above mentioned. Such statement will not only aid our project but will tend to eliminate some unfriendliness towards the service which we believe has grown out of a total misapprehension of the position of the government in matters of this nature.

Yours respectfully,
FRED W. BARBER,
N. V. FOSTER,
GEO. SMITH,
Directors of Orchard Mesa Irrigation District, Mesa, Colorado.

Washington, Oct. 12, 1908.
Mr. Howard Morris,
Attorney for Directors of Orchard, Mesa Irrigation District, Colby-Abbo Building, Milwaukee, Wis.

Sir:--I am in receipt of your letter of the 8th instant in regard to the relations between the Orchard Mesa Irrigation District in Colorado and the Grand Valley Project of the Reclamation service.

You refer to an announcement of the Project Engineer under date of March 1, to the effect that the Reclamation service has not decided to abandon any part of the Orchard Mesa and that in any event such farmers as own lands on the lower part of the Orchard Mesa and desire to do so will be allowed to purchase water from the proposed government canal.

This announcement was made sometime prior to the contract entered into between the Reclamation service and the Orchard Construction company for the benefit of the Orchard Mesa district.

Since the execution of that contract the conditions have changed materially and the Reclamation service will necessarily avoid taking any action which will

Daily News, Oct 20 1908

interfere with the carrying out of the plans as contemplated by said contract.

The policy of the government in all matters relating to the operations under the Reclamation act has been to avoid interference with private enterprise as far as may be possible without detriment to the interests of the projects. In such a case as the present where an irrigation district is actively proceeding with a view to the construction of a system for the irrigation of lands within the district, the project of the Reclamation service will not be so developed as to conflict therewith. The plans, of the Reclamation project have not yet been definitely outlined so that the exact area to be irrigated thereunder cannot now be defined and must remain to some extent undetermined until the surveys and investigations have been reviewed by a board of engineers and have been presented to me with appropriate recommendations by the director of the Reclamation service.

Very respectfully,

JAMES R. GARFIELD.

Daily News, October 24 1908

ASSOCIATION APPROVES THE HIGHLINE CONTRACT

Aupperle Returns From Washington and Makes Report--Election Will be Held

----

FULL TERMS OF CONTRACT SECURED

Secretary D. W. Aupperle returned to Grand Junction yesterday from Washington where he went to secure the signature of Secretary Garfield and his report was made to the Water Users' association which met in special session last evening. The committee approved the report of Mr. Aupperle and the contract was signed and approved.

The president and secretary of the association were authorized to call a special election for the purpose of allowing the shareholders to ratify the action and the election will be called at once. As soon as this formality and minor matters are dispensed with work of construction on the canal will commence.

Mr. Aupperle thinks that actual work will commence during the month of December, as the sum of $25,000 has been set aside by the Reclamation department for immediate use after the formalities are dispensed with and $100,000 is available immediately after the first of January. The subscriptions to the fund raised in the valley may be called in at any time and those who are to pay their assessment in work are anxious to have it commence as early as possible. The route that the canal will take yet remains to be definitely decided, but it is believed that this is also only a matter of form.

Those most interested in the project are greatly surprised that any of the sum to be furnished by the department is available before the first of the year, but due to the very efficient work and efforts of Mr. Aupperle and the willingness of the department that has been accomplished.

"Our people have cause for congratulation" said Mr. Aupperle last night, when discussing the matter. "The great interest manifested by Secretary Garfield and others in the Reclamation department is the principal cause of such prompt action in the matter and we all owe them a debt of gratitude. The first thing Mr. Garfield said to me was: 'I hear you people have raised $125.000 by subscription and I'm mighty glad of it.' We spent part of three days going over the matter and everything was thoroughly went into." The contract as signed and approved is as follows:

Now, therefore, the parties hereto agree that:

1. The association shall, as hereinafter provided, make available for use by the United States in the construction of the said Grand Valley project the money and labor so subscribed, and the Secretary of the Interior shall allot the sum of one hundred and twenty five thousand dollars ($125,000) from the Reclamation fund established under said act, such allotment to be utilized for the construction of the said project during the remainder of the present year and during the calendar year of 1909 to an extent equal to the amounts of said subscriptions made available to the United States by the association in money and work for such construction.

2. The amounts subscribed in cash to the fund so contributed by the association shall be collected by the association at such times and in such amounts as may be designated by the director of the Reclamation service and shall be made available for use in payments of obligations incurred in said construction within 30 days from notice by the director of the Reclamation service that the same are required. Such moneys when collected shall be deposited in such bank or banks at Grand Junction as may be selected by the association when countersigned by the special fiscal agent of the United States Reclamation service assigned to the project. Such funds shall remain on deposit at the risk of the association until the same shall have been drawn from the bank or banks on checks thus issued; and the treasurer of the association shall be required to give a bond to the association in the sum of $25,000 for the faithful performance of his duties in connection with the handling of these funds.

3. The association may give to the persons paying moneys to the treasurer for the fund contributed, a receipt or certificate showing such payment and stating that when the moneys have been applied in payment of the cost of construction of said project, the payor or his designated assignee will be entitled to receive from the association, co-operation certificates to the amount of the moneys paid by him.

4. The association shall contract or place orders for such work and materials as may be requested by the project engineer of the Grand Valley project and when such work has been performed or the materials delivered to the Reclamation service, the project engineer will prepare and approve accounts therefore in favor of the association showing by whom the work was done or the materials furnished.

5. When these accounts have been submitted to the association and checks therefor have been delivered to the proper payees, the project engineer will register co-operation certificates in total amount equal to the total of such accounts and the association will deliver such co-operation certificates to the persons who have paid in moneys upon the calls issued by the association.

6. Such co-operation certificates shall not be redeemed in money by the United States, but will, in accordance with the rules and regulations established by the Secretary of the Interior be accepted and applied at their face value in reduction of the installments of water right charges hereafter to accrue against the lands within the project, when public notice thereof shall have been issued in accordance with Section 4 of the Reclamation act.

7. Such certificates may be redeemed by the association by special

Daily News, October 24 1908

assessment or assessments made against all the lands under the project to create a fund for such purpose, or they may be taken up by the association with funds which may come into its hands in pursuance of its first assessment made for the collection of the charges announced in said public notice issued by the Secretary of the Interior.

8. Of the allotments heretofore made by the Secretary of the Interior for said project, amounting to the sum of one hundred thousand ($100,000) dollars, to be utilized prior to and during the calendar year 1909, for the acquisition of the necessary property, rights of way, and other purposes in connection with said project, the unexpended balance shall remain available in addition to the allotment of one hundred and twenty-five thousand dollars ($125,000) herein provided for.

9. For the work to be performed in pursuance of subscriptions payable in work, specific contracts will hereafter be entered into between the parties hereto, providing for the character of the work and the method of making settlements therefor in co-operation certificates:

10. Only those who are or may become members of said association, under the provisions of its articles of incorporation and by-laws, shall be accepted as applicants for rights to the use of water available by means of said proposed irrigation works, and no application for water right shall be accepted by the United States without a certificate signed by the secretary of the association to the effect that the applicant for such water right has duly subscribed for stock of the association for the lands described therein, and that all assessments levied against said stock have been fully paid up to the date of the applications.

11. After the year 1909 the construction of the project shall be continued by expenditures from the reclamation fund to the extent that money may be found available for that purpose by the Secretary of the Interior. In making final plans for the system the Secretary of the Interior, so far as he may find it practicable and feasible, shall make provision for supplying water to all the lands in the Grand Valley not otherwise provided for, on condition that a sufficient acreage is assured to the project under conditions which are practical for the economical application of water.

12. All work done or material furnished in pursuance of this agreement shall upon acceptance become the property of the United States, subject to the terms of the Reclamation act of June 17, 1902, and to the rules and regulations established in pursuance thereof.

13. No member of or delegate to congress, officer, agent or employe of the United States is or shall be admitted to any share or part of this contract or to any benefits to arise therefrom, and sections 37339, 37340, 37341, 37342. revised statutes of the United States, so far as applicable, enter into and are part of this agreement.

Daily News, Nov 18 and 21 1908

IMPRACTICABLE TO TAKE WATER OVER THE DIVIDE

"The Water users of the Grand Valley need not loose any sleep over the reported plan of a number of capitalists on the Eastern Slope attempting to secure large filings on Grand river water. At the best the Eastern Slope men could not get more than 20 per cent of the water that flows in the Grand. With proper storage the flow of the Grand river at its lowest stage is sufficient to supply a million acres of land."

This is the statement of Reclamation Engineer E. E. Sands of the Grand Valley project in regard to a flashy story printed in the Rocky Mountain News and clipped and republished by a local evening daily yesterday purporting that the waters of the Grand were in danger of being stolen by Eastern Slope men

"The Central Colorado Power company has already begun construction of a number of large reservoirs partially solving the problem of storage for all of the irrigation projects below Glenwood Springs. Others can be built if necessary.

The principal reason for any particular alarm over any movement on the part of the Eastern Slope men in an effort to obtain filings to such an amount of water as to seriously affect Western Slope projects can be based on the fact that to obtain water from the Grand and its tributaries and to get it over the mountain the water would have to be taken so high up on the Continental Divide that but a small per cent of the river's actual flow in the lower regions could be secured. The river would have to be tapped very close to the snow and water line.

The Tennessee Pass project is one of these projects which has been frequently talked of. The promoters of the project plan to take the waters of the Grand and its tributaries through the Tennessee Pass in to the Arkansas valley. They would have to secure their water from the Eagle river. They could not secure ten per cent of the full flow of that river and get it over the divide with any reasonable amount of tunneling.

"I do not believe that over ten per cent of the flow of the Roaring Fork river could be successfully taken over to the Eastern Slope. Twenty five per cent of the Fraser river might be tunnelled and ditched over into the other side of the mountains and perhaps ten per cent of the normal flow of the Blue river. None of the Western branches of the Grand can ever be tapped by eastern Slope projects. I believe that not over 20 per cent of the headwaters of the Grand river can be taken to the eastern slope. The enormous expense involved in such a scheme is another reason for belief that it will not be attempted in any very extensive fashion."

There is a suit before Judge Shackelford of the district court to adjudicate the water rights of the Grand. This suit was brought by the Grand Valley

Daily News, Nov 18 and 21 1908

Irrigation company and a number of other ditch companies early in November and was filed by Attorney Carnahan and Griffith. It is believed that when the full rights all of the present water owners of the Grand are fully adjudicated it will be found that there is but little water left in the Grand river for further filing.

A greater part of the rights on these are some filings on the part of Eastern Slope men that antedate some of the recent Western Slope filings. The filings made by the Government for the high line canal are among these latter.

The story in the Rocky Mountains states that papers are now being drawn by a syndicate of Denver lawyers attacking the water rights of the following companies:

Grand Valley canal, Irrigating 100,000 acres of land.

United States Government (Gunnison project) irrigating 100,000 acres

High Line Canal, Irrigating 45,000 acres.

Orchard Mesa canal, irrigating 12,000 acres.

Bonita Valley canal, irrigating 4,000 acres.

Central Colorado Power company, capital $12,000,000.

DeBeque Power Company.

Involved on the Eastern Slope are:

Poudre Valley canal (Skyline) taking 400 cubic feet a second.

Henrylyn canal, taking 700 cubic feet a second.

SPECIAL HIGH LINE ELECTION WILL BE HELD DECEMBER 5

M.O. DELAPLAIN ELECTED PRESIDENT TO SUCCEED L.M. MILLER

The special election of the Grand Valley Water Users' association to formally ratify the contract entered into between the association and the government for the construction of the High Line canal will be held on December 5. At this same election the proposition of issuing interest bearing certificates in return for work and money advanced by the members of the association and others for use on the project during the present and coming year will also be voted upon. Both of these propositions will certainly be approved.

Receiving the resignation of President L M. Miller of the association with great regret, which was afterwards expressed in a formal set of resolutions drawn up by the board, the directors of the Water Users' association, which met yesterday morning. elected M. O. Delaplain, president of the association. Mr. Delaplain then resigned as treasurer and W. S. Wallace was elected to take this position.

With the resignation of L. M. Miller from the Board of Directors was also the resignation of John T. Beaman. To fill these vacancies D. W. Aupperle and L. A. Cosler were elected to the Board of Directors.

Daily News, Nov 18 and 21 1908

The annual meeting of the board will be held on the second Tuesday in January, 1909. Routine matters were taken up and discussed and bills due were audited and ordered paid. Secretary Aupperle reported that his last trip to Washington had cost the association $161.00 and turned back a balance of $89.00 which had been appropriated for the expenses of the trip.

NATIONAL PARK SYSTEM OUTSIDE LAW'S SANCTION

ATTORNEY IN FENCING CAN ATTACK PARK POLICY OF GOVERNMENT

NO RIGHT TO SET ASIDE ANY OF THE LAND OF THE PEO-PLE.

Declare it has Exceeded Powers in Taking Up the Gunnison Tunnel Pro-ject.

All the reclamation projects of the government for the irrigation of millions of acres of arid land are illegal, all the national parks set aside by the government like Yellowstone and the Yosemite, are illegal and all the forest reserves set side for the conservation of valuable trees are illegal--in a nutshell, these were the sweeping charges made by Attorney Ethelbert Ward in arguments before the federal court yesterday in suits brought against Fred Light of Aspen, Sheriff Zimmerman of Glenwood Springs and Representative Joseph Luxon of Rifle, based on violations of the federal law in regard to the public domain

If Mr. Ward's arguments are sustained by Judge Lewis and ultimately by the supreme court, where they will undoubtedly be taken, the entire forest reserve system of the forestry bureau will be overturned and the constitutionality of the work of the reclamation depart-

Daily News, Nov 18 and 21 1908

ment and the national parks question. No longer could the government extend a helping hand to the almost desert sections of the West. Such a project as the Gunnison tunnel, which will convert thousands of arid acres into fertile farms and orchards, would be illegal.

According to Mr. Ward, the government has only the power to sell and condemn public lands by proper legal proceedings, within the different states. It cannot set aside a certain proportion of the public domain in Colorado, for instance, for a forest reserve. It can, however, he argues, sell that portion to the state of Colorado and let the state government look after the preservation of the forest on the land.

To say that cattlemen must not allow their cattle to stray on public land is interference with a state's rights by the government, according to Mr. Ward. If the people of Colorado want forest reserves, then let the state give them. But the government is given no power to make the people have reserves against their will.

"Congress can exercise no powers," declared Mr. Ward "other than those granted to it by the constitution, either expressly or through implication. There is no implication in the constitution of power by which congress can reclaim arid lands. Nor is there any implication in the constitution of power by which congress can set aside forest reserves in the different states. The constitution does not grant congress the legislative power over the states."

He read decisions from the supreme court which held that public land can be improved by congress or set aside for some purpose if such an action is necessary for the public welfare. Land can always be condemned by the government for post offices or for forts and military posts, the decisions declared. One decision read was rendered in the suit which attacked the government's right to place monuments on the field of Gettysburg or to make certain other improvement to commemorate, the great battle. The supreme court ruled that the constitution could be stretched a point in order to legalize a movement of so much vital importance. The placing of the monuments the court declared, was in a broad sense a public necessity and benefit.

In reply to Mr. Ward's arguments, former Special Assistant Attorney General Ernest Kaebel, who is appearing for the government, will argue that if the placing of monuments on a battlefield is of such public benefit as to justify stretching of the constitution, surely the preservation of the trees of the nation for posterity is of sufficient public benefit to stretch it even further. Mr. Kaebel is relying chiefly on section IV of the constitution which gives congress the power to "dispose of public lands and to make the necessary rules and regulations" for it. The arguments were continued until tomorrow.

Daily Sentinel, Daily News, Denver News, Nov 21-22 1908

To Mr. Sands Entirely Competent Authority

With due respect for the technical engineering knowledge of Mr. E. E. Sands the project engineer for the reclamation service in this valley, the Sentinel very pointedly disagrees with him upon the possibility of danger from the deflection of the headwaters of the Grand river to the eastern slope.

No one will take any exceptions whatever to the knowledge that Mr. Sands possesses upon matters relating to the government service: but he has not demonstrated that he is a Sir Anthony Absolute upon all other knowledge of the wiles and schemes of mankind.

The lands of the eastern slope of the Rocky Mountains in Colorado need water. They are proportionately valuable as they are watered. There is insufficient water from the rivers and streams of the eastern slope to irrigate those great bodies of land. Men owning same are determined that water shall be provided in some manner: they have, therefore, turned in the direction where sufficient waters are available and propose to secure those waters at any cost if the law does not step in and interfere.

The combined press of Denver has stated that the Henrylyn organization has millions at their command for the purpose of irrigating great bodies of land. That press has also stated that the organizers of that and other companies have taken, already, more than the initial steps to divert the headwaters of the Grand and other western slope streams to their purposes.

Under the circumstances, we think it is presumptuous upon the part of Mr. Sands to attempt to ridicule schemes of such vast import, with millions of money back of them and with men of acknowledged ability and shrewdness at their head, and lull our people into a security that is not warranted.

Engineers of far greater experience than even Mr. Sands claims for himself have studied that project for years and have placed upon it the seal of their approval and have staked their reputation upon it, so that, to make the sweeping assertion (as has been made by Mr. Sands) that this important scheme to divert the waters is impracticable is certainly far fetched upon his part, to say the least.

If there was not some immediate danger, the Grand Valley Irrigation Company would not proceed, as they have in the past week or so, to protect the water necessary for their use by court procedure. This was not deemed necessary before the possibility of a vast struggle to obtain water became imminent. This golden product of the hills is absolutely necessary to reclaim the vast areas of government lands, and to get it vast sums of money are to be spent in arid America.

The Sentinel believes that this is the time for the people to arouse themselves and protect their rights. It may seem to some, as it has to Mr. Sands, extremely indefinite when there should be a probability of our losing our waters but to those who have undertaken to appropriate these waters there is nothing indefinite to them; and, without we do take care of our own rights, we may awaken some morning and find that they have been stolen from us.

Not one among the number of those engaged in this colossal enterprise either believes in its impossible execution or are at all abashed by what other engineers may say of the project. We hardly imagine that Mr. Sands will maintain that there are not just as ca-

Daily Sentinel, Daily News, Denver News, Nov 21-22 1908

pable engineers without the government service as there are in it. Anyhow, we cannot understand why it is necessary at all for Mr. Sands to place the seal of his disapproval upon any movement to protect the waters of our half of the state.

We do not apprehend that it is the purpose of the reclamation service to enter controversially upon all such matters, or that subordinate engineers should speak for it upon matters that do not come within the province of the government work in which they are specifically engaged.

Let us, by all means care for our rights while yet there is time.

'RECLAMATION RECORD' PUBLISHES NEWS ON HIGH LINE

Formal announcement of the appointment of the board of engineers to consider the plans for the Grand Valley project of the Federal Reclamation Service was made in the last issue of the Reclamation Record published for the month of November.

The board consists of W. H. Sanders D. C. Henry, I. W. McConnell and J. H. Qulnton. The Record states that this board will meet in Grand Junction on January 5, 1909. but this date has been changed since the publication of The Record and will probably take place during the middle of next month. The Grand Valley project is reported to be one per cent completed. The entire irrigable area of the project having been mapt (sic) on a scale of 100 feet to the inch with ten foot contours. Sixteen miles of the main canal are reported located and a number of the most important structures designed.

The same issue of The Record has the following report on the Uncompahgre project:

The total advance in the construction of the Gunnison tunnel during October amounted to 251 linear feet. Owing to excessive inflow of water, no work was done in the river portal during the month. The river portal crew completed the excavation connecting the tunnel to its headworks. All of the machinery necessary for concrete lining in this portal was delivered in October. The flow of water in Cedar Creek portal has averaged 7.3 second-feet throughout the month, and had increased to about 8 second-feet at the close of the month, About 275 linear feet of arch and sides of Class 4 tunnel were concreted during the month. The seasoning of South Canal was continued throughout the month as was the operation of the Montrose and Delta canal. There are 39,600 acres of cultivated lands on the Uncompahgre project, 15,600 acres of which are under the canals now operated by the Reclamation Service.

Daily Sentinel, Daily News, Denver News, Nov 21-22 1908

WATER FOR ALL IN THE GRAND, SAYS U. S, EXPERT

Engineer in Charge of High Line Project Reassures Western Slope.

[SPECIAL TO THE NEWS] GRAND JUNCTION, Colo., Nov. 20—Alarmed by the reports that companies are being organized on the eastern slope to take water from the Grand river across the divide, water users and land owners in the Grand valley are waking up to take notice of where they stand in the matter.

At the first intimation of trouble the Grand Valley Irrigation company, one of the largest water users at the present time, began suit seeking the final adjudication of the water rights in the Grand, and to obtain a definite settlement once and for all of the rights of all concerned. A referee has been appointed to hear the evidence, and it is hoped that when he gives his decision the chaos now existing in the filings will be cleared up.

The High Line canal, the government project now under construction. will be by all odds the largest water user on the slope. E.E. Sands, the project engineer in charge of the work, has issued a statement which, if true, is calculated to allay the fears which are entertained that the eastern slope speculators are about to appropriate a large share of the Grand's water.

He is of the opinion that it will be found impracticable from an engineering standpoint to tunnel the divide and tap the sources and tributaries of the Grand low enough down to obtain any appreciable amount of the flow. Only 20 per cent of the water at best, he believes, can be taken to the other slope. The expense, and the obstacles in the way of coming lower to take more water, he says, makes it out of the question.

And, even granting the scheme were feasible, Sands states that the filings on the water, all of which are on record in his office, by the users on the western slope antedate those of prospective users across the divide. The courts will amply protect Grand valley land owners should a situation ever arise to require it.

Daily News, Dec 1 1908

ATTEMPT TO BLOCK ORCHARD MESA WORK

T. C. Henry and Son Obtain Temporary Injunction Restraining Orchard Construction Co. From Disposing of Bonds

----

CONSTRUCTION WILL NOT BE DELAYED

Suit Result of Henry's Quarrel With Magenheimers Over Water Rights

----

Like a thunderbolt out of a clear sky came the announcement here yesterday morning that an injunction had been granted against The Orchard Construction company and C. C. and G. J. Magenheimer of the Chicago capitalists who are about to build the Orchard Mesa Irrigation works, restraining them from disposing of the bonds or any money which they may receive from the district in compensation for their services, pending the suit which has been filed by T. C. Henry and son of Denver, regarding a dispute over certain water rights connected with the project.

Judge Greely Whitford issued the temporary injunction in Denver Saturday and a copy of it was served on the officers of the Orchard Mesa District yesterday. They are not all directly a party to the suit however and, Attorney S. M. Logan stated last night that the whole matter was merely the result of a quarrel which the Henrys' have with the Magenheimers and will in no way affect the Orchard Mesa undertaking.

Preparations have been completed for the immediate commencement of the work and it is not thought that this legal tangle will cause any interruption. Mr. Greer and Mr. Barber, officers of the

Daily News, Dec 1 1908

Orchard Mesa District returned from Chicago yesterday. They stated that the agreement with the Magenheimers by which they were to take the bonds in payment for the work had finally been signed. The City National Bank of Chicago has been selected as the trustee to hold the bonds and to turn them over to the Orchard Construction company as the work progresses.

Plans and specifications for the work are all complete and bids for the subcontracts will be advertised for within a few days and immediately upon the awarding of the contract work will commence.

The following is the copy of the court's order:

George C. Henry. T. C. Henry and E.C. Henry. plaintiffs.

The Grand Mesas Land and Canal and Power company, The Orchard Construction company, Woodland Construction company, C. C. Magenheimer, Ira D. Buck, G. J. Magenheimer and E. G. Eberhardt, defendants.

THE PEOPLE OF THE STATE OF COLORADO.

To defendants above named and their agents, servants, employees, representatives and confederates:
SEND GREETINGS:

At this day come the said plaintiffs by their attorneys and the said plaintiffs having filed their complaint with the district court against the above named defendants praying for a restraining order against the said defendants requiring them to refrain from certain acts in said complaint and hereinafter more speparticularly (sic) mentioned.

It is therefor ordered by the said court that until further notice of said court you and your servants, agents, attorneys, employees, and all other persons acting under your control, authority or direction of yon, do absolutely refrain from and desist from selling, assigning, transferring paying out or otherwise disposing of any bonds or cash received by you or any of you for or on account of the contract between the Orchard Mesa Irrigation District of Mesa county, Colorado and the defendant, The Orchard Construction company or pursuant thereto or in connection thereof, and from removing the same or any part of them from the jurisdiction of the court and from doing or attempting to do any of the acts herein specifically mentioned or any acts looking to connected with or designs for the doing of the same until a hearing upon said application for the writ of injunction and further order of the court in the premises.

Witness whereof the Honorable.

Greely W. Whitford, one of the judges of the District Court of the Second Judicial District of the state of Colorado with the seal thereof affixed at my office in the city of Denver this 28th day of November A.D. 1908

Percy W. Clay
Clerk.

T. C. Henry has been a sort of nemesis to this valley. It was most unfortunate that those who were working to build a ditch on Orchard Mesa were not fully apprised of Henry's ways before engaging with him in any enterprise.

Daily News, Dec 1 1908

RIGHT OF WAY FOR HIGH LINE IS ADJUSTED

Railroad Officials and Engineer Sands Inspect Route

As a result of a meeting and trip of inspection yesterday between Project Engineer E. E. Sands of the Reclamation Service and General Manager Ridgeway and Chief Engineer Gwyn of the Denver and Rio Grande there is every prospect that the present negotiations over the High Line canal right of way along the tracks of the Rio Grande above Palisade will be satisfactorily settled within the next two weeks.

Engineer Sands and the railroad officials left Grand Junction yesterday morning in a special car and traveled along the entire route of the proposed canal right of way. The officials of the railroad were given a complete statement as to the plans of the government at this point and at the conclusion of the inspection the railroad men were apparently willing to do everything in their power to assist the government in regard to the building of the ditch,

At Newcastle the trip closed, Ridgeway going to Denver in his special and Chief Engineer Gwyn and Engineer Sands returning to Grand Junction over the Midland route.

"I will go to Denver within a few days," stated Engineer Sands yesterday afternoon, "and will then probably close an agreement with the Rio Grande in regard to the use of a part of their right of way for the canal. I found the railroad men anxious to be very fair in the matter and I have every belief that the matter will be satisfactorily settled".

Daily News, Dec 3 and 10, Denver Post, Dec 31 1908, Daily News, Jan 19 1909

HIGH LINE DITCH CONSULTING BOARD MEETS

Plans and Estimates of Work to be Gone Over

IN SESSION TODAY

Contracts Will be Let as Soon as Findings Are Approved

Commencing this morning and continuing for at least the next seven days and probably ten, the United States Reclamation engineers, J. H. Quinton of Los Angeles, W. H. Sanders of Boise, Idaho, and Will E. Gerig of Manila, Philippine Islands, will sit in Grand Junction as an examining board of engineers to pass upon the plans for the Grand Valley project. Supervising Engineer I. W. McConnell will also be a member of the board. Project Engineer E. E. Sands will attend all of the meetings and present the maps and reports as prepared by his office. Besides looking over the maps and reports, the estimates of cost and material, the board will make a personal inspection of the ground through which the canal is to run.

After completing their examination they will report to Director Newell of the Reclamation Service, who will in turn file their report to Secretary Garfield. It is expected that Secretary Garfield will have passed upon the findings of the project board by Christmas day and work will then commence as soon as contracts can be let.

Yesterday Project Engineer Sands returned from Denver where he and Supervising Engineer McConnell had been in consultation with General Managers Vallery of the Midland and Ridgway of the Rio Grande in relation to the canal right of way along the railroad tracks above Palisade. No trouble has been experienced by the engineers' in coming to an agreement with the railroad men and they have every hope of having the negotiations successfully concluded.

Referring to the work of the project board Engineer Sands declared yesterday that he hoped to be through with the examination within a week.

"The board will probably be employed at least a week in the examination" he said. “Mr. Quinton is expected tonight from Los Angeles and Mr. Gerig is already here. We will take a trip over the proposed right of way of the canal and examine all of the plans in the Reclamation office."

Daily News, Dec 3 and 10, Denver Post, Dec 31 1908, Daily News, Jan 19 1909

STATE ENGINEER FEARS U. S. WILL GOBBLE UP WATER

T. W. Jaycox in Biennial Report Describes Government Encroachment.

URGES STATE TO FIGHT

Recommends Surveys of Rio Grande for Defense of Rights

In the biennial report of State Engineer T. W. Jaycox a recommendation is made that the seventeenth general assembly appropriate an adequate sum for the work of collecting data upon the surface flow of the Rio Grande river, and other physical characteristics of the stream, which information is needed to defend the state's rights against the aggressions of the reclamation service.

The action of the federal bureau in denying applications for rights of way for irrigation systems In the San Luis valley makes it incumbent upon the state to take some action to protect the settlers within the borders, the state engineer thinks, and he views the San Luis condition as the first of others that will be caused by increasing activity of the bureau. For that reason he believes the state should prepare itself now for a conflict with the bureau, a battle which is inevitable and imminent.

'°I believe as a matter of prudence," writes Engineer Jaycox, "that the legislature should provide an adequate sum for the collection of data upon the surface flow of the Rio Grande and its tributaries in Colorado and in Texas even to El Paso; loss of underground waters; data regarding deforestation of watersheds of the Rio Grande and its tributaries in New Mexico, and whether or not a private land scheme may profit from the building of reservoirs by the reclamation service."

The last phrase is a reference to the federal irrigation project at Engle, New Mexico, where it is planned to construct a reservoir and a diversion dam that will cost $8,000.000, and will reclaim 110,000 acres in New Mexico and 7,000 acres in Texas. It is this project which has caused the federal bureau officers to issue orders against the use of the flood waters of the Rio Grande by the settlers of the San Luis valley. The land to be reclaimed in New Mexico is all new land, and privately owned. It is not a part of the public domain. The Texas land is also free of government authority, for the reason that the Lone Star state retained all of her lands as a condition of her admission to the union.

Another reason for activity on the part of the legislature is that a test case attacking the powers of the reclamation bureau officers and raising the question of constitutionality of the entire act is now being tried in San Francisco. It may be necessary to begin proceedings in this state in the federal court to get a ruling that will revoke the orders of the bureau chiefs.

Daily News, Dec 3 and 10, Denver Post, Dec 31 1908, Daily News, Jan 19 1909

NOT PLANNING TO FIGHT GOVERNMENT

The Mesa County Land Holders' Protective Association, which was recently formed at Palisade, with J. H. Sayles, president, and A. M. Echerkneck, secretary, is not, as has been understood, fighting the early completion of the High Line canal project, declares R. H. Bancroft of Palisade.

"There is a misapprehension in regard to the organization," declared Bancroft yesterday. "It was organized about 30 days ago by the men that own land which will be crossed by the High Line canal. Its articles of association expressly state that it has not been formed to hinder in any way the work of building the canal. All the members desire is a just and equitable settlement with the government, and they hope to get this easier by means of an organization than as individuals. One meeting with Engineers McConnell and Sands has already taken place and another will occur shortly."

WATER USERS ELECT DIRECTORS

The Water Users association, held an uneventful meeting yesterday and elected a new board of directors for the next year. Other than that no important business took place there were no particular contests and the following is the composition of the new board:

W.S. Wallace, H. W. Blakeslee, D. W. Aupperle, M. O. Delaplain, Wm. Campbell, James Smith, F. F. Harriett, Lester Jaynes, Daniel Lynch, J.H. Hunter, John Hynes.

They will meet at 10:00 o'clock Saturday to elect officers.

Denver News, Dec 16 1908

Colorado Needs Irrigation Legislation

I.W. M'Connell –
Able government engineer makes suggestions – the News invites discussion

On Friday evening, December 10, I. W. McConnell, chief engineer of the government reclamation service in this section, spoke on "Needed Irrigation Legislation" before the Grand Junction Chamber of Commerce. The News feels that Mr. McConnell's suggestions are of such importance as to justify their very serious consideration by the people of this state. Those who are interested in irrigation are invited to discuss Mr. McConnell's proposals through the columns of this paper. Communications should be made as brief as possible.

Mr. McConnell's address follows;

COLORADO is today a state of opportunity. The development of its possibilities in precious minerals, coal, clay products, cement-making materials, building stone, factories and farms hold out alluring promise not only to the promoter but also to the laborer, the artisan and the farmer who come after to build homes and enjoy the fullness of the land.

Of all these great resources that one which is most enduring and which is now and will continue to be the greatest single source of wealth is the matchless fertility of the soil under irrigation.

There is today a great fund of money looking for investment in irrigation enterprises. The volume of this demand may be inferred from the statement that the United States reclamation officials of the state examine an average of six applications per week for rights of way upon the public domain for ditches and reservoirs. Not all of these are feasible and many of them are of small magnitude, but many enterprises located upon private lands do not come to the attention of federal officers at all.

Day of Cheap Water Practically at an End

Now, the first requirements of irrigated farming are land and water. Of land there is an abundance, but in spite of the great floods which run every year across the state line, the day of cheap water as we understand that term is practically at an end in Colorado. With the exception of a few streams in the northwestern part of the state the low water flow of all the rivers has been filed upon and will be appropriated within the next four or five years by works already under construction. Upon many streams of the eastern slope the limit of reliable storage possibilities has also been reached, and in some cases it has been passed.

There are in contemplation at the present time at least six plans, which have the indorsement of reputable business men and civic bodies, proposing the diversion of waters from the western slope, across the Continental divide, for the irrigation of lands upon the eastern slope.

The western slope does not abound in good reservoir sites, meaning by that term localities where waters may be impounded at a low cost. Consequently, we may anticipate a progressively increasing price for water rights and for lands which have a reliable irrigation supply,

Denver News, Dec 16 1908

with a corresponding increase in litigation and uncertainty.

Now we hold that the first requisite of successful investment is reliability. We hold further that the good name and the final prosperity of the state depend less upon the amount of money which exploiters of its resources make than upon the value rendered to those people who arrive within our borders for the purpose of redeeming the lands, building homes and adding to the wealth produced.

A water right on paper is readily manufactured. A water right which will actually furnish the amount of water which the gaudy document that represents it says it will furnish, is an expensive thing to find, a more expensive thing to develop, or a most expensive thing to maintain. There is no more popular game of chance known in Colorado at the present time than the purchase of water rights, and the funds of worthy mechanics, professional men, widows and orphans are being spent in great volume in this form of investment.

The authority for the worth of these rights varies from the opinion of the pastor of the purchaser's church to the opinion of the lawyer or the engineer who will give an opinion for the smallest fees.

Now in mining that kind of development is called "wild catting." Its harmful results will be in no way alleviated by the fact that it goes on in the new fields of irrigation and power development and we hold that the great state of Colorado can not afford to disappoint its actual settlers, home builders and investors for the small immediate benefits which will follow the feverish exertions of its exploiting-promoters, who hope to put something over before the excitement is past.

Some Legislation Which Is Required

We hold that properly designed laws will weed out at least the most fraudulent and the worst designed of the schemes, and that such a law will add value to those enterprises of actual merit, of which the state has many. Under present laws we believe that unsound development is encouraged. We recognize the contention of our state's rights friends and agree that these matters should be remedied by the state and not by the United States.

We suggest for the consideration of the business men of this community who are interested in sound development, the following modifications of the law which need not disturb vested rights, and which will in some measure at least, throw the protection of the state about the small investor:

First--Relieve the office of state engineer of political control by making his term of office ten years, subject to removal by the governor for incompetency or malfeasance. Place all his assistants under a rigid civil service rule. Place all water commissioners under the state engineer and upon state pay.

Second--Filing on water. Require the state engineer to review all filings and to reject those which apply to streams already over appropriated. For reservoir filings, require a certain definite progress on the work and require all filings to lapse automatically if requirements are not observed.

Third—When adjudications by the courts are asked on streams which run in more than one district, require by ample notice all appropriations on the stream to intervene regardless of districts.

Denver News, Dec 16 1908

Fourth--Require all irrigation construction to be subject to approval and inspection by the state engineer.

Fifth--Require the state engineer to examine irrigation plans, and reject those which propose the utilization of only the select portions of those projects which by reasonable expense can be enlarged to make a greater and more complete use of water and land.

Sixth--To guard against arbitrary decisions by state engineer, provide for an appeal to a board of review composed of state officials.

Seventh--Provide that promoters upon completion of their filings shall apply to the state engineer for a statement as to water supply and land available under the proposed development. This statement shall set forth the possibilities of the project and no sale of stock, or assignment of an interest, or sale of water right or land shall be made until the purchaser shall have been given the statement of the state engineer concerning the proposed development,

As to Water Rights

We print on this page the speech recently made by Mr. I.W. McConnell, chief engineer of the reclamation service, at the banquet in Grand Junction.

And we hope that all persons interested in the irrigation question in Colorado will carefully read this speech, and see whether they have any objections or confirmations to offer. Particularly do we request all irrigation engineers to send us their suggestions, whether these follow the plan outlined by Mr. McConnell or not.

Mr. McConnell is entitled to be heard with respect when he speaks on the irrigation question. He is one of the foremost civil engineers in this section of the country, and he is a broad-gauge thinker and close observer as well. We know that Mr. McConnell tells the truth when he says that there is a vast amount of "wildcatting" just now in the irrigation business. Every purveyor of the wildcat mining stocks of ten years since--who has managed to avoid state's prison without leaving the state between suns--is now selling irrigation stock. And the public is buying. It is buying just as it bought the worthless mining stocks in the old days, because it has no means of distinguishing the good from the bad, and the increasing prosperity of the state inclines the investing public to consider everything good.

Whether Mr. McConnell's plan for stopping this wildcatting is feasible or not The News will not undertake to say without careful examination. We want all the expert advice we can get. If you have any send it in.

Daily News, Jan 6, Jan 21 and Feb 2 1909

INJUNCTION HAS DELAYED WORK ON ORCHARD MESA

Magenheimers Expect to Quash Case at Hearing Feb. 5

Tile suit instituted by the Henrys of Denver against the Orchard Construction Co., which has the contract to build the Orchard Mesa ditch and irrigating plant, has at least one result which has undoubtedly been a source of gratification to the instigators of the suit, it has tied up temporarily at least all work on the ditch. The restraining order of the court which was granted several weeks ago has prevented the work from being commenced promptly on the first of this year as was planned and until the hearing which is set for Feb. 5. absolutely nothing can be done. At that time Henry, through his attorney, will ask the court to make the restraining order permanent. The Orchard Construction Co., through Attorneys Hale Dawson and Wright of Denver, will represent the Magenheimers who seek to have the injunction dissolved. On the ground that regardless of merits of the case the court's order is delaying the construction of an important and public utility, the Magenheimers will ask that they be allowed to proceed with the work pending ultimate settlement of the suit. Messrs. Smith and Barbour returned yesterday from Chicago where they had been conferring with the Magenheimers regarding the status of affairs. Barbour stated that there was every prospect that after February 5th work could start with a rush. The members of the Orchard district who are anxious for the water are confident that matters will be so adjusted that the ditch can be built as originally planned. The Magenheimer brothers will be here early next month in order to be present at the hearing which probably be held in Denver.

Daily News, Jan 6, Jan 21 and Feb 2 1909

"TRANSPARENCIES" TO ADVERTISE VALLEY AT THE ALASKA-YUKON FAIR

Reclamation Service to Do Work Free

Grand Junction and the Grand valley are to be given a big boost at the Alaska-Yukon exposition next summer by the federal government and furthermore it will not cost the men of the Grand valley a single cent for the advertising.

This is the valley's reward for entering the panoramic display proposition which was put up to the Chamber of Commerce several months ago by Statistician J. C. Blanchard of the Reclamation service, who will be in charge of this department of the government's display at the big Seattle show. Secretary Aupperle of the Chamber of Commerce received a letter from Mr. Blanchard yesterday outlining the matter.

It seems that the number of applications for space on the panorama proposition had dwindled down to two, one of which was the Grand Junction Chamber of Commerce. The original number hoped for was four. This number was secured at one time but later two were withdrawn and this caused the government to drop the scheme, Blanchard, however, felt that the Reclamation service should secure its display and so has arranged for a number of monster transparencies which will be placed in the government building. This will be supplemented by two lectures each day illustrated by colored lantern slides.

"These lectures proved to be very effective at Jamestown", he writes, "and I have no doubt but that they will be equally so at Seattle. We can achieve excellent results in the way of the promotion of settlement by their use. I shall personally see that the matter of securing transparencies of the Grand valley are properly displayed and that illustrated scenes of the Grand valley are in the lantern slides. You can count on my abiding interest and active co-operation in promoting your district."

Sincerely.

J. C. BLANCHARD, Statistician

Daily News, Jan 6, Jan 21 and Feb 2 1909

SETTLEMENT IS NOW REACHED BY WATER USERS

Basis for Damages for Canal Right of Way is Fixed

After a protracted session lasting over three days the Water Users association and the Mesa County Land Owners Protective association yesterday afternoon agreed upon a contract for the payment of land to be taken up by the High Line canal right of way in the Palisade district.

This afternoon at four o'clock Secretary Aupperle of the Water Users will leave Grand Junction for Washington with the contract in his pocket and with the intention of camping on the steps of the capitol until the High Line canal contract is formally signed by Secretary Garfield.

The contract between the Palisade men and the Water Users association provides for the payment of ten per cent of the agreed upon valuation of the individual properties to be paid down within sixty days and the balance optional with Secretary Garfield of the Secretary of the Interior. It is believed that the Reclamation service will approve of the contract and if they do so the last obstacle of moment will be removed from the path of the secretary and he will be free to sign the contract calling for the construction of the canal. The cause of the long discussion between the local men was brought about by the difference in opinion as to the time when the money for the right of way was to be paid. Some held out for an immediate payment of the full amount, others wanted ten per cent down and full payment within two years. The compromise as above noted was finally agreed upon. The signing of the agreement took place at four o'clock in the office of the Chamber of Commerce where the session was held.

Daily News, Feb 13 and 14 1908

SIGNING OF HIGH LINE CONTRACT ONLY AWAITS ARRIVAL OF PAPERS

Waiting for the necessary papers to arrive from Denver prevented Secretary Garfield from yesterday placing his signature to the High Line project papers. That they will be signed either tomorrow or Thursday without any hitch is assured beyond any possibility of a doubt. As announced exclusively yesterday morning the last obstacle which stood in the way has been removed and aside from the formality of the signing the preliminary steps in the giant undertaking are now completed.

The following dispatch was received last night from the Daily News correspondent, Mr. Sheldon Cline:

"The signing of the Grand Valley Project papers awaits the arrival of the revised contract from the D. & R. G. which is expected to get here by Monday. Secretary Garfield would have signed the papers today had these other papers been on hand. As it is, it will probably receive his signature Monday afternoon or possibly Tuesday."

The following dispatch was sent out by the Associated Press last night from Washington, relative to the matter. By means of its own special correspondent the Daily News received practically this same dispatch 24 hours ahead of the regular Press Service.

WASHINGTON, Feb. 13.--That the contract for the construction of the Grand Valley irrigation project |n Colorado, which has been withheld because of what Secretary Garfield believed to be an objectionable clause, will be signed officially within a few days, is the belief here. As the canal would parallel the Rio Grande road at several points the attorneys inserted a clause obligating the government to assume responsibility for the damage to the right of way and injuries of the workmen. Garfield objected and at a conference today with George Gould and President Jeffries and D. W. Aupperle of Grand Junction, representing the farmers, an agreement was reached.

Daily News, Feb 13 and 14 1908

SUBPOENA SECRETARY GARFIELD IN MAGENHEIMER-HENRY CASE

DENVER, Feb. 9.--The suit instituted by T. C. Henry and son against the Orchard Mesa Construction Co. and the Magenheimer Brothers of Chicago begun before Judge Whiteford today. Probably the entire week will be spent in arguing the question of a change of venue which is what the defendants are fighting for. Attorney Clyde Dawson representing the Henrys spent the entire session today in argument against the change. It was announced today that Secretary Garfield would be subpoenaed as a witness.

Daily News, Feb 13 and 14 1908

PROSPECT IS GOOD FOR NATIONAL IRRIGATION CONGRESS HERE IN 1910

Engineer I. W. McConnell Urges Chamber. of Commerce to go After This Coveted Honor

----

SUPERVISING ENGINEER I. W. McCONNELL OF THE RECLAMATION SERVICE HAS COME TO THE FRONT, IN A LETTER TO SECRETARY MAHONEY OF THE CHAMBER 0F COMMERCE ENDORSING AN IDEA WHICH THE DAILY NEWS STRENUOUSLY FOUGHT FOR LAST SUMMER, IN ADVOCATING THAT THE WESTERN SLOPE MEN GET IN LINE TO SECURE THE 18TH NATIONAL IRRIGATION CONGRESS FOR GRAND JUNCTION IN THE FALL OF 1910

MR. McCONNELL SUGGESTS THAT THE WESTERN SLOPE CHAMBERS OF COMMERCE GO INTO THE FIGHT TO SECURE THE BIG MEETING FOR GRAND JUNCTION WITH A VIM. HE DECLARES THAT HE BELIEVES THAT WITH THE OPENING OF THE WORK ON THE HIGH LINE AND THE COMPLETION OF THE WORK ON THE GUNNISON TUNNEL AND THE OPENING OF THE UNCOMPAHGRE VALLEY THE WESTERN SLOPE WILL BE THE CENTER OF INTEREST IN THE IRRIGATION WORLD

The suggestion is one that should be seriously considered by every Chamber of Commerce in the Grand Valley. Through the agitation of the Daily News for this same project last summer the Chamber of Commerce at Green River, Utah, passed resolutions endorsing Grand Junction as the meeting place for the big convention and urged the Grand Junction men to go after it.

After some consideration of the project the Chamber of Commerce decided that it would be advisable to delay going after the meeting until next year. This makes Engineer McConnell's letter peculiarly pertinent at this time. It is as follows:

Department of the Interior
United States Reclamation Service
Denver, Feb 12 1909

Secretary Chamber of Commerce
Grand Junction, Colorado

Dear Sir,

The Seventeenth National Irrigation Congress will be held at Spokane, Washington, August 9-14, 1909. This convention attracts a very large number of visitors from all sections, mainly men of prominence who are interested in the promotion of enterprises which tend to settle the States of the arid region. At the Sixteenth Congress held at Albuquerque, New Mexico, last September there were present over 1000 delegates with a large representation of unofficial visitors. Cer-

Daily News, Feb 13 and 14 1908

tain communities added very appreciably to the value of the Congress by sending either delegates or boosters who distributed special literature and howled for their special communities in season and out of season. The Seventeenth Congress will be a sort of overflow for the Alaska-Yukon-Pacific exposition which will be going on at Seattle at the same time.

It is my judgement that the towns of the Western Slope, particularly those interested in the development of the Uncompahgre and Grand Valley projects can afford to organize with view to chartering a special car and sending a delegation to advertise the communities interested. Such an effort would reach in a personal way a very large number of people interested in fruit growing sections as the inquiries concerning congress in the various districts which I visit indicate the intention of large numbers of people to take this opportunity to inspect the fruit lands and other opportunities of the Pacific Northwest. While such an outcome would be difficult to bring about I should not be surprised if a systematic effort in that direction, might land tile Eighteenth National Congress in Grand Junction or some other city of the Western Slope. Such a result, however, would require a persistent and somewhat expensive campaign which should be inaugurated at once.

In view of the early opening of the Uncompahgre project and the proposed development in the Grand Valley I deem it wise to call these matters to your attention in order that a systematic effort looking toward the advertising possibilities in connection with this congress may be put forward if deemed advisable.

Very respectfully,
I.W. McCONNELL,
Supervising Engineer.

Daily News, Feb 13 and 14 1908

GARFIELD TO SIGN HIGH LINE CONTRACT

GEORGE GOULD TAKES HAND IN RAILROAD RIGHT OF WAY FIGHT AND MAKES PROPER AGREEMENT

Aupperle Visits New York Magnate and Clears Away Final Obstacle to Secretary's Formal Action

----

THE HIGH LINE CANAL CONTRACT WILL BE SIGNED BY SECRETARY OF THE INTERIOR RUDOLPH GARFIELD TODAY OR MONDAY. A FINAL AND SATISFACTORY AGREEMENT IN REGARD TO THE RIGHT OF WAY OF THE CANAL ON THE RAILROAD PROPERTY ABOVE PALISADE WAS REACHED YESTERDAY AFTERNOON BETWEEN GEORGE GOULD, PRESIDENT JEFFERY AND SECRETARY D. W. AUPPERLE OF THE WATER USERS ASSOCIATION IN NEW YORK.

THIS REMOVED THE LAST OBSTACLE IN THE PATH OF SECRETARY GARFIELD AND UPON RECEIVING MR. AUPPERLE'S REPORT HE DECLARED TO MR. SHELDON CLINE, THE DAILY NEWS' SPECIAL CORRESPONDENT IN WASHINGTON, THAT HE WOULD SIGN THE CONTRACT AT ONCE.

THE DISPATCH WAS SENT TO THE MEETING OF THE DIRECTORS OF THE CHAMBER OF COMMERCE BY THE DAILY NEWS LAST EVENING AND WAS RECEIVED WITH CHEERS.

THE DISPATCH IS AS FOLLOWS:

BY SHELDON CLINE

Secretary Aupperle of the Grand Junction Water Users' association has returned from New York where he saw President Jeffery and George G. Gould about the demands made by the railroad in relation to the right of way of the main canal of the Grand Valley project which Secretary Garfield could not accede to. Gould and Jeffery took the matter up at once with their Denver officials by wire and today Aupperle received word from Jeffery saying that a satisfactory adjustment had been reached. This removes the last obstacle to Garfield's signing the contract with the Water Users' association. Nothing could be done today it was a legal holiday, tomorrow unless a rush of business in the Department of the Interior forces the matter over until Monday.

Daily News, Feb 13 and 14 1908

The News dispatch, which up to midnight tonight is the only one in this city from Washington in regard to the High Line matter lifts a heavy load from off the mind of President M. O. Delaplain of the Water Users' association and will allow Project Engineer E. E. Sands to begin at once planning steps toward immediate construction. Authority to advertise for bids on construction will probably be received by Engineer Sands within the week. The bids will be advertised for ten days and actual dirt should be moving on the canal immediately after the contracts are let.

The difficulty in the way of an earlier settlement of the High Line matter was the objections of the government to signing the contract until the matter of the details of the adjustment of payments for the right of way of the canal through the Palisade district was settled with the Mesa County Land Owners Protective league and the right of way difficulty with the Denver and Rio Grande was cleared away.

The first of these obstacles was removed after a three day session with the land holders late last month. The railroad adjustment was not completed until yesterday. It was a source of worry to the Water Users which they concealed from the general public but it is not going beyond the bounds of the strictest truth to say that President Delaplain has been greatly worried over the High Line matter since Secretary Aupperle left for Washington.

The receipt of Mr. Cline's dispatch clears away all doubts as to the certainty that the canal is assured the Grand Valley.

Speaking of it last evening President Delaplain said:

"This is the word that I have been longing to receive for the last week. It is certainly great news and will be a splendid story for The Daily News to give to the people of the valley tomorrow."

The dispatch was then read aloud to the members of the Board of Directors of the Chamber of Commerce and cheers rang out when the reading was completed. "I expected this from a telegram I received from Engineer McConnell today," said Project Engineer E. E. Sands last night.

"That means that the canal will surely be built. I never was doubtful but still the uncertainty over the matter was uncomfortable."

Daily News, Denver Republican, Feb 17 1909

GARFIELD BALKS AGAIN ON HIGH LINE CONTRACT – DELAY IN SIGNING ONLY LIKELY TO BE TEMPORARY

The signing of the High Line contract still awaits the pleasure of Secretary Garfield. It was thought that when the demands of the D. & R. G. over the Right of Way question were modified last week the final obstacle in the way of an immediate approval of the project had been removed. It was announced at that time that Secretary Garfield would sign the contract as soon as the modified contract from the railroad arrived. This did not reach Washington yesterday as had been expected. In the meantime a dispatch received last night from Mr. Sheldon Cline the special correspondent of The Daily News at Washington states that the secretary considers excessive the price which the Palisade land owners ask for land which will be appropriated for the canal's right of way. That despite his objections, Secretary Garfield will approve the contract as it now stands is practically certain. The following is the dispatch in full which was received by The Daily News last night:

Washington, D. C., Feb. 16, 1909.

The Daily News, Grand Junction:

The signing of the Grand Valley project papers is being delayed by the failure of the revised railroad contract to arrive and because Secretary Garfield thinks that the price asked by the Palisade land owners for the rights of way is excessive. He has asked that the Improvement association reopen the negotiations with the Palisade owners. Secretary Aupperle has told Garfield this would be useless until the contract with the association was signed. He urges Garfield to sign the contract then resort to condemnation proceedings for a right o£ way. He is hopeful that the secretary will take this view.

SHELDON CLINE

Daily News, Denver Republican, Feb 17 1909

TO RECOMMEND RIGHT OF WAY FOR HIGH LINE

DIRECTOR OF RECLAMATION SERVICE REMOVES LAST OBSTACLE.

LAND HERETOFORE WORTHLESS BECOMES VALUABLE ON PROSPECT OF DITCH.

Fifty-Five Thousand Acres of Promising Soil to Be Reclaimed By Federal Project

----

Director F.H. Newell of the National Reclamation service yesterday signified his intention to recommend to the secretary of the interior the contract prepared between the government and the Denver & Rio Grande Railroad company granting a right of way for the High Line canal of the Grand Valley project through the Grand canon below Palisade.

This assurance practically removes the last obstacle to the completion of a government project which will ultimately reclaim 55,000 acres of the most promising land in the United States. Land identically like it at Palisade, in the same neighborhood, has sold recently for $4,000 per acre and land along the proposed route of the canal, heretofore absolutely worthless, has recently advanced to $125 per acre on the mere prospect of the early completion of the government ditch.

The completion of the canal, which will cost $2,500,000, according .to government estimates, is made possible through the waiving by the Denver & Rio Grande company of the indemnity clause of the contract. The canon of the Grand is very narrow for most of the distance along the route of the proposed canal, and nearly all the available space is occupied by the railroad right of way. For part of the distance the ditch will run between the railroad and the river, and in others between the railroad and the mountain. The average width of the available space between the hills and the river is scarcely more than 100 feet, and the railroad company felt that it was running grave risk of damage if the construction of the canal should prove faulty, or if it should be carelessly managed after construction.

But rather than see the project held up, the Rio Grande people consented to the waiving of the indemnity clause which had been originally part of the contract taken to Washington, and Director Newell lost no time in sanctioning the bargain. Upon his recommendation depends almost entirely the appending of the signature of Secretary James A. Garfield of the interior, who has long been friendly to the project. It has been the desire of the Grand valley water users and of the Denver & Rio Grande alike for a long time to see the plans for the project perfected during the present administration, and this now seems to have been accomplished, at the eleventh hour, by the action of the railroad company in waiving the indemnity clause.

D.W. Aupperle, secretary of the Grand Valley Water Users' association, went to Washington two weeks ago to take up the matter of having the contract and his quest has been attended with success.which insured the opening of yet another empire in the western part of Colorado.

Daily News, Denver Republican, Feb 17 1909

Land Values Leap at Mere Prospect of Government's Big Irrigation Enterprise

Point on Grand River, four miles east of Palisade, where water is to be taken out for government's Grand Valley project.

Daily News, Feb 18-19 1909

GEORGE GOULD TAKES HAND IN RAILROAD RIGHT OF WAY FIGHT AND MAKES PROPER AGREEMENT

Aupperle Visits New York Magnate and Clears Away Final Obstacle to Secretary's Formal Action

----

THE HIGH LINE CANAL CONTRACT WILL BE SIGNED BY SECRETARY OF THE INTERIOR RUDOLPH GARFIELD TODAY OR MONDAY.

A FINAL AND SATISFACTORY AGREEMENT IN REGARD TO THE RIGHT OF WAY OF THE CANAL ON THE RAILROAD PROPERTY ABOVE PALISADE WAS REACHED YESTERDAY AFTERNOON BETWEEN GEORGE GOULD, PRESIDENT JEFFERY AND SECRETARY D. W. AUPPERLE OF THE WATER USERS ASSOCIATION IN NEW YORK.

THIS REMOVED THE LAST OBSTACLE IN THE PATH OF SECRETARY GARFIELD AND UPON RECEIVING MR. AUPPERLE'S REPORT HE DECLARED TO MR. SHELDON CLINE, THE DAILY NEWS' SPECIAL CORRESPONDENT IN WASHINGTON, THAT HE WOULD SIGN THE CONTRACT AT ONCE.

THE DISPATCH WAS SENT TO THE MEETING OF THE DIRECTORS OF THE CHAMBER OF COMMERCE BY THE DAILY NEWS LAST EVENING AND WAS RECEIVED WITH CHEERS.

THE DISPATCH IS AS FOLLOWS;

BY SHELDON CLINE.

Secretary Aupperle of the Grand Junction Water Users' association has returned from New York where he saw President Jeffery and George G. Gould about the demands made by the railroad in relation to the right of way of the main canal of the Grand Valley project which Secretary Garfield could not accede to. Gould and Jeffery took the matter up at once with their Denver officials by wire and today Aupperle received word from Jeffery saying that a satisfactory adjustment had been reached. This removes the last obstacle to Garfield's signing the contract with the Water Users' association. Nothing could be done today, it is a legal holiday, but the secretary assures me that the contract will be signed tomorrow unless a rush of business in the Department of the Interior forces the matter over until Monday.

Daily News, Feb 18-19 1909

The News dispatch, which up to midnight tonight is the only one in this city from Washington in regard to the High Line matter lifts a heavy load from off the mind of President M. O. Delaplain of the Water Users' association and will allow Project Engineer E. E. Sands to begin at once planning steps toward immediate construction. Authority to advertise for bids on construction will probably be received by Engineer Sands within the week. The bids will be advertised for ten days and actual dirt should be moving on the canal immediately after the Contracts are let.

The difficulty in the way of an earlier settlement of the High Line matter was the objections of the government to signing the contract until the matter of the details of the adjustment of payments for the right of way of the canal through the Palisade district were settled with the Mesa County Land Owners Protective league and the right of way difficulty with the Denver and Rio Grande was cleared away.

The first of these obstacles was removed after a three day session with the land holders late last month. The railroad adjustment was not completed until yesterday. It was a source of worry to the Water Users which they concealed from the general public but it is not going beyond the bounds of the strictest truth to say that President Delaplain has been greatly worried over the High Line matter since Secretary Aupperle left for Washington.

The receipt of Mr. Cline's dispatch clears away all doubts as to the certainty that the canal is assured the Grand Valley. Speaking of it last evening President Delaplain said"

"This is the word that I have been longing to hear for the last week. It is certainly great news and will be a splendid story for the Daily news to give to the people of the valley tomorrow."

The dispatch was read aloud to the members of the board of directors of the Chamber of Commerce and cheers rang out when the reading was completed. "I expected this from a telegram I received from Engineer McConnell today," said Project Engineer E. E. Sands last night.

"That means that the canal will surely be built. I never was doubtful but still the uncertainty over the matter was uncomfortable."

Daily News, Feb 18-19 1909

GARFIELD'S SUDDEN ABSENCE FROM WASHINGTON PREVENTS PROGRESS ON HIGH LINE DITCH CONTRACT

The status of the High Line canal contract remains unchanged. A telegram received last night by The Daily News from Sheldon Cline in Washington, stated that Secretary Garfield had been called to New York by the serious illness of his son and consequently the signing of the project papers was in abeyance until his return.

The telegram was as follows:

Washington, D. C., Feb. 17, 1909

The Daily News,

Grand Junction, Colo.

Because Secretary Garfield was called to New York by the illness of his son no progress was made today in the signing of the Grand Valley project papers.

SHELDON CLINE.

Daily News, Feb 21 1909

SIGNED!

GARFIELD FORMALLY APPROVES CONSTRUCTION OF HIGH LINE

Contract With Water Users Association is Accepted by Government---Four Million Dollar Ditch Now a Certainty

----

LAST PRELIMINARY CONCLUDED – CONSTRUCTION WORK WILL COMMENCE WITHOUT DELAY

Washington, D. C., Feb. 20, 1909.

M. O. Delaplain,
Grand Junction, Colo.
Garfield signed High Line Contract at 5:47 this evening.

D. W. AUPPERLE.

After negotiations which has extended over a period of many months the last preliminary in the great High Line project was completed, the last possible obstacle cleared away for the construction of the greatest single enterprise this section of Colorado has ever known, when Secretary Garfield late yesterday afternoon affixed his signature to the contract and put the formal approval of the Department of the Interior on the project papers. Yesterday's action means that the Federal government is now on record with a promise to spend nearly four million dollars in the Grand Valley during the next five years. It means that one hundred thousand acres of now arid land will be converted into the finest fruit land in Colorado.

Daily News, Feb 21 1909

M. O. Delaplain president of the Water Users' association received a telegram from Secretary Aupperle at eight o'clock last night bringing the glad tidings.

The News special correspondent in Washington, Mr. Sheldon Cline sent the following dispatch.

WASHINGTON, Feb. 20.--The Grand Valley project papers were signed this afternoon by Secretary Garfield after two weeks of delay during which time the claims of the D. & R. G. railroad and the Palisade land owners damages for right of way have been under consideration. The railroad modified their demands and the new papers reached the secretary several days ago.

The delay had been due to the fact that Garfield thought the prices for right of way asked by the Palisade land owners was too high. Secretary Aupperle's argument persuaded him to sign the contract and leave the right of way prices for future adjustment. Secretary Aupperle starts for home Sunday. He secured Pinchot's promise to visit Grand Junction between March 15 and 25.

Daily News, Feb 21 1909

Secretary James R. Garfield

Daily News, Feb 21 1909

ENGINEERS WHO ARE ENTRUSTED WITH WORK OF BUILDING CANAL

I.W. McConnell

E.E. Sands

Engineer I. W. McConnell was in Denver last night when he received a telegram from Washington telling of the signing of the contract. Earlier in the day he had received a letter mailed in Washington on Wednesday, which forecasted today's action. It is probable that Mr. McConnell will arrive in Grand Junction early in the week when fuller advices from Washington have arrived.

"It is what we all hoped and expected," said Project Engineer E. E. Sands' last night when shown the telegram containing the news of Secretary Garfield's action. "Whether the digging will start at once I cannot tell yet. Of course it will be necessary to advertise for bids for the work and that would take several weeks. Then too, it is quite likely that negotiations with the Palisade land owners over whose land the canal is surveyed will have to be completed first and that alone might take a month. Anyway, we are now sure of the canal and that is the im

Daily News, Feb 21 1909

portant thing. With Garfield's signature on that contract there is nothing that can interfere with the project. Whether it is two weeks or eight before the dirt actually begins to fly is a comparative minor matter."

ESTABLISHED 1882. GRAND JUNCTION, COLORADO, SUNDAY MORNING, FEBRUARY 21, 1909.

SIGNED!

GARFIELD FORMALLY APPROVES CONSTRUCTION OF HIGH LINE

ENGINEERS WHO ARE ENTRUSTED WITH WORK OF BUILDING CANAL

I. W. McConnell.

E. E. Sands.

Contract With Water Users Association is Accepted by Government — Four Million Dollar Ditch Now a Certainty

LAST PRELIMINARY CONCLUDED ... CONSTRUCTION WORK WILL COMMENCE WITHOUT DELAY

Washington, D. C., Feb, 1909.

M. O. Delaplain,
Grand Junction, Colo.
Garfield signed High Line contract at 5:47 this evening.
D. W. AUPPERLE

PAGE FACSIMILE: *Signed!*

Daily Sentinel, Feb 25, Daily News and Daily Sentinel, Feb 27 1909

Mr. Project Engineer Sands is a very touchy individual. In the story that was given the public the day after the signing of the canal contract, The Sentinel endeavored to be as correct in its statements as possible. Perhaps we did not speak as frequently of Mr. Project Engineer Sands as he thinks he should have been quoted. Several times The Sentinel has spoken of the High Line canal and Mr. Sands has gone out of his way to make denial. Now, Mr. Sands does not frighten us in the least, nor are we impressed by his silly attempts to discredit The Sentinel. We propose to say what we regard as the truth on matters concerned with the High Line canal, because we are entitled to do so for the years of work that we have done in its behalf. The construction of the High Line canal is not a political job, as Mr. Sands seems to so regard it, and The Sentinel is a democratic paper. Mr. Sands is not yet big enough to place The Sentinel under ban, because if he thinks so he will be mightily fooled. In the construction work we propose to be treated fairly, whether Mr. Sands thinks we ought to be or not. The Sentinel dislikes to say this, but it has been invited, and there is more to come if necessary. The Sentinel never asks anything but what is fair and right.

KNOCKING the High Line canal work, knifing the Chamber of Commerce, jeering at the Law Enforcement element in Grand Junction, lying about the Charter matter, sneering at the efforts of the lyceum course and covertly attacking the "church people" of the city, is a fair record for the Speer sheet to make in less than six weeks. Now we will listen for the smashing of the editorial typewriter while the recently self-appointed engineer of the High Line canal sets about it to tell you how strong he stands with the plain people.

Upon several occasions we have heard Mr. Project Engineer Sands blow off steam which would drown a siren whistle attached to a flat boat in a Louisiana swamp. Perhaps he did not rise to the pressure of which George Smith told in the legislature, wherein the whistle was the larger part of the steamboat, and, when it blew, landed the flat boat high and dry on the "sands."

Daily Herald, Feb 23, Daily Sentinel, Feb 26 1909

SHOULD NOT EXPECT HIGH LINE FOR YEARS

Daily Newspapers Give One Erroneous Impression Regarding Early Completion of High Line Project

----

Anent their desire to beat each other to the news, it is common for both the News and the Sentinel to make statements that are without foundation in fact. Such is the charge against the Sentinel of yesterday with regard to its speculation as to the manner and time of construction of the High Line project.

For the information of those who want the facts we would say that the government work will all be handled under force account for the present, no bids being advertised for. The water users' association may handle its portion of the money in that manner, tho we are unable to verify this report at this time. It is quite unlikely that the government will take up the actual construction work before spring and possibly not as soon as that, depending entirely upon the terms of the contract which was signed in Washington and which will be brot back by Secretary Aupperle.

The Sentinel also states that the government officials will divide the canal site into small sections and let contracts thereon. This, too, is without foundation, as that paper has not interviewed any government officials upon the subject and therefore has no knowledge of the plans of the government officials in charge of the High Line work. That paper also guest that the cost of moving some of the dirt would be as high as sixty-five cents a yard. History does not record wherein any dirt on any government project has ever cost even as much as two-thirds of that amount to move, and the work on this project is not as difficult as in many other places, among them being the South canal at Montrose.

The Sentinel also calls attention to the fact that "In the meantime the local officials will organize their staff and accomplish whatever preliminary work may be necessary." Engineer Sands assures The Herald that the "local staff is pretty well organized at this time, so it is not probable that any extensive changes will have to be made in this.

The first work will likely be done on the tunnel thru Mount Lincoln, which will be something like 7,000 feet long. Instead of requiring a year for construction, as stated in the Sentinel, of this important portion of the work, it is likely that it will require more than three years. As the location of a canal above the big tunnel is still undecided, the statement as to there being two smaller tunnels to construct is all buncombe. There is no

Daily Herald, Feb 23, Daily Sentinel, Feb 26 1909

definite information at hand to show that Engineer Carroll will be in charge of the tunnel work He is not a civil service man, therefore it is easy to conjecture all kinds of things in connection with his name. No superintendent of this branch of the work has been selected, nor is it likely that one will be for some time to come.

No agreement has been entered into or even tacitly understood as between the land owners in the Palisade district and the government with regard to the right-of-way. The government has not intimated that it will stand the cost of improvements of anything else with regard to the right-of-way, tho we can not deny that the Sentinel made a good guess when the statement appeared that this would be 130 feet wide. In all filings on homestead entries after Oct. 2, 1S88, there is a release clause which makes it possible for the government to select its right-of-way at any time for canal purposes without damages. Just what terms of settlement will be devised for those having established their residence prior thereto can not be stated at this time, as Engineer Sands refused to be quoted upon the subject. Again we must take issue with the Sentinel for the reason that there will be no headgate at Cameo. The site of the headgate has not, as yet, been selected, but it will be above the mouth of Plateau creek somewhere, tho its ultimate location is very indefinite. The canal will be approximately fifty-six miles in length and will not leave any great amount of $3,000,000 unexpended, if it is really completed for that money. If the work is done inside of five years it will cause many men who have had experience with government construction to miss their guess.

EX.SOLDIER AND OLDTIMER SUCCUMBS.

Tom Charleston, an ex-soldier and member of the local G. A. R. post, died at his home near the sugar factory Monday evening about 7 o'clock, from the effects of a paralytic stroke. He was a Negro slave before the war, and during the war he was in the Union army and part of that time was the mess cook.

He came to Grand Junction or western Colorado with the first pack trains and was a resident of Grand Junction long before it was on the map.

The funeral arrangements-have not yet been made.

SECTION OF GOVERNMENT WORK

This is the Class of Work that will be Followed out in Building the High Line Construction

Government Camp and Power Station

Daily Herald, Feb 23, Daily Sentinel, Feb 26 1909

Scene to Be Witnessed When the Construction of High Line Canal Starts

SERIES OF CONCRETE DROPS
A Part of the South Canal, Which Will Convey the Gunnison Water to the Smaller Channels fir Distribution to the Water Users Under Montrose Project

SECTION OF TUNNEL
Shows Timbered Portion Before Concreting

CARTOON FACSIMILE: *When the Construction of High Line Canal Starts*

Daily News Feb 26 and 27, Daily Sentinel Feb 27 1909

AUPPERLE BRINGS GOOD NEWS FROM WASHINGTON

"Grand Valley Project is Best Ever," Says Garfield

Pen Which Signed Contract Now Cherished Souvenir

"I look to the people of Grand Junction to make the Grand Valley project the finest irrigation project in the world. Tell my good friends in Grand Junction this.'"

This was the message carried from Secretary Garfield of the Department of the Interior to Grand Junction by D.W. Aupperle who arrived here early Thursday morning. Garfield made the statement just after handing Mr. Aupperle the pen and holder with which he had signed the High Line canal contract. '

Formal ratification of the contract will probably take place next week when a mass meeting of the members of the Chamber of Commerce and the Water Users' association will be called and appropriate resolutions drawn up, passed and sent to Secretary Garfield and the men in the Reclamation Service thanking them for their interest in the Grand Valley project which had so much to do with the three years of negotiations which have been carried on by Mr. Aupperle in the interests of the High Line.

It was late Saturday afternoon last week when Mr. Aupperle and Director Newell of the Reclamation Service started on the short journey between Newell's office and Garfield's to witness the formal signing of the contract. Besides these two men and the secretary, Judge Parker and Mr. Brown, secretaries to Garfield were in the room. The contract was handed to Mr. Garfield. He took a final glance over its provisions and then picking up a pen signed the paper which meant so much to the people of the Western Slope.

Turning to Mr. Aupperle he said: "I understand you want this pen," and handed the instrument to him. Mr. Aupperle has it with him now and will keep it as a treasured souvenir for the long fight, for the success of the canal.

The fate of the High Line, as far as immediate action went, hung in the balance for many days this month. At one time, on February 16, to be exact, Garfield addressed a letter to Aupperle formally stating that as he understood the situation in the Grand Valley at present in regard to right of way matters, he could not sign the contract.

It developed later that at that time he had fully determined not to sign but to recommend the project to the favorable consideration of his successor. Upon receipt of this letter Secretary Aupperle immediately indited a lengthy reply going over the entire situation in the Grand Valley at the present time and explaining how he believed that it would be impossible to secure better right of way terms until the government had formally entered into the contract. This letter apparently changed the tide of affairs and secretary Garfield made a new appointment with Mr. Aupperle and the matter was

Daily News Feb 26 and 27, Daily Sentinel Feb 27 1909

again gone over. As a result of this consultation and the arrival of a favorable right of way contract from the Denver and Rio Grande he put his signature to the document. While in Washington Mr. Aupperle was unable to make a definite date with Pinchot in regard to his appearance in Grand Junction next month. He did secure Pinchot's promise to be here however and the matter of the date will be arranged through Denver.

MANY RECLAMATION PROJECTS SOON DONE

CHICAGO, Feb. 26.--Several great government reclamation projects in various western states will be completed this year. The works include storage systems as well as river distribution systems. Among the largest of the projects nearing completion are those of the Roosevelt dam project in Arizona and the Uncompahgre tunnel project in western Colorado, where a tunnel has been bored through the mountain for a distance of two miles in order to divert the water of the Gunnison river into the Uncompahgre valley. This is the greatest underground waterway in the world. Several other large projects are to be completed in Idaho and Wyoming.

This year's construction will bring into use water-controlling systems the aggregate cost of which has been more than $18,000,000 and which will irrigate 800,000 acres of land that has hitherto been arid and worthless.

It sounds good to our ears to be given the title of "Project Engineer Bunting." For and in behalf of this city, dear to us in every possible way, we have fought many battles, planned many projects, been maligned by very cheap individuals, berated and vilified; but still our purpose to assist in making it the one and only great city of western Colorado has not abated in the least.

Our services are at the command of Mr. Project Engineer Sands for any purpose whatever which will advance the High Line canal project. Having been created assistant project engineer by the Yellow, we are ready to make good our title. Any way, they should have a "cherubic" individual in their construction corps to offset the mightiness of his excellency, Mr. Project Engineer Sands.

THE PROPOSAL that the people of Grand Junction meet together some day next week and in mass meeting pass resolutions of thanks to Secretary Garfield of the Department of the Interior for his deep personal interest in the welfare of the Grand Valley project is a very good one and one that will undoubtedly be acted upon. Only by steady team work between the people and the Department of the Interior, which has existed in the Grand Valley up to date with one notable journalistic exception, can the High Line be constructed swiftly and without friction

Daily News, Apr 10 1909, Daily Sentinel, Feb 27 1909

QUARTER MILLION YEAR'S FUND

Government to Spend Amount Equal to Subscriptions

Formal notification has been received by D. W. Aupperle from Project Engineer Sands of the Reclamation Service that as soon as the Grand Valley Water Users' association collect $10,000 of their subscribed High Line fund work will then be started on the Palisade tunnel with government funds. Just as fast as the Water Users' association collect their funds the government will match their collections with the same amount, and on that basis the work will be done this year. The necessary certificates are being prepared and will be sent front Washington.

In formally notifying Secretary Aupperle of the government's intentions Engineer E. E. Sands wrote the following letter;

Grand Junction, Colo., April 7, '09.

Mr. D. W. Aupperle, Secretary,
Grand Valley Water Users' Ass'n.,
Grand Junction, Colo.

Dear Sir:

Cooperative certificates are now being prepared for the Grand Valley project.

There has been added to the agreement regarding payment of interest, forming part of the form, the following:

"Interest on this agreement will cease after 90 days notice in accordance with resolution of the board of directors, adopted.. ...... "

It will therefore be necessary for the board to hold a meeting and adopt a resolution prescribing the manner of publication of the fact that interest is to be discontinued on outstanding coupons.

Please let me know as soon as the board has taken action on this matter in order that the Washington office may be advised at once and the proper date filled in. I also wish to inform the Water Users' association that I am authorized to begin the construction of tunnel No. 3 with government funds and to continue such construction at a rate which will not utilize the funds more rapidly than the rate at which funds are collected for cooperative work by the Water Users' association. As soon as you have collected $10,000 I will expend a similar amount on the tunnel at once. In order that I may be properly safeguarded in these expenditures, I will request that you send me a signed statement as soon as you have collected $10,000 and have it deposited for work under the cooperative plan, and that you also send me statements in the future indicating the increment of the fund.

Very respectfully,
E. E. SANDS,
Engineer.

Daily News, Apr 10 1909, Daily Sentinel, Feb 27 1909

HIGH LINE CANAL MASS MEETING TOMORROW

Tomorrow evening a mass meeting to which all the citizens of Grand Junction and the entire valley are invited will be held at which the formal notification of the signing of the High Line contract will be made. At this meeting a vote of thanks will be tendered to Secretary Garfield who is now on the eve of his retirement for all that he has done for the Grand Valley. A vote of thanks will also be extended to the reclamation service and Engineers McConnell and Sands for their share in the work. It will not be known until today just where the meeting will be held. Nor can the list of speakers be announced as yet. It was hoped that Mr. McConnell could arrange to be here but he has been compelled to refuse the invitation. Two men from Fruita will be among the outside speakers.

Not the Ability but the Actions of Mr. Sands Involved

It is perhaps the greatest possible exhibition of les majesty to presume to criticise a petty official of the, government; but The Sentinel does not think so.

Mr. Project Engineer Sands invited all that was said of him by The Sentinel, and we have no apologies to make for what was said.

This paper never uttered a criticism of Mr. Sands until Mr. Sands, who has a well developed desire to see his name in print, went out of his way to discredit what news matter in relation to the High Line canal was printed in this publication.

It is the very general opinion here that Mr. Sands has a largely developed case of the swell head and that his clothes will hardly contain him. He is pompous in his relations with men, not at all warranted by anything which he has accomplished or in comport with his position.

The Sentinel is not attempting to force Mr. Sands to do anything. The reverse is the case evidently, as Mr. Sands immediately rushed into print, following the very conservative statement which appeared in this paper, following the signing of the contract by Secretary Garfield, and which was true.

Nor is it the first instance that Mr. Sands has attempted to discredit statements made in connection with the High Line project which have appeared in The Sentinel.

Daily News, Apr 10 1909, Daily Sentinel, Feb 27 1909

The Sentinel regards Mr. Sands connection with the High Line project simply as an incident. He is here because he happens to be connected with the reclamation service in Colorado under Mr. McConnell. He cannot be charged with having had any connection whatever with the securing for this Valley of this greatly desired irrigation project.

The question of the ability of Mr. Sands in connection with the work of the High Line project is entirely extraneous to the criticism offered of Mr. Sands. In fact, there was no question raised whatever upon the ability or capacity of Mr. Sands, save in the comment of the Yellow.

What was said in The Sentinel was simply in answer to Mr. Sands criticism of news matter contained in this publication.

With the very highest marks of courtesy upon every occasion have all matters in relation to the canal project been spoken of by The Sentinel. It has treated Mr. Sands most courteously, never anticipating or seeking information from him which was not of a public nature. But most studiously has Mr. Sands endeavored to close avenues of information upon all possible occasions to this paper.

It is not even necessary to attempt to make answer to the Willie in control temporarily of the Yellow. It might not be inapropos to say to him and of him in relation to all the rot published in behalf of Mr. Sands: that he would better "tie his bull out of doors."

The single purpose that animates the Yellow is to possibly secure some rejected credit to itself, in. the very long and rambling defense, totally foreign to the subject matter in controversy between Mr. Sands and The Sentinel, which gives its editor opportunity to fill space.

Daily News, Feb 27 1909

PROJECT Engineer E. E. Sands has received his orders. Engineer in Chief, Gloomy Isaac of the Speer sheet, has taken over formal charge of the High Line construction work and hereafter Supervising Engineer McConnell, Director of the Reclamation Service Newell, Secretary Garfield, and after March 4th. Secretary Ballinger of the Department of the Interior will please report promptly at 7 a. m. each blithsome morning and get their orders.

Project Engineer Bunting of the Grand Valley project "hates" to take over this work of building the High Line but with sadness and a heavy heart be finds it necessary. Sands and McConnell have been running things along the High Line now for over a year and yet not once have they darkened the door of the Speer organ's office and with humble mien asked, "Oh Chief, what is your will?" and at the same time slipping Project Engineer Bunting two columns and a half of legal advertising with a hope that the returns therefrom might sweeten the Speer organ's unsanitary tongue.

It seems that Project Engineer Bunting fears that Engineer Sands is about to construct the High Line with gross impartiality in regard to the Evening Sentinel. Hercules joined two rivers and ran them through the Augean stables in an effort to cleanse that noisome sink. Project Engineer Bunting evidently fears that Sands plans to sneak the High Line through the editorial offices of the Speer organ in another well meant effort in this line. In any event, however, Engineer Sands has got his orders. Hereafter the headquarters for High Line construction will be at the Main street joint where hangs the Speer banner.

Just what the Speer sheet is attempting to force Sands to do is somewhat problematical but the Daily News can assure its readers at the outset, from our personal acquaintanceship with the gentleman in question, that. he's not of the breed that can be forced by a shower of the billingsgate or by any species of intimidation. Men like McConnell and Sands who have been able to handle from a thousand to six thousand laborers and direct one of the mightiest battles against nature ever projected on the North American continent are not going to be bulldozed by the Speer sheet's snarls into submitting their engineering plans for the approval or disapproval of the editorial cherub that presides in the Main street hang out.

The Speer sheet has again declared its unalterable determination to speak the truth in regard to the High Line. If the Speer organ has any charge to prefer against Engineer E. E. Sands no doubt Mr. Sands and the department would be glad to consider them. If the Speer organ, angered because of its failure to extract a sack of loot from the Reclamation officials, desires to. formulate an indictment against the work that Supervising Engineer McConnell and Project Engineer Sands have completed up to date, why doesn't the Speer sheet stop howling "smoke" and point to fire? If there "is more to come," as the Speer sheet charges, let it come now.

The Daily News has absolute confidence in the ability of Supervising Engineer McConnell and Project Engineer E. E. Sands to complete the work of the High Line in the same splendid way that the Gunnison tunnel is about to be completed. The Daily News is also confident that the High Line will be built quite irrespective of the floundering attempts of the Speer sheet to dislodge Engineer Sands, and that it will be built ably and with all speed possible. The Speer organ has fought the High Line in the past and evidently is serving notice that it will either run the High Line now or ruin it if it can. Fortunately for the people of Colorado and the Western Slope, the Speer sheet is but a fly speck on the map in the contemplation of the Reclamation officials and regardless of its whining the work will be carried on, as Reclamation work is always done, with the highest efficiency. In the meantime, however, as a local amusement we commend our readers to the promised serial, "The High Line As I Would Build It," by Project Engineer Bunting of the Speer organ.

WASHINGTON PRETURBED AT SERIOUSNESS OF HIGH LINE SITUATION AT JUNCTION

Roosevelt at Once Weakens to Engineer Bunting's Demands—Taft Affected—Wall Street Jarred--Treasury Vaults Opened

----

BY SKELTON PINE.

WASHINGTON, Feb. 26.---Consternation reigned in administration circles last night when word was received here of the serious state of affairs existing in Grand Junction, Colo. Late editions of all the papers gave Editor Bunting's complete editorial concerning the High Line canal, and it was the universal belief that something must he done at once to pacify Mr. Bunting.

President Roosevelt was observed to turn pale as he was handed the news by our special correspondent. Refusing at that time to discuss the matter, the president immediately called together a special session of the cabinet, in spite of the fact that a diplomatic reception had been scheduled for last night at the White House. It was perceived, however, that the serious items of the situation in western Colorado brooked of no delay.

The cabinet members hurried into their rooms, and for some time remained closeted there with the president. After a conference lasting about an hour, a messenger was hurriedly dispatched for President-Elect Taft, who had been advised of the seriousness of the situation, and was waiting in his club. After Mr. Taft's arrival the conference continued, until after 2 o'clock this morning. Not since the Spanish war has official Washington been so perturbed.

At the adjournment of the conference President Roosevelt gave out to the visiting newspaper men the following statement;

"I want you to assure the country that everything possible will be done to pacify Mr. Bunting. I cannot tell you how I deplore the fact that he has been driven to take this extreme step, which

Daily News, Feb 27 1909

he has taken, I am sure, only after grave consideration, and a full regard for its effects on Wall Street and the entire nation.

We are now waiting word from Mr. Bunting in regard to the overtures tor peace we have made. I don't mind telling you that we have offered him full charge of the construction work of the High Line canal. To tell the truth, it was my impression that he had always been in charge. Gold is being rushed to the subtreasury in New York, which it is hoped will allay the panicky feeling there. If only Mr. Bunting will accept our terms, which are most generous, we shall yet weather the storm."

Secretary Garfield seemed especially worried. "It was evidently a mistake on the part of Mr. Sands," explained Mr. Garfield to the newspaper men. "His instructions were explicit, that Mr. Bunting was in full charge of the construction work, and that in no case should any steps be taken without consultation with him. Mr. Sands has always shown exceptional ability in his work, and I am sure this awkward situation was not caused by a deliberate action on his part to refuse to obey Mr. Bunting's orders, but merely did not understand my instructions, in which Mr. Bunting was given full charge."

President-Elect Taft refused to be interviewed. "I hope the situation will be cleared by the time my term begins," he said. "I had hoped," he added, "that my administration would begin under the most favorable conditions. And now this blow is almost more than I can bear." And the big man turned away to hide his tears.

There is general sympathy with Mr. Taft, as it is realized that he is in nowise to blame. It is expected that the occasion will be utilized by the enemies o£ Mr. Roosevelt, however, as the subject for a bitter attack. Mr. Bunting's well known modesty assures the people of the capital that the step he has taken was necessary; it is felt here that his prerogatives have been most grossly overlooked, and that the government cannot too quickly offer all reparation for his wounded feelings.

School Notes

At the general teachers' meeting Tuesday evening the teachers were given lists of books for supplementary reading. This class of work is especially desirable to add variety and interest until the close of the school year. The importance of breathing exercises in the daily physical drill was brought to the teachers' attention. The appearance of a few beanshooters on the school grounds furnished the occasion for some remarks on kindness to animals. The superintendent stated that in visiting one of the rooms he was pleased to note in a prominent position upon the blackboard these words: Be kind to dumb animals.

As a digression from the routine of school work we give the following from an eastern paper.

A Girl's Essay on Boys

Boys are men that have not got as big as their papas, and girls are women that will be ladies by and by. Man was made before woman. When God looked at Adam He said to Himself: "Well, I think I can do better if I try again." And he made Eve. God liked Eve so much better than Adam that there have been more women than men.

Daily News, Feb 27 1909

Boys are a trouble. They wear out everything but soap. If I had my way half the world would be girls and the rest dolls. My papa is so nice that I think he must have been a little girl when he was a little boy.

Man was made, and on the seventh day he rested. Woman was then made, and she has never rested since.

Miss Helen A. Dewey, principal of the Whittier school at the general teachers' meeting gave a report of the Colorado State Teachers' association which met in Denver in the Christmas vacation. Miss Dewey was the only one of our corps who attended this meeting. While speaking, Miss Dewey had the close attention of the teachers. She said that the thing which meant

the most to her in the meeting was a report made by the secretary of the Denver Chamber of Commerce. He had been requested by the management of the S. T. A. to state in what respects our public school system is deficient,

He said that, preferring to give facts rather than empty theories, he visited several business men in Denver and without exception each one said that poor hand-writing is the greatest defect. When asked, how are they in arithmetic? The business men said, "fine". This seems to be a general criticism of nearly all school systems. In view of these statements it is worth while to give special attention to this subject in our public schools. The members of our board of education have decided to employ a special teacher of penmanship in our city schools next year. In this way the best possible attention will be given to this subject, and we look for a steady and marked improvement in the handwriting of our boys and girls.

Daily News, April 1 and 17 1909

T.C. HENRY SEVERELY GRILLED IN LETTER TO SECRETARY BALLINGER

D.W. Aupperle Informs Department of Interior That People Want Government

----

Secretary Ballinger of the department of the Interior now knows, if he was ever in any doubt, just what the people of the valley think of T. C. Henry and any proposition he may have for the private construction of the High Line canal. Ever since ex-Secretary Garfield formally signed the Grand Valley project contract, Henry has been restive. Recently The Daily News received from Henry a long file of letters relating to a correspondence which Henry had had with Director Newell of the department stating that he was preparing to build a gravity high line here for $1,000,000. As a result of this correspondence Secretary Ballinger wrote to Secretary Aupperle last week asking him how the people felt about the government going ahead with the work. Secretary Aupperle immediately wrote to Secretary Ballinger and also to Director Newell of the reclamation service plainly telling them that the people wanted nothing to do with the Henry project.

Work on the High Line cannot commence until Secretary Ballinger has received the copy of the formal contract with the government by the Water Users' association which was approved here last week. He will probably receive these papers this week.

Secretary Aupperle's letter to Secretary Ballinger is as follows:

Grand Junction, Colo., Mar. 29, 1909. Hon. R. A. Ballinger,
Washington, D.C.,
My Dear Sir:

We are in receipt of yours of March 23rd. In reply we wish to say that we infer that the parties termed "various individuals or associations" refer to one, T. C. Henry of Denver.

We are aware of the fact that he has recently made an attempt to keep up his record as a "trouble breeder", but we wish to say that the people here have paid absolutely no attention to the matter.

The files in your office will show in November, 1907, Secretary Garfield submitted to us a proposition to either withdraw entirely from this project and leave the matter open for private construction or to take up the work on the basis proposed by him in the statement submitted to us. This proposition was thoroughly canvassed by our local people, was given wide publicity and was submitted to the people here directly and indirectly interested in the matter and at mass meetings held in Grand Junction and Fruits the government construction proposition was unanimously and enthusiastically adopted.

The sentiment of our people has not changed in the least degree since that time. The statement we think is fully substantiated by the action of our people in subscribing a fund to assist in getting construction by the government started at the earliest possible date. Also by the expression given by the people at various times in public meetings and still further and more emphatically by a unanimous vote of the shareholders of our association in an election held December 5, 1908 to vote on a contract submitted at that time by Secretary Garfield and also by their expression at an election held on the 27th inst., at which election they voted on the question of approval of the regular form of contract between the government and our association and at which time there was again a unanimous expression, there not being a single vote against the government contract.

A copy of the report of the judges of this last election made to the directors on the 27th is forwarded today to Director Newell, together with the contract duly signed showing the number of votes cast and how they were cast.

We understand fully, we believe, your position on this matter and if these official expressions are not sufficient as a matter of record, we will gladly call another mass meeting of our interested people to submit the question but we can give you positive assurance of the result of such meeting, for our people are not only well satisfied with the Reclamation Service, but they are most heartily opposed to any agitation from the Denver source. Our people are not looking for any private construction proposition and if they were, we feel that we can speak for our people in giving assurance that they would give no consideration whatever under any conditions to Mr. T. C. Henry.

We would make this suggestion however. We believe that it is his purpose to stir up contention if possible and is trying to create the impression that he could build a gravity system at a much less cost than that estimated by the government engineers. He knows very well, as we know, that he could not build the kind of system the government will build at any such prices as he is proposing and his record is well enough known to leave no doubt that such a system as he would construct would not meet with our approval. If he is in position to do good, substantial construction work on such basis as he intimates, let him take the government plans and bid on them and if he can do the work cheaper than the government can, let him have the contract and make a profit on the work. We would not object to that, because we would feel certain of getting substantial work if he were doing it in accordance with the government plans and specifications

We are impressed with the thought that his chief purpose in attempting to stir up matters at this time is the fact that he knows full well that the government construction is a settled fact and this gives him a safe opportunity to break in with an attempt to annoy some members of the Reclamation Service and if possible to cause a few of our people who may not be fully advised to think that the work might be done at a smaller cost to the land holders,

Sincerely yours.
D. W. AUPPERLE.

Daily News, April 1 and 17 1909

HIGH LINE MONEY IS READY — WORK BEGINS

TEN THOUSAND IS TODAY GIVEN GOVERNMENT

Expect Dirt Will Fly Within Next Ten Days

Some time today Secretary Aupperle will file a formal notification with Engineer E.E. Sands of the Reclamation Service informing him that the Water Users' association has on hand its first $10,000 and that this amount has been deposited with the local banks to the credit of the Reclamation Service.

When Engineer Sands receives this notification he will at once set in motion the machinery of the service preparatory to an immediate opening of the work on the High Line canal. It is highly probable that work may be begun next week, but this is largely dependent upon the timely arrival of certain equipment and materials.

It is certain, however, that before the thirtieth of April rolls around, "dirt will be flying" on the High Line.

The initial $10,000 however, is not by any means the end of the activity of the Water Users' association. It is their aim to secure at least $20,000 within the next few weeks. Twenty thousand dollars means an additional twenty thousand from the government and as all of the work is paid half in cash and half in Water certificates it follows that if the Water Users can collect $20,000 they can be assured of $80,000 worth of work being done on the canal between now and early next fall when the second call for High Line subscriptions funds will probably be made.

Daily News, Mar 8, April 22, May 2, Daily Sentinel, May 11 1909

BECAUSE Project Engineer Sands of the reclamation service was compelled upon several occasions to publicly deny misleading and inaccurate statements, to avoid the use of the shorter and uglier word, published in the local Sunset Whine, he has called down upon his head a stream of low abuse and scurrilous personal epithets which are so large a feature in the literary style of the editorial writer of the organ of the local Speer annex.

Public opinion in Grand Junction has no part in this discreditable assault. The people of Grand Junction are heartily in sympathy with Mr. Sands and his fellow officials in the reclamation service and the great work they are doing for Grand Junction and the Grand Valley.

Project Engineer Sands came to Grand Junction, an officer in the reclamation service of the federal government. He has gone about his work quietly, efficiently, without bombast or abuse. He has won the high respect and confidence of every man with whom he has ever come into personal contact. Because he refused to accept the domination of the editor of the Sunset Whine he has been foully attacked in language that would be shameful in the mouth of a wharf rat.

The Daily News cannot believe that such an assault will be long tolerated by the people of this community. It has no basis except in the wounded egotism of one man who has yet to prove to the people of Grand Junction that he has any call upon the reclamation service to be treated with any more deference or consideration than any other citizen of the valley.

ADVERTISE FOR BIDS FOR WORK ON HIGH LINE

The Grand Valley Water Users' Association this morning invite bids for three miles of excavation work on the High Line canal one mile north of the present Grand Valley canal and between Grand Junction and Fruita.

The section of three miles will be sub divided into small sections in order that small contractors may have an opportunity to bid on the work. The excavation work will involve the removal of about 121,000 cubic yards of dirt. All of the work will be done under the direction and supervision of the U. S. Reclamation Engineers. The bids will be opened on May 4th and dirt should be moving on the work by the end of that week.

Contractors who do not know the territory involved in the bid will be furnished with a guide whereby they will be enabled to personally examine the section. The right to reject any and all bids is of course reserved by the Water Users' Association.

Bids were opened yesterday at the Reclamation Office for supplies and material for the formation of the first construction camp on the High Lin at a point about two miles above Palisade where tunnel No. 3 will be dug. The camp may be located on the railroad right of way or on leased grounds. It will be established as soon as the parties awarded the contract can get the supplies on the ground. The awards will be announced today.

Daily News, Mar 8, April 22, May 2, Daily Sentinel, May 11 1909

HIGH LINE WORK HELD PENDING NEGOTIATIONS

Secretary Aupperle of the Water Users' association is still waiting for official confirmation from Secretary Ballinger of the contracts provisionally let Tuesday for the construction of the first three miles of High Line ditch. As soon as that is received the gangs will start at once. Yesterday some of the men working on the camp near Palisade were let off and at the same time a rumor was current of some hitch in proceedings. Mr. Aupperle was unwilling to admit that there was any significance in these rumors.

"The work on the digging north of town will start as soon as we get Secretary Ballinger's confirmation," he said, "and as regards any possible trouble at Palisade, the Government officials are handling that direct and the adjustment which they are making are their private affairs. It is manifestly absurd to suppose that an injunction suit could be brought against the government, as I was told it had been rumored."

DIRT TO FLY ON CANAL SOON

High Line and Orchard Mesa Both Ready to Start

Bids For Construction to be Opened Monday and Tuesday

The sub-contracts for the construction of the Orchard Mesa ditch will be let in Chicago tomorrow. At that time the bids will be opened, the contracts awarded and the last step prior to actual work of construction will be completed. Already contractors from Denver and Pueblo have been at Palisade and Cameo, looking over the ground and getting in readiness to start work as soon as the signal was given.

A camp is to be built across the river from Palisade and a force of probably two hundred men put on the work almost immediately.

C. C. Magenheimer will arrive here Friday and will personally superintend the start of the project.

Coincident with this, work, the dirt is scheduled to fly on the High Line project during the next ten days. The bids for the first three miles of excavation, on that portion of the canal north of town will be opened here Tuesday, and the contracts awarded at that time.

The work will probably commence one week from tomorrow. It has purposely been divided into small sections in order that no contract will be more than can be handled by local firms. With the money now available and what is expected to be subscribed and received from the government during the summer, the work once commenced will be continued without interruption.

OPPOSITION TO HIGH LINE IN GRAND VALLEY CAUSED DELAY

BY SHELDON CLINE

WASHINGTON, D. C.. May 10. —It was stated to The Daily News correspondent today that opposition in the Grand Valley to the High Line canal was responsible for the suspension of work on that project. The conflict of interests became so strong that Secretary Ballinger decided to suspend operations until he could make a thorough investigation of the entire question. Details regarding this opposition have not been made public.

Daily News, May 9 and 13, Denver Republican, May 13 1909

THE Grand Junction High Line canal should never be constructed by private capital. Some excitable souls, seeing cyclones in every cloud, frost in every breeze and depression in every rumor have apparently worked up a report to the effect that private capital was knocking at the door of the department of the interior and was apparently attempting to influence Secretary Ballinger to in some way abrogate the government contract to build the canal and hand the valley's great project over to the tender mercies of the private capitalist.

The Daily News doubts if there be a bakers dozen men in between the Palisades and Mack who would countenance the thought of turning over the great ditch to private parties. The men of the valley are sick of the very thought of letting the canal go into the hands of any man or set of men. They are fervently almost passionately determined that the government and only the government should build the "High Line". The people have confidence in the Reclamation Service; they have had the splendid illustration of the potency of its workings in the boring of the Gunnison tunnel and the construction of that great ditch system. They have set their hearts upon having Uncle Sam do the work. They KNOW that if the Reclamation Service builds the ditch it will be built and be built right and quickly. Their experience with private capital in connection with the High Line has not been of that character that is calculated to inculcate respect and confidence. Whether T. C. Henry or Thomas Walsh are anxious to build the canal or not The Daily News has no certain information but The Daily News does KNOW that the people of the valley, the owners of the High Line land, the constructive men of the valley want the government, and only the government, to build the big ditch.

NO MYSTERY IN CANAL'S TIE-UP

There is no great amount of mystery concerning the delay of suspension of work on the High Line canal, in spite of the effort of some people to involve the whole matter in deep mystery. There is every reason to believe that the suspension has come about through the efforts of T. C. Henry and his very eminent and distinguished attorney, a man who has been greatly honored by the state in years past and who has a great deal of influence down at Washington..

The Sentinel has all along believed that T. C. Henry was the cause of the suspension and facts that have developed in the last few days plainly indicate that such is the case and a great many people who have made investigations share the same opinion with The Sentinel.

It is really foolish to attempt to leave the impression that "all is shrouded in mystery" and only heroic and mighty efforts will unravel the mystery and get at the cause of the suspension.

T. C. Henry still wants to build the canal and he is leaving no stone unturned to get the government to abandon the project.

Daily News, May 9 and 13, Denver Republican, May 13 1909

TO ASK WHY WORK STOPPED ON BIG DITCH

SECRETARY BALLINGER TO BE SEEN ABOUT HIGH LINE PROJECT.

T.C. HENRY THINKS CHICAGO CONTRACTOR MAY HAVE KNOCKED.

Interior Department Answers Senator Teller That Money Available Now Is Insufficient.

GRAND JUNCTION, Colo., May 14 — (Special.)--State Senator DeLong and, Secretary D. W. Aupperle, of the Water Users' association, started tonight for Washington to interview Secretary Ballinger regarding the sudden cessation of work on the government high line canal project. They hope to convince him that there is an overwhelming sentiment in favor of the project being completed by the government and to get him to issue an order for the immediate resumption.

Authentic private information from Washington indicated that it was opposition from some unknown source here in the Grand valley which caused the secretary to send his telegram stopping the work. In this connection it was said that ex-Senator Teller had gone to Ballinger as the intermediary between the government and the private parties interested, and in reply to this senator Teller tonight issued the following signed statement:

"I received a letter from certain lawyers asking me to carry out the contract made by Garfield with the Grand Valley Water Users' association. I called on Ballinger and he said he could furnish $125,000, as Garfield had agreed to that amount, provided the people could furnish a similar sum. Ballinger could not say whether he could furnish any more money.

Later Benton Cannon, of Grand Junction, wrote me about the project and I again went and saw Ballinger and he said he could not say what could be done beyond the $125,000. He didn't have any more money than that now and might or might not have it in 1910. He said the project would cost $2,500,000.

"I asked him if he was willing to allow private parties to build the canal, and he said that if he was perfectly satisfied they had the means and were capable of completing it, he was entirely willing to let them go on with the work. I also interviewed the chief engineer on the subject and got practically the same information. Ballinger promised to furnish me with a statement, but I have not yet received it. I have no interest in the project and I am not trying to block any irrigation project in the state. Certain people wrote me for information and I tried to get it for them."

Senator Teller positively refused to disclose in whose interest he had been working.

Due primarily to the fact that T. C. Henry of Denver, a few months ago offered to build the canal for $2,000,000, he was the first person to be suspected in connection with the present complication. He stated that ever since disposing of his water filings to C. C. Magenheimer of Chicago, which took place two

Daily News, May 9 and 13, Denver Republican, May 13 1909

months ago, he had had nothing to do with the Grand Valley project He accused Magenheimer of being the one now implicated. Magenheimer is constructing a million dollar irrigation ditch here on Orchard mesa and has several times professed willingness to build the high line if given a chance. He refused to make any statement regarding Henry's charges.

The Water Users' association have a petition with more than 1,000 signatures, asking the government to build the project and protesting against letting the contract to private individuals. These people represent 90 per cent of the land owners along the line of the canal.

Daily News, May 9, Denver Republican, May 16 1909

WEST SLOPE WROTH OVER HALTING OF HIGH LINE DITCH

Water Users Send Committee to Washington to Protest.to Ballinger

'GOVERNMENT MUST BUILD'

Valley Will Never Consent to Private Capital Completing Canal to Be Ultimatum

SPECIAL TO THE NEWS

GRAND JUNCTION, Colo., May l4. The High Line canal, begun a few weeks ago by the government, after seven years of waiting, and which, when completed, will do more to exploit the fruit growing resources of the Grand valley than all the other irrigation schemes of the last ten years put together, the project which has been the golden dream of every land owner and every well-wisher of the valley, has suddenly been held up by a peremptory telegram from Secretary of the Interior Ballinger.

Without the least intimation of the real hitch, the local reclamation officials a week ago last Monday received the order to suspend all operations. So disastrous and unexpected was this turn of affairs that it was a week before the officials of the Water Users' association, with whom Project Engineer Sands had confided the order, made the bad news public. During that week every known source which might reveal the cause for the order was investigated without the slightest hint of the reason being discovered.

Secretary Ballinger, in response to queries, merely stated that it was his desire to more thoroughly familiarize himself with the details of the project.

When Secretary Garfield, as one of the last official acts of his term of office, signed the contract with the Water Users' association on behalf of the government, it was believed that the last obstacle had been removed. The government was pledged to spend an amount equal to that raised by the water users, and it was believed it could not withdraw from its bargain if it wished to. One hundred and twenty-five thousand dollars was pledged here, and $20,000 was immediately deposited in a local bank, a similar amount of cash being deposited by the government. Bids on sub-contracts were advertised and the contracts let. It was in the midst of this that the stop order came, and the question which arose was, "Where are we?" Authentic private information was received from Washington that opposition in the Grand valley had been responsible for Secretary Ballinger's order, and the first thing to be done was to find "'the nigger in the woodpile."

Want Government to Build.

Ninety-five per cent of the people of the valley are eager for the government to build the canal. This is not an idle estimate, but the result of a careful canvass made by the Water Users' association and proved by petitions now on file in their office, containing thousands of signatures, asking the government to do the work, and stating that they were opposed

Daily News, May 9, Denver Republican, May 16 1909

to having the project undertaken by private capital.

There are, however, many Colorado capitalists, backed by some men in the East, who have professed to stand ready to build the canal themselves if given the opportunity, and have several times in the past blocked the government proposals.

Chief among these men are T. C. Henry & Son of Denver. The reclamation service offered to build the canal seven years ago, the first project to be undertaken, but gave it up then chiefly, so it is said, because of Henry's representation that he could build it cheaper and that the people favored private enterprise. He has figured in all the negotiations since then.

Henry Blames Others.

Henry has issued a statement denying that he is responsible for the present delay and stating that C. C. Magenheimer, a Chicago capitalist who is now building the $1,000,000 Orchard Mesa project here and with (too damaged to read)

formerly associated. and Orson Adams, president of the Mesa County bank of this city, are in the deal to block the government project and build it themselves, and that it was their influence, in conjunction with that of former Senator Teller, that caused Ballinger to issue the order of suspension.

Neither Adams, who returned yesterday from Washington, nor Magenheimer, who is in Chicago, will discuss the situation. The latter qualified his refusal by a statement that he would not talk at present. Several men here supposed to be in his confidence, who were in Chicago last week when the matter first became known, stated tonight that Magenheimer told them that he had nothing to do with the present complication and didn't know who was in it, but that he was ready, as he always has been, to construct the canal if the government gave it up.

The history of the seven years' negotiations which preceded the signing of the final contract by Secretary Garfield last February throws a strong side light on the actions and motives of the leading figures in the drama now being enacted. It was a long battle between the Water Users' association headed by Secretary D. W. Aupperle and backed by the rank and file of the land owners along the proposed route of the canal, and T. C. Henry and a few others, in conjunction with important water filings on the Grand river which Henry claims superseded all other. Henry won in the earlier stages of the conflict, but last summer the reclamation service became convinced that the people wanted the government to build the canal, and at once commenced preparing to undertake the project.

The final survey was made, estimates prepared and preliminary contracts entered into. Henry had been eliminated from these negotiations, and meanwhile had formed a combination with the Magenheimer Brothers in Chicago in the proposal to build another irrigation works here on Orchard Mesa, to cost $1,000,000. Henry was to trade his water rights for his share in the company, and the entire proposition became involved in the High Line project, due to a conflict over water rights. The Magenheimer Brothers parted company with Henry and closed the contract with the Orchard Mesa district to build the ditch. Henry secured an injunction in Denver and tied the proposition up, and then startled the valley with a formal offer to the government to build the High Line canal, according to the government's specifications, for $2,000,000: as against the government estimate of $2,500,000.

Petitions containing 10,000 signatures, and asking the government to build the project were sent to Washington in charge of a local delegation, and after several weeks of lobbying Garfield signed the contract. At the same time the Henry-Magenheimer suit was compromised.

On the night the suit was settled Magenheimer wired his friends here: "Henry is out of the Grand valley forever. It has cost us a lot of money, but we have bought all his claims to water rights on the Grand river."

Henry now claims that from that day he has never concerned himself with any of the projects here.

Former Senator Teller tonight issued the following signed statement from his office in Denver: "I received a letter from certain lawyers asking me if I could find out from Secretary Ballinger whether he was going to carry out the contract made between Garfield and the Grand Valley Water Users' association. I called on Ballinger and he said he could furnish $125,000, and Garfield had promised that amount, provided the people would furnish a similar amount. He could not say whether he could furnish any more money or not.

"Later, Benton Cannon, now in Grand Junction, wrote me about the project, and I again went and saw Ballinger, and he said he could not say what could be done beyond the $125,000. He didn't have more than that now, and he might and he might not have more in 1910. He said that it would cost $2,500,0(0.

"I asked him if he was willing to allow private parties to build the canal. He said that if he was perfectly satisfied that they had the means and were capable of completing it he was entirely willing to let them go on with the work. I also interviewed the chief engineer on the subject and received practically the same information. Secretary Ballinger promised to furnish me with a statement, but I have not yet received it.

"I have no interest in the project, and I am not going to block any irrigation project in the state. Certain people wrote me for information and I tried to procure it for them."

Keeps "Interests" Secret.

Teller positively refused to disclose the identity of the interests which had asked him to get the information. Benton Cannon, the man referred to in Teller's statement, also has issued a statement to the effect that his letter to Teller was merely a general request for information regarding what the government intended to do and that he is well aware that there is an overwhelming sentiment in favor of government construction. Cannon is engaged in the real estate business here.

State Senator Horace DeLong and Secretary D. W. Aupperle of the Water Users' association started for Washington tonight to have a personal interview with Secretary Ballinger. They will try to convince the secretary that the Grand valley will never consent to the work being undertaken by private capital, and will urge the secretary to recommence the work at once.

Daily News, May 9, Denver Republican, May 16 1909

PEOPLE ANXIOUS FOR GOVERNMENT CANAL WORK

GRAND JUNCTION, Colo., May 15-- (Special) —The statement of former Senator Teller, made public this morning, concerning the high line canal situation, containing the significant fact that Secretary Ballinger was doubtful of the government's financial ability to carry on the work and professing his willingness to allow the project to be undertaken by private capital, was a leading subject of conversation here today. There is an overwhelming sentiment in favor of the government doing the work, and the intimation that it may abandon it has caused a sensation.

In response to repeated requests for a statement regarding his alleged connection with the sudden cessation of work on the canal C. C. Magenheimer of Chicago today telegraphed that he had nothing to do with the government stopping the work.

Magenheimer Loyal Booster of Grand Junction

REPUDIATES CHARGE OF ATTEMPTING TO THWART BUILDING OF HIGH LINE

Cooperates With Government in facilitating Work:- Ready However, to Bid on Project Should Uncle Sam Relinquish it

DE LONG AND AUPPERLE AT CHICAGO MEETING

Senator DeLong and Secretary Aupperle arrived in Chicago yesterday and spent the day in conference with President Magenheimer of the Orchard Construction Co. regarding the present situation relating to the High Line canal project.

The matter was gone over in detail especially as related to the alleged connection between the Magenheimers and the sudden order of cessation of work sent out by Secretary Ballinger two weeks ago. Mr. Magenheimer repudiated the accusation attributed to T. C. Henry, that the former was at the bottom of the present tangle. He realized, he told his interviewers as well as anyone else how anxious the landowners here were for the government to do the work and he had not and would not put any obstacle in the way. However, should the government be obliged to abandon the project he was willing as he had been in the past to bid on the project and carry it through.

At the close of the conference Mr. Magenheimer wrote a statement of his position in the form of a letter to Senator DeLong, which he telegraphed to the Daily News.

Senator DeLong and Secretary Aupperle, left for Washington last night and will have their conference with Secretary Ballinger tomorrow... Mr. Magenheimer's letter is as follows:

CHICAGO, Ill., May 17.--"Hon. Horace T. DeLong, Grand Junction, Colorado. Dear Sir: Referring to our interview of this date and answering your questions as to whether or not there are any differences between us and the United States reclamation service and the proposed grand Valley high line canal project, and as to whether we know anything regarding the secretary of the interior stopping the work on the proposed high line project, and as to whether we know anything regarding the secretary of the interior stopping the work on the proposed high line project and as to whether or not we wish to defeat the building of the high line canal by the government.

There are absolutely no differences between the reclamation service and ourselves as to the building of the orchard Mesa irrigation district system, which we are now constructing, and which has been approved by the Chief engineer of the United States reclamation service. On May 16, 1908, we made a contract with the secretary of the interior, which settled all differences. Said contract has been published many times in your local papers. We have no knowledge whatsoever as to why the secretary ordered the work stopped on the High Line Canal project.

We do not wish to defeat the building of the High Line Canal by the government.

The Grand Mesas Land, Canal and Power company made a filing of 1,200 second feet of water in the Grand river in the fall of 1907 for the purpose of supplying both the orchard Mesa irrigation district and the High Line district with water when we found that the government intended to undertake the building of the High Line canal, we, at once, stepped aside, and in fact put 750 second feet of water, at the disposal of the secretary of the interior which was necessary to be acquired by him to enable him to rapidly proceed with the building of the High Line canal project. The people of the Grand Valley, and the officers of the reclamation service fully understood that we do not wish to interfere with the government's building the High Line canal, but they also understand that if the government does not wish to proceed with the High Line canal project, that we have at all times, stood ready to submit a proposition to build the High Line canal, which is our present possession. We do not, however, desire to take up the project in any way that will interfere with the purpose of the government or not be of substantial advantage to the irrigation district and the people of the Grand Valley

Yours very truly, — C.C. MAGENHEIMER

Orchard Construction Co.

ADAMS DECLARES HENRY NOT TRUTHFUL

Henry Attempted to Get Local Man's Support But Failed

Charging that news reports sent out from the Daily News office by the Denver News correspondent in Grand Junction, R. C. Mulligan, news editor of the Daily News, had been so worded as to besmirch the character and efforts of Mr. Orson Adams, appeared last evening in the Whimper.

The statements made by the Whlmp are, as usual absolutely false. To place the position of Mr. Adams in connection with the High Line canal fairly before the people the News makes this statement, believing that the article which appeared in the Whlmper, while ostensible pretending to clear Mr. Adams of any imputations in regard to the holding up of the work, really makes it appear that he might have some hand in the matter.

In Denver last week Mr. T. C. Henry was quoted by a well known newspaper man of that city with the statement that he had knowledge of the persons who were concerned in the holding up of the High Line. '"These men", he is quoted with saying, "are the Magenheimers of Chicago and Orson Adams of Grand Junction." This report was promptly sent to the Daily News but Mr. Henry's alleged remarks were regarded so lightly by the publisher of the News that publication of them was withheld. Henry's statement however were known among Denver newspaper men who were working on the High Line story.

Mr. Henry has made statements of a like character before. In the letter files of the News today can be found a letter to the publisher of the News from Henry stating that Mr. Adams was against the government's construction of the High Line. The News at that time regarded the matter so lightly as to make no comment on it or to refer to it in any way.

On Friday night the Denver News called their correspondent here, Mr. Mulligan and asked that he give them what he understood Henry to have said in connection with the High Line story. This was done. A charge that in so doing their correspondent was besmirching Mr. Adams is ridiculous.

'The statement made by Mr. Henry is absolutely false," stated Mr. Adams last evening to the News. "I am very anxious that the government construct the High Line canal. I was in favor of the government building the ditch as far back as 1902 when Henry took it over and failed to accomplish anything. I believe that the work of the reclamation service in building these canals has made the sale of irrigation securities in the east much easier than formerly. While I was in Washington I went to see secretary Ballinger with Senator Guggenheim and then saw Mr. Davis of the reclamation service. At that time I did not know that the work had been held up but was so informed at the reclamation offices and expressed my regret. My only purpose in going to the offices of the reclamation service was to get exact information as to the status of the work. I have in the past twelve months been approached by different parties, among them Henry, attempting to interest me in propositions in connection with the building of the canal by private parties. I have turned these matters down absolutely. I believe that fully 75 per cent of the people of the valley if not more desire the work done by the government. I sincerely hope that the government will see its way clear to do the work at once. I'm certainly not afraid of letting the people judge between me and Mr. Henry in regard to the matter of our personal integrity."

Daily News, May 22 1909

HIGH LINE NOT ABANDONED IS OFFICIAL WORD

Senator Patterson Promised by Ballinger There Shall Be No Needless Delay Upon Project.

RED TAPE IS SNARLED UP, SAYS SECRETARY'S LETTER

Time Required to Determine Regularity of the Steps Already Taken,

----

Enforced delay to straighten out legal tangles, not abandonment is the position of the government regarding the Grand valley High Line canal project, according to a letter from R. A. Ballinger, secretary of the interior, received by Senator Patterson yesterday in response to an inquiry of the latter. Because of the general interest in the matter the correspondence is given herewith:

Senator Patterson's Letter.

"Hon. R. A. Ballinger, Secretary of the Interior, Washington, D. C.--Dear Mr. Secretary: You are doubtless familiar with what is known as the Grand valley canal project in western Colorado. It is an enterprise of profound importance to the entire state, but particularly to the section through which the canal is to be constructed. The reclamation service signed a contract last February, I think it was, with the Grand Valley Water Users' association, and shortly afterward work, under government supervision, was actively commenced and was prosecuted until about two weeks ago, when, upon peremptory orders from Washington, all work was stopped.

"You can well understand that the stopping of the work has occasioned a great deal of uneasiness throughout Colorado, but particularly in the section the canal was directly to affect, and surmises of many kinds are being indulged in as to the reason for the government's peremptory action.

"There is no question about the vast importance and significance of the work that was undertaken. It, doubtless, when completed, would have brought more annual wealth and population to the section of the state involved than any other government enterprise of the kind. I feel a very deep concern about the enterprise, although I have no interest in it near or remote, directly or indirectly, except such interest as every loyal Coloradan should have for the welfare of the state.

"I would like very much, indeed, to have from you the reasons for the stoppage of the work, whether it is merely suspended or a permanent stoppage, and what are the reasons that moved you to issue or approve the order.

"I would not encroach upon your time with these inquiries were not the subject of such vital importance, and I sincerely hope you will favor me with as early and as full a statement as you feel the welfare of the service will permit you to make. With sincere best wishes. I am truly yours, T.M. PATTERSON.

"Denver, Colo., May 14, 1909

Daily News, May 22 1909

No Unnecessary Delay.

"Hon. T. M. Patterson, Denver, Colo.

"My Dear Senator--I am in receipt of your Enquiry of the 14th inst., in reference to the Grand valley irrigation project, and in answer I beg to say that certain questions have arisen with reference to the regularity of the proceedings already had. I have directed the temporary suspension of work pending a disposal of these matters that are now under consideration. When the result is determined, I will he in position to advise you as to the government's attitude respecting the project.

"I do not intend that there shall be any unnecessary delay on the government's part in disposing of these matters. I realize the importance of speedy action. Sincerely yours,

"R. A. BALLINGER, Secretary.

Washington, D. C., May 17, 1909.

The High Line Canal.

An Enterprise That Must Not Be Lost–Rudolph Spreckels' New Philanthropy—The Ex-Confederate Moderator

ELSEWHERE in this issue will be found the correspondence between Senator Patterson and Secretary of the Interior Ballinger on the subject of the Highline canal in the Grand valley. The News deems the subject important enough, and the danger serious enough to call for a full statement of the case.

The proposed Highline canal is a reclamation service project for the watering of 60,000 acres of land on the right bank of the Grand river, below Grand Junction. Complete surveys and estimates of the cost have been made, and it is known that no physical difficulties stand in the way of the scheme The plan is to take the water from the river a little distance above Palisade, and carry it along the edge of the cliffs that border that town, down onto the broader mesas below. The estimated cost is something over $2,000,000. Work was begun on the project about a month ago, and was suddenly stopped a couple of weeks later. No reason has been given for the stoppage, and it is hard to believe that a good reason exists. There are two theories abroad in the Grand valley as to the bedrock cause of the stop order. One is that certain parties who claim water rights in the Grand are tying up the job through political influence until these claims are bought out. The other is that certain capitalists want the job of putting water on the land themselves. What lends color to this latter theory is the fact that the stop order came shortly after the demonstration that the fruit crop of the Grand valley could be saved from frost by orchard heaters.

The News does not pretend to know which of these theories is true, or whether either is true. What it does know is that the stopping of work on the highline canal is a detriment to the whole state;

and that the permanent cessation of that work would be nothing short of a calamity.

For the matter is in such shape that if the federal government does not put through the scheme it cannot be put through at all, at least, for very many years to come. When the orchardists of the upper slope of Palisade secured title to their lands, the government reserved the right to construct a ditch through those orchards. That right is not transferable; it is a piece of sovereignty pertaining to the government alone. And nothing short of an incontestable right will ever get a canal through the narrow, neck that must be traversed to get the water to the waiting mesas. Palisade orchards are worth $2,000 per acre, often more; and men with holdings like that do not take kindly to having ditches cross their land. An attempt of a private corporation to get a right of way would be the signal for endless lawsuits, for criminations and recriminations, for a fight that would set back the progress of the valley, instead, of hastening-it. And there is no way around. The canal must cross this particular neck of land, or stop where it begins.

The only party on earth, then, who can build a canal to water these 60,000 acres of splendid land, a canal where the water will flow without being pumped, is the United States government.

Could Not Vote Bonds

But the water can be pumped and a private company can build a pumping plant. Technically, yes, but practically, no. The cost of pumping the needed amount of water up the rise would be great; though not too great to pay if the pumping company could secure title to the lands to be irrigated. But this is exactly what the company could not do. Two-thirds of the area to be watered consists of homesteads which have been filed on, but not proven up. This means that two-thirds of the land to be watered is in such shape that neither the government nor the settlers can give clear title to it. You cannot vote bonds on land held in that way. You cannot do anything with it. A pumping plant would cost more than a gravity canal, both to construct and maintain. For a private company to put up such a plant, it would need not less than $3000,000 of ready cash. It would have to spend this sum to get ready to sell water to customers who could give no security. Do you happen to know of any corporation which does business in that fashion? We do not.

From whatever side you approach the matter you come to the same plainly evident fact: THE UNITED STATES RECLAMATION SERVICE IS THE ONLY ORGANIZATION WHICH, UNDER EXISTING CONDITIONS, CAN BUILD THE HIGHLINE CANAL. The News believes most potently in the efficiency of private enterprise as against public enterprise where the two can be fairly set one against the other. But here it is public enterprise or nothing--and we are not long in making our choice.

The Highline canal must be built; and the whole state must get busy to see that it is built with the smallest possible delay. THIS IS A STATE MATTER. It affects the Grand valley most nearly; but it affects us all. For every three houses built in the Grand valley, one house is built in Denver. For every three families that move into the Uncompahgre country this summer, one family will make its home in Denver. The whole state must join in the demand for the Highline ca-

Daily News, May 22 1909

nal, and Denver must lead the way. The News calls on every commercial organization throughout the state, on every live newspaper, on every Colorado congressman and senator, to take up the fight for the Highline canal. And take it up now.

Daily News, May 18, Daily Sentinel, May 24 1909

ARE READY TO BUILD THE HIGH LINE CANAL

IF GOVERNMENT CANNOT TAKE UP PROJECT AT ONCE

ORCHARD Mesa Company Announces Its Readiness to Make Proposition to the People---Guarantee Completion in 1911.

THE Orchard Mesa Construction Company stands ready to build the Grand Valley High Line canal, under the district plan, and have water ready for the lauds under the canal for the season of 1911, providing that the government cannot take up the early construction of the big project and it is the desire of the people that this company take up the project.

This afternoon in a special dispatch to The Sentinel the above most important announcement is made, the message being sent by Mr. George Smith of this city, who is now in Chicago.

It has been rumored for some time that, in case the government did not proceed with the work, one or more big private concerns stood ready to make propositions to the people, but the identity of all these concerns has until today been kept secret.

The Orchard Mesa Construction Company, backed by some of the largest financial interests of the country, has the contract for building the great Orchard Mesa project near the city.

The following is the telegram received by The Sentinel this afternoon:

Chicago. Ill., May 24.

The Daily Sentinel, Grand Junction, Colo.

After seeing the article in the Denver News of May 21 on the High Line canal, in which Secretary Ballinger told Messrs. DeLong and Aupperle that there was no doubt about the government's right to build the canal and still more doubt about having the funds for its early construction, I saw the Magenheimers and their associates, who are now building the Orchard Mesa canal, and they stated that if the government could not promise early construction, they were willing to submit a proposition to build the High Line canal under the district plan, to furnish water for the season of 1911.

GEORGE SMITH

Daily News, May 18, Daily Sentinel, May 24 1909

LAWYERS TO REPRESENT PALISADE GO EAST

F. C. Goudy of Denver and D.D. Potter to See Ballinger

THEIR MISSION SECRET

Frank Goudy of Denver, probably the greatest irrigation lawyer in the State of Colorado, and Attorney D. D. Potter of Palisade, have left for Washington in the interests of the Mesa County Protective association of Palisade to appeal to Secretary Ballinger for a "'just and more equitable" proffer of damages for the injury which it is claimed will be done to the orchards of the Palisade district if the High Line canal is constructed along its present lines.

The Mesa County Land Owners and Protective Association is made up of the land owners about Palisade who are affected by the Grand Valley project. All of these men own land across which the High Line canal is surveyed. The damage that will be done to their land has been variously estimated from $75,000 to $200,00. Some of the land involved sells at high as $2,000 per acre. Other sections are not nearly as valuable. For the last two months J.R. Alexander of the Reclamation Service has been circulating among the ranchmen offering them the damage the government saw fit to give. In most cases these proffers have been refused. Suits were in prospect at the date of the holding up of the entire Grand Valley project by Secretary Ballinger.

The Mesa County Land Owners and Protective association was formed last December. W. J. Sanford is president, Ray Lundy secretary and treasurer. Attorney D. D. Potter is chairman of the board of directors. It is estimated that the combined wealth of the men in the association is close to $800,000 and these men declare that they stand ready to take their case to the United States Supreme court.

A few days ago the association addressed a letter to Secretary Ballinger asking that they receive a rehearing in regard to the damages against their property caused by the construction of the High Line canal. They then decided to follow their letter up with a personal visit to Secretary Ballinger. Officers of the association in Palisade emphatically declare that they were in no ways responsible for the holding up of the project by the Secretary.

Daily News, May 25 1909, Daily News, unknown date.

WATER IN HIGH LINE IN 1911 IS MAGENHEIMER'S OFFER

TENTATIVE OFFER IS SUBMITTED TO BALLINGER

If Uncle Sam Consents Private Capital is Ready to Do It

BALLINGER STILL SILENT

C. C. Magenheimer has submitted a detailed proposition to Secretary Ballinger of the Department of the Interior offering to build the High Line canal and assure water in the ditch by 1911. He makes this offer based on the understanding that the Reclamation Service has not at present sufficient funds in their treasury to go ahead with the immediate construction of the line. Secretary Ballinger has not yet replied to Mr. Magenheimer's offer. Until he does so Mr. Magenheimer will not make public the details of his offer.

Last evening The Daily News received the following dispatch from Mr. George Smith, now in Chicago consulting with Mr. Magenheimer in regard to Orchard Mesa construction work.

His dispatch is as follows:

CHICAGO, May 24.--Mr. Magenheimer has written Secretary Ballinger stating that if the Reclamation Service does not have funds available for immediate construction of the High Line they desired his permission to submit a proposition to the people of the Grand Valley to build a High Line canal under the district plan They will agree to furnish water for the season of 1911. Until the Magenheimers receive a reply from Secretary Ballinger they do not desire to make their proposition public.

Yesterday afternoon Mr. Smith sent the following to The News relative to the High Line canal:

After seeing the article in the Denver News of May 21st on the High Line canal in which Secretary Ballinger told DeLong and Aupperle that there was doubt about the government's right to build the canal and still more doubt about having funds for early construction. I saw the Magenheimers and their associates who are now building the Orchard Mesa canal and they stated that if the government could not promise early construction they were willing to submit a proposition to build the High Line canal under the district plan to furnish water for the season of 1911.

GEORGE SMITH.

AUPPERLE AND DELONG ON JOB READY FOR ANY EMERGENCY

BY SHELDON CLINE.

WASHINGTON, D. C., May 24--Aupperle and DeLong are daily working on the project of gathering data preparing to meet any emergency that may arise. Ballinger has not given out any new information.

Daily Sentinel, May 25 1909

PEOPLE WILL TOLERATE NO ATTEMPT TO BUILD CANAL BY PRIVATE CAPITAL

CHAMBER of Commerce, Water Users Association and Citizens Pass Resolutions and Send Telegrams to Effect That High Line Canal Must be Built Only by Government---If Government Dropped Project Magenheimers Would Completely Control Situation---Important Letter From Aupperle---Gov. Shafroth Declares He Will Exert All His Influence in Behalf of Project---Resolutions Thanking Patterson Adopted--A Day of Important and Vigorous Action.

That Grand Junction and the Grand Valley, that the citizens of the city and the citizens of the valley are determined that the government and not private capital shall build the Grand Valley High Line canal; that the citizens of the city and the valley are willing to wait for the government to build this project so dear to their hearts, and that the interference of private parties with propositions to build the canal will not be tolerated for a single moment, was made plain today at a meeting of the board of directors of the Chamber of Commerce, with a number of other prominent citizens in attendance; and has also been made plain by expressions received from every section of the valley since the announcement in yesterday's Sentinel to the effect that the Magenheimers of Chicago and the Orchard Mesa Construction Company have a proposition to make to the people looking toward the building of the High Line canal by this company and to have it completed by 1911.

The meeting this morning was called on short notice and followed numerous informal consultations among the members and among business men and citizens generally. It was an important meeting, called for a most important purpose. The meeting opened at 10.30 a.m.

There-were present the following gentlemen: President C. E. Cherrington, Secretary Mahoney, Directors M. O. Delaplain, S. B. Hutchinson, W. H. Buthorn, W. M, Campbell, J. H. Rankin and Walter Walker; Messrs. Edwin Price, A. R. Sampliner, I. N. Bunting; George A. Currie, Karl A. Bickel, T. M. Jones, John F. Moore and Smith.

Immediately upon the calling of the meeting to order President Cherrington announced that the meeting had been called to consider an all-important matter — the question of whether or not any efforts to have the High Line canal built by private ownership shall be tolerated and to take action that may seem expedient in view of the telegram sent to this City yesterday by George Smith to the effect that the Magenheimers wanted to submit a proposition if the government announced that it could not take up the work very shortly

The first thing .given consideration was a letter Mr. Delaplain, as president of the Water Users' association, received this morning from Secretary Aupperle, who with Senator DeLong, is in Washington.

The letter from Mr. Aupperle was in the nature of a report as to what he and Senator DeLong had accomplished to date He told of stopping at Denver and seeing Senator Teller, Senator Patterson, and several other gentlemen. He stated that Senator Teller gave him a letter to Secretary Ballinger urging in the strongest terms that the government build the canal. Mr. Aupperle said that the letter was as powerful and satisfactory as could be asked. Mr. Aupperle told of the stop in Chicago and the talk with the Magenheimers. He then detailed the work done up to last Friday in Washington. He stated that Senator Guggenheim, Senator Hughes, Congressman Martin, all of Colorado, were working hand in hand with him and Sen. DeLong but at that time Congressman Taylor had not been seen. Mr. Taylor's friendship for the canal and the fact that he is working for the government construction of same are certain.

Therefore there is not the slightest foundation for some attempts made here to make the people believe that politics is cutting any figure in the matter. The Democrats and Republican officials are alike in their fight for the government construction-of the canal, and the influence of each side is about equal.

Mr. Aupperle stated that Mr. Davis of the Reclamation office was giving them every assistance. He told of the early conferences with Secretary Ballinger, the details of which have been announced here. The letter was on the whole very encouraging.

Among other things Mr. Aupperle stated that Crawford Hill had promised Senator DeLong the support of the Denver Republican in the matter. Senator Patterson some time ago promised his personal influence and that of his papers, the Times and the News and he has made that promise good.

John F. Moore, manager of the Grand Junction Fruit Growers' Association, was called upon. In a brief talk Mr. Moore emphatically declared that the people must not tolerate the idea of having private capital build the canal; that the government should and must build it. He declared that he believed that there would he a great depreciation in High Line land values if the government dropped the project and it was turned over to private ownership. Mr. Moore's remarks were very forceful and convincing.

A. R. Sampliner was then called upon, and this gentleman was equally as emphatic as Mr. Moore in declaring that the only thing the people of the valley must stand for is the government construction of the canal. He declared that he was ever opposed to private capital taking up the big project. He presented a number of very strong arguments showing why the government and not private capital must build the canal if the best interests of the people are conserved.

MR. SAMPLINER POINTED OUT THE FACT THAT 1T SEEMS CERTAIN, ACCORDING TO THE TERMS OF THE CONTRACT BETWEEN THE GOVERNMENT AND THE ORCHARD MESA CONSTRUCTION COMPANY, CONCERNING WATER RIGHTS, AND ACCORDING TO THE RIGHTS UNDER WATER FILINGS, THAT. IN CASE THE GOVERNMENT DOES NOT BUILD THE CANAL, THE WATER RIGHTS WILL REVERT TO THE MAGENHEIMERS, WHO ARE THE GENTLEMEN BACK OF THE ORCHARD MESA CONSTRUCTION COMPANY.

Daily Sentinel, May 25 1909

This brought out a startling phase, new to most of the gentlemen present, of the proposition. It would mean that should the government not build the canal, the Magenheimors would be the only men with whom the people could do business, as they would control the water rights and, as a result, the private company could make whatever demands it desired.

Mr. Sampliner also pointed out the fact that, while under government construction, the land owners would pay for the land and water in annual payments WITHOUT INTEREST, making the whole project cost probably 2 ½ million dollars, if a private concern built the big ditch these payments WOULD DRAW INTEREST, and interest on such a big amount for a number of years would almost DOUBLE THE COST OF THE PROJECT TO THE LAND OWNERS by the time the indebtedness was all paid off.

This and many other pointed arguments were presented by Mr. Sampiner. Both he and Mr. Moore agreed that the people should wait on the government to build this canal, even if it will take some longer time to get it. Both gentlemen also declared that a government-constructed canal meant a perfect canal; a privately constructed canal meant a canal just sufficiently good to permit the contractors to get their money and meet the terms of the contract.

George A. Currie was then called upon. He is for government construction first, last and all the time, He pointed out the undesirable features of standing for a bond issue by a private concern for the construction of the big ditch. He thought no time should be lost in notifying the officials at Washington that this city and valley will not for a moment hear to any proposition for the private construction of the canal, and he thought that local men and outside men who had interfered with a private proposition, such as that of the Magenheimers and George Smith, should be censured by the Chamber of Commerce; that such action on the part of these gentlemen was "an unfriendly butt-in." Mr. Currie thought that the feature mentioned by Mr. Sampliner concerning the Magenheimers would have to control the water rights in case the government does not build the canal, should be pointed out, clearly to Mr. Ballinger as a condition dangerous to the best interests of the city and valley.

Then Mr. Currie referred to the editorial appearing in the Denver News at the direction of Thomas M. Patterson, the owner of the Denver News and Times, the same editorial being reproduced in The Sentinel yesterday, and to other articles in the interest of the High Line canal and government construction of the same appearing in Mr. Patterson's papers. Mr. Currie declared that ex-Senator Patterson deserved the thanks of the people of this section for his interest in behalf of the project and his influence, which is very powerful at Washington.

Mr. Currie then presented the resolution given below, which was adopted by a unanimous vote of the board of directors:

Resolved, That the Grand Junction Chamber of Commerce, on behalf of the entire people of Mesa county, express to the Hon. Thomas M. Patterson their appreciation of the interest he has shown in our High Line canal project, and for valuable assistance he is rendering us through his papers---the Rocky Mountain News and Denver Times--in our efforts to have said project constructed through the United States reclamation service. We heartily endorse his editorials as ably presenting the views of our people and the reasons why the government should construct this canal.

Resolved, That these resolutions be printed in our local papers and that a copy be sent to Hen. Thomas M. Patterson by the president.

Then Secretary Mahoney read the following resolutions, passed at a recent special meeting of the Chamber of Commerce and sent to Secretary Ballinger, but which have not heretofore been made public, and which plainly indicate that the Chamber of Commerce and the people want the canal built by the government and not by private capital. The resolutions follow:

Whereas, Word has been received by the Chamber of Commerce of Grand Junction that the interior department has ordered the reclamation service to discontinue the work of construction on the Grand Valley project, commonly known as the High Line canal; and,

Whereas, No reason is known and no reason has been given why the work of construction of the canal should be discontinued at this time; and,

Whereas, The people of the Grand valley have complied with all the requirements of the interior department necessary to secure the aid of the Reclamation service in building this canal, by forming a water users' association, according to the rules of the department: and, Whereas, The government, through the interior department, has entered into a contract with the Grand Valley Water Users' Association to build this canal; and,

Whereas. The Grand Valley Water Users' Association has complied with all of the terms of this contract and has met all the other requirements of the interior department and the reclamation service, in order that the work of building this canal might be taken up and completed by the government; and,

Whereas, After a careful canvass made by the Chamber of Commerce, we have been unable to learn of a single person in the Grand valley who is not heartily in favor of the construction of this canal by the government instead of by private capital; therefore, be it

Resolved, That we, the board of directors of the Grand Junction Chamber of Commerce, are most heartily in favor of the building of this canal by the government in accordance with the terms of its contract with the Grand Valley Water Users' Association, and are absolutely opposed to its construction by private capital. It was announced that the Water Users' Association had passed similar resolutions and had sent them to Washington.

Edwin Price suggested that the Chamber of Commerce pass another resolution stating that in view of the offer of Magenheimer the Chamber of Commerce wanted to again go on record as being absolutely pledged and loyal to the government construction of the canal and would not consider ANY private proposition.

I. N. Bunting was then called upon and he endorsed a statement made by Mr. Currie to the effect that some people here having lands under the proposed canal were simply anxious for the project to be built whether by the government or by private capital, just so that they could sell out the lands at the earliest possible moment for heavy profits. Mr. Bunting declared that he believed no persons, other than these, wanted any proposition other than that of the government considered for a moment. He declared that this restless spirit among these several land holders should be eased as quickly as possible, and to let

Daily Sentinel, May 25 1909

the desire to have the government build the canal and the desire to have the cost as little as possible prevail instead of a desire to get the canal as quickly as possible, no matter who built it or at what cost. Mr. Bunting then stated

(Continued on page eight)

Page 8 not found.

Daily News, May 20 and 26, Daily Sentinel, May 8 1909

PRIVATE PARTIES WILL NEVER CONSTRUCT BIG PROJECT

UNANIMOUS OUTBURST OF LOYALTY TO PRESENT PLAN ATTESTED BY RESOLUTIONS —MANY TELEGRAMS ASKING BALLINGER TO CARRY OUT CONTRACT

Chamber of Commerce and Water Users Association Pass Resolutions, Wire Congressman and Receive Many Assurances of Support From Patterson, Shafroth and Others

MAGENHEIMER DOES NOT OPPOSE IT, AUPPERLE HOPEFUL OF RESUMPTION

PRIVATE CAPITAL WILL NOT BUILD THE HIGH LINE CANAL. IN A MESSAGE RECEIVED LAST EVENING BY K. A. BICKEL FROM C. C. MAGENHEIMER OF CHICAGO MR. MAGENHEIMER STATES THAT HE WILL NOT ATTEMPT TO INTERFERE WITH THE GOVERNMENT CONSTRUCTION OF THE GRAND VALLEY PROJECT IF THE PEOPLE OF THE GRAND VALLEY DO NOT WISH IT.

FROM THE TENOR OF YESTERDAY'S PROCEEDINGS AT THE SPECIAL MEETINGS OF THE CHAMBER OF COMMERCE AND THE GRAND VALLEY WATER USERS' ASSOCIATION IT IS PLAIN THAT THE PEOPLE OF THE VALLEY DO NOT DESIRE THAT ANY ONE SAVE THE UNITED STATES RECLAMATION SERVICE CONSTRUCT THE BIG DITCH. MEANWHILE EVERYTHING IS LOOKING FAVORABLE FOR THE RESUMPTION OF WORK ON THE GRAND VALLEY PROJECT IN WASHINGTON. YESTERDAY AFTERNOON SECRETARY MAHONEY RECEIVED A DISPATCH FROM SECRETARY AUPPERLE OF THE WATER USERS' ASSOCIATION IN WASHINGTON STATING:

"EVERYTHING LOOKS FAVORABLE HERE FOR THE CANAL WILL HAVE SOMETHING DEFINITE I BELIEVE SOON."

Sheldon Cline, The Daily News special representative in Washington who is thoroughly familiar with the entire High Line situation, wires The News that the telegrams to Secretary Ballinger from the Chamber of Commerce and the Water Users' association declaring that the people of the Grand Valley were unalterably opposed to the construction of the High Line ditch by any other agency

Daily News, May 20 and 26, Daily Sentinel, May 8 1909

-

PROTESTS REACHED WASHINGTON TOO LATE FOR CONSIDERATION

BY SHELDON CLINE.

WASHINGTON, May 25.--D.W. Aupperle and H. T. DeLong received the dispatches from the Chamber of Commerce and the Water Users' association late this afternoon as did Secretary Ballinger and Senator Guggenheim. They came too late for consideration with the Department of the Interior today but will be taken up tomorrow. Nothing definite was accomplished today but I expect to be able to wire you something of a more definite nature tomorrow. Attorney Frank Goudy and D. D. Potter were in conference with Secretary Ballinger today. They leave Washington tomorrow. They did not oppose government construction of the canal in any way. I secured this direct and from authoritative sources.

than the government reached Washington too late for the consideration of Secretary Ballinger, but that he would take up the dispatches early today. Definite word from Washington will be wired The News tomorrow evening.

Immediately upon receipt of the message from Mr. George Smith to The Daily News last evening stating that Mr. Magenheimer had made representations to Secretary Ballinger expressing a willingness to build the canal, a dispatch was sent to Magenheimer as follows:

"People of valley protesting against private capital building canal. How do you expect to bond unplated government lands and what do you estimate your canal would cost land owners?

"K. A. BICKEL,
"Daily News."

To this Mr. Magenheimer replied:
Karl A. Bickel,

Publisher Daily News

Answering your telegram of the twenty-fifth will say that the legislature of Colorado at last session passed amendment to irrigation law which enables government and state lands to be included in district by contract so as can be bonded. I most positively decline, as I have always done, to interfere with the government building the High Line or make any proposition to build the High Line without the consent of the government and the wishes of the people of the Grand Valley.

C. C. MAGENHEIMER.

From Mr. Magenheimer's dispatch it is inferred that he has not yet made any proposition to the government, but merely requested permission to do so. As he states that he will not do so unless backed by the wishes of the people of the valley he will in all probability not submit any. Later that evening, Governor Shafroth wired the following telegram to Secretary Ballinger urgently requesting Secretary Ballinger to exert every effort to have the Reclamation Service placed in a position as to enable them to resume work within the near future. The telegram was as follows:

"The people of the Grand Valley through the Chamber of Commerce of Grand Junction and the Grand Valley Water Users' association have gone on record as absolutely opposed to the construction of the Grand Valley project by private capital and are as one in voicing the sentiment that they are relying upon the government to build this great enterprise. Unless this work is performed by the government the landowners under the proposed irrigation system would be absolutely at the mercy of a single contractor. The citizens of the Grand Valley are convinced that the government, and only the government can construct, economically and rightly, the canal they desire. The citizens of Grand Junction urge you to do all that you can justly to see that this splendid enterprise is constructed by the government."

Immediately upon the general reading of the Daily News yesterday containing the dispatch from Mr. George Smith stating that Mr. Magenheimer had requested permission of Secretary Ballinger to lay before him a proposition to build the High Line canal, President M. O. Delaplain of the Water Users association met with President Cherrington of the Chamber of Commerce and meetings of the organizations to formulate a protest was at once arranged. At 10:30 a. m. the directors of the Chamber of Commerce met with Messrs. John F. Moore, George Currie, A. R. Sampliner, Edwin Price, I. N. Bunting, K. A. Bickel and after a short discussion telegrams were drawn up and sent to Secretary Ballinger, Senator Guggenheim, Senator Hughes, their congressional delegation, ex-Congressmen Haggot and Bonygne, Senator DeLong and Secretary Aupperle and telephonic communication was secured with Governor Shafroth and ex-Senator Thomas M. Patterson.

George Currie introduced the following resolution in regard to the splendid work Senator Patterson had been doing for the people of the Grand Valley through his papers the Times and Rocky Mountain News and the resolution was ordered sent to the senator. It read as follows:

Resolved, That the Grand Junction Chamber of Commerce, on behalf of the. entire people of Mesa county, express to the Hon. Thomas M. Patterson their appreciation of the interests he has shown in our High Line canal project, and for valuable assistance he is rendering us through his papers the Rocky Mountain News and Denver Times in our efforts to have said project constructed through the United States Reclamation Service. We heartily endorse his editorials as ably presenting the views of our people and the reason why the government should construct this canal.

Resolved; That these resolutions be printed in our local papers and copies sent to Hon. Thomas M. Patterson by the president:

"We could afford to wait five years for the construction of the High Line by the government rather than let it be taken up by private capitalists now," declared John D. Moore at the meeting. of the Chamber of Commerce. "I believe that the land under the ditch would immediately depreciate heavily in value the moment that the announcement was made that the government was not going to do the work, but that the ditch was to be built by private capital. We were shunted off the track eight years ago by private capitalists when everything was shaped up for government construction and we still are without the canal. I believe that we would have water in the High Line today if it had not been for this fact. We must stick by the government now, and have them stick by us."

"I always wanted the High Line canal built by the government," declared A. R. Sampliner. "If the government does not build the ditch we must realize clearly that the water rights, which make the High Line possible will revert to the Magenheimers and that the Magenheimers will have complete control of the situation. They will be able to dictate what terms they choose. The government offers to build the canal for $2,500,000 without interest. Private parties will ask

Daily News, May 20 and 26, Daily Sentinel, May 8 1909

at least $2,500,000 with interest which means fully five million dollars before the work is paid for.The loss of this great sum of money means much to the valley. It should be kept here. I would like to see the Chamber of Commerce absolutely put itself on record for the government construction of the High Line regardless of the essence of time."

"I feel like the others that have spoken," declared George Currie. "We all know that a government bond is worth 100 cents on a dollar but that an irrigation bond by private parties on a project of this size is worth about 90 cents. The people of the valley have to pay the difference. I'm willing to let my High Line land lay for five years yet if I am assured that the government will build the canal. We ought to let it be known that we regard it as an unfriendly act if any one attempts to butt in on this proposition now.

Mr. Currie then introduced his Patterson resolution. President M. O. Delaplain read a lengthy letter from Mr. Aupperle sent from Washington on May 21, expressing confidence that the government would soon satisfactorily arrange the High Line matter. Messrs. I. N. Bunting, C. E. Cherrington, W. A. Campbell, S. B. Hutchinson and others spoke. A telegram from Congressman at Large Taylor was read by I. N. Bunting, expressing a desire that Governor Shafroth be asked to come to the assistance of the people of the Grand Valley in the High Line tangle and the proposition met with favor. Mr. Bunting was authorized to act for the Chamber of Commerce in conferring with the governor over the telephone. In the conversation with the governor he assured Mr. Bunting that he would do all in his power and asked that full information be sent him. Postmaster Edwin Price, Secretary Mahoney and I. N. Bunting were delegated to write out a statement for the governor. Messrs. Mahoney, Bunting and Bickel were then placed on a committee to draft telegrams to the Washington congressional delegation and others in reference to the project.

A stiff resolution from the Chamber of Commerce declaring that it could not find any person in the valley favoring private construction of the High Line and asserting that it stood opposed to the construction of the ditch by all parties save that of the government was read and approved. The following dispatches were then sent to Washington and Chicago:

Hon. R. A. Ballinger, Sec'y of Interior, Washington, D. C.

The Chamber of Commerce of Grand Junction at a meeting held today again reaffirms all its previous actions for the building of the High Line canal by the government. We will not entertain any proposition for private building of this canal, believing such a proposition to be vast injury to the land owners as well as the community at large. We urge you to go on with the contract as prepared with Sec'y. Garfield.

CHAMBER OF COMMERCE,
T. F. Mahony, Sec'y.

D. W. Aupperle, Washington, D. C.

If the government does not build canal, water rights revert to Magenheimer's. We would be at his mercy. Urge this point with Ballinger. Wired all Colorado Congressmen today.

T. F. MAHONY.

Senator Simon Guggenheim,
Washington, D. C.

Chamber of Commerce and people of the Grand Valley are a unit in favor of government construction of the High

Line canal. Are absolutely opposed to construction by any private enterprise. Will Colorado delegation so inform Secretary Ballinger

CHAMBER OF COMMERCE
T F. Mahoney, Sec'y.

This same message was sent to Senator Hughes and the congressmen.

George Smith,
Woodland Construction Co.,
First Nat'l. Bank Bldg.,
Chicago Ill.

Chamber of Commerce in meeting with citizens of city today passed resolutions declaring any attempt to interfere with government construction of canal "an unfriendly act". Public aroused and angry.

KARL A. BICKEL.

A resolution was then passed by the directors thanking the local papers for their stand and their support in this matter, and asking them to carry out their present policies in the matter.

A RESOLUTION WAS THEN PASSED TO THE EFFECT THAT THE CHAMBER OF COMMERCE AND THE PEOPLE OF THIS CITY WILL REGARD AS AN UNFRIENDLY ACT THE INTERFERENCE OF ANY PERSON, PERSONS OR COMPANY WITH THE PLAN TO HAVE THE GOVERNMENT BUILD THE CANAL.

Messrs. Bickel, Bunting and Mahoney were then instructed to prepare telegrams for all members of the Colorado delegation in the house, and senate, urging them to insist on the government construction of the canal. These messages will go out tonight.

Mr. Bunting was then instructed by a resolution to call up Governor Shafroth for the Chamber of Commerce and urge him to take action in behalf of the High Line.

Mr. Mahoney was instructed by resolution to call up former Senator Patterson, read to him the Currie resolution, thank him for his efforts in the canal's behalf and ask him to continue to use his influence for the government construction of the big ditch.

Secretary Mahoney will also ask Senator Patterson to enlist, if possible, the influence of the Denver commercial organizations, as the matter is of state importance. Secretary Mahoney will talk to Mr. Patterson some time this afternoon.

THIS AFTERNOON, AT ABOUT 3 O'CLOCK, MR. BUNTING CALLED UP GOVERNOR SHAFROTH OVER THE TELEPHONE. He read to the governor the message he had received from Congressman Taylor, and then informed him that the Chamber of Commerce was anxious that he take action in the matter if he cared to do so, and if he couldn't go to Washington, to send an appeal or statement to Secretary Ballinger.

The governor said in part:

I AM PERFECTLY WILLING TO ACT IN THIS MATTER. I KNOW IT IS A GREAT ENTERPRISE AND I WANT TO HELP IT ALONG ALL I CAN.

"I am not thoroughly familiar with the situation as it stands today, but I know what the project means to that section AND WILL CERTAINLY BE MIGHTY GLAD TO DO ALL IN MY POWER TO URGE THE GOVERNMENT TO RESUME WORK ON THE HIGH LINE CANAL and to do my part toward showing Mr. Ballinger that YOU PEOPLE WANT THE GOVERNMENT TO BUILD THE CANAL. You gall count on my most hearty and loyal support.

Daily News, May 20 and 26, Daily Sentinel, May 8 1909

"As soon as a statement reaches me from you, I will at once send a letter to Mr. Ballinger urging action along the line your people desire."

Mr. Bunting and Mr. Mahoney got a statement of the situation out to the governor tonight.

The governor's statements are certainly most satisfactory and show that he stands ready to do all he can in the interest of the big project.

"This afternoon there was a special meeting of the board of the Grand Valley Water Users' association and the following resolution was offered by Mr. Campbell and was adopted by a unanimous vote:

Resolved, That we consider the offer of private capital to construct the High Line canal, as made through George Smith, as an unfriendly act and against the best interests of all the people of the valley;

Resolved, That we favor the construction of the canal, first, last and all the time, by the government and are unalterably opposed to the construction by any private enterprise.

"We express a hope that the government will be able to construct this project without undue delay, but favor government construction in preference to that by private capital even though such construction might not be completed as rapidly as that promised by private capital.

Then the following message was ordered sent to Secretary Ballinger:

Mr. R. A. Ballinger
Secretary of the Interior,
Washington, D.C.

At meeting of the Water Users' Association, called to consider offer of private capital to construct the Grand Valley project, a resolution was unanimously adopted declaring that we favor first, last, and all the time, government construction

----

AUPPERLE AND DELONG TODAY INTERVIEW BALLINGER

BY SHELDON CLINE.

WASHINGTON, D. C., May 19. Secretary Aupperle and Senator DeLong arrived here this afternoon and have an engagement with Secretary Ballinger and Senator Guggenheim for tomorrow morning. They have not as yet located the trouble but are hopeful of straightening it out with little difficulty.

The Daily News special Washington correspondent, Mr. Sheldon Cline will be in close touch with the momentous events which are likely to transpire during the next few days in Washington in relation to the High Line canal and through his despatches to the News, the earliest possible information will be known here. His dispatches will serve the added purpose of enabling our "esteemed contemporary" the Evening Whimper, to keep track of what is going on.

Daily News, May 20 and 26, Daily Sentinel, May 8 1909

From the present look of things in the reclamation service, we are not to have as smooth sailing with our High Line canal project as we anticipated. Richard Achilles Ballinger, secretary of the interior, is a man entirely different from his predecessor, James Rudolph Garfield. Ballinger for a time was general land office commissioner under Garfield, placed there at the special invitation of the president. He found some conditions in the reclamation service of the government that did not suit him then, but which he could not change, owing to his position under Garfield. He is now in a position to make some changes, and we believe this is, perhaps, the reason that work is not proceeding on the High Line project like we should like to have it. If an article in the last Saturday Evening Post can be believed, written by one of the most brilliant writers upon men at the capital, Ballinger is going to tear up things quite a little in his department. So, we may look out for squalls. Ballinger is the subject of the article in the Post, and it is well worth reading as a description of the kind of man we have to deal with for favors on the High Line canal project.

Denver Republican, May 21 and 24 1909

EXPLAINS WHY WORK STOPPED ON HIGH LINE

TO DECIDE VALIDITY OF CONTRACT WITH WATER USERS.

BALLINGER TELLS COLORADO QUESTION IS UP TO ATTORNEY GENERAL.

If Opposed to Law Work Will be Delayed Till Reclamation Fund Accumulation.

WASHINGTON, May 20.--(Special.)-Messrs. DeLong and Aupperle, of the Grand Valley Water Users' association, called with Senator Guggenheim this morning on Secretary Ballinger and discussed with him the questions involved in the stoppage of work upon the proposed Grand valley high line reclamation project.

Secretary Ballinger explained that his order directing that work should be stopped was issued, first, because he had some doubts as to the legal authority of the interior department to make the contract with the Water Users' association under which work was commenced and, second, because the reclamation service did not have sufficient funds on hand or in prospective receipt to carry the work to completion.

The contract with the Water Users' association contemplated the payment by the members of the association at this time of $100,000 in labor and material, as against a like amount to be furnished by the reclamation service, the association members to be reimbursed by credits upon future charges for water rights.

Secretary Ballinger stated that he was not certain that such a contract could properly be made under the terms of the reclamation act, and until he could be satisfied on that point he preferred to have work suspended upon the enterprise. He is having the legal phase of the matter carefully examined by the law officers of his department and before determining finally he will probably submit the question to the attorney general.

If it should be determined that the contract is legal, the plan of co-operation between the government and the Water Users' association may be followed and this will overcome in part the objection as to lack of funds. If the contract, according to the law officers of the interior department and the department of justice, is not legal, the project will have to await construction until adequate funds can be provided from the reclamation fund.

PUBLIC SATISFIED THAT NO KNOCKERS AT WORK

GRAND JUNCTION, Colo., May 20.-

(Special.)—A report was received here tonight from the local delegation now in conference with Secretary Ballinger in Washington regarding the cessation of work on the High Line canal, which tends to confirm the earlier rumor that the stop order was issued in order to permit the department to more fully investigate the project and not at the instigation of outside parties.

D. D. Potter, a local attorney, and F.C. Goudy of Denver, started for Washington tonight to represent the Mesa County Protective association, which is the organization of the Palisade land owners along the ditch who have so far refused to accept the government's indemnity offer for right of way. The association indignantly denies the charge that it has been instrumental in blocking the project.

PETITION FOR BIG DAMAGES

PALISADES FRUIT GROWERS TAKE CANAL CASE TO WASHINGTON.

ASK PAYMENT FOR ALL LANDS TAKEN

Contestants Not Benefited by Big Irrigation Project on the Western Slope.

In a petition from the fruit land owners of Palisades, Colo., to be presented in Washington today, the secretary of the interior will be asked to render a decision that will have a far-reaching effect in determining the rights of the reclamation service in running ditches across patented lands, and which will save thousands of dollars to the fruit owners if their contention is upheld.

This contention, in short, is that while the federal statutes of 1890 provide that the government shall have the right to run a ditch across homestead or pre-emption lands in the future and shall take a right of way for that purpose, the statute intended that the government should take the land by right of eminent domain. Under the law of eminent domain the government must pay the land owners for the value of the property taken and for all improvements upon it. But the reclamation service has held that under the statute of 1890 the land needed by the government for ditches or canals shall be taken without payment and thus the owners shall receive only the value of the improvements placed there by them. The fruit land owners of Palisades claim that their property will be damaged $500,000 by the government canal, but they do not wish to stand in the way of the project at all. They will merely ask the secretary of the interior to pay for the value of the land actually taken for the right of way, and for the improvements. They do not ask for damages. The sum they have agreed to take is $156,000, while the reclamation service has offered about $70,000.

WILL DESTROY VALUE OF TRACTS FOR HOMES

Frank C. Goudy of the well known Denver law firm of Goudy & Twitchell, and D. D. Potter, a fruit land owner of Palisades are now in Washington and will present their petition today and ask for a hearing, which has been promised them.

The fruit land owners of Palisades, who are acting through their association, the Mesa County Irrigation District Land Owners' Protective association, have nothing to gain from the government canal. It is to benefit the owners of land 40 miles further down the river on the mesa back of Grand Junction. Those whose lands are to be reclaimed, who are band-

Denver Republican, May 21 and 24 1909

ed together under the name of the Water Users' association, secured the consent of the Palisades orchard owners before enlisting the aid of the reclamation service in the canal project. They recommended to the government that the money asked by the orchard owners be paid them, representing that the offer of the right of way for $156,000 was entirely just and reasonable. Especially as the government canal will destroy the value of the orchard tracts as homes.

The physical geography of the land in question is peculiar. The Grand river passes the palisades, the immense rock dyke from which the town takes its name, at right angles. when it sweeps around in a gradual curve until it flows parallel with them. The land between the river and the palisades is very valuable fruit lands, divided into two strips. From the palisades the ground slopes steeply down for a short distance, and then changes to a more gradual slope, which runs to the river.

Reclamation Service Wants Gravity Supply

The gradual slope was first taken up by settlers. By using a pumping station and raising the water from the Grand river 40 feet, a ditch running along the intersecting line of the two slopes was filled, and from this the long slope was irrigated. But experiment showed that the short, steep slope was more valuable as fruit land. Again a pumping station was constructed at the intersection of the rock dyke and the short slope. Water was thus raised 80 feet and run through a ditch at the very foot of the palisades and laterals irrigated the narrow strip. This strip extends down the valley for 10 miles and is divided into five, 10, and 20 acre tracts.

When the owners of the land on the Grand Junction mesa wished to get water it was necessary to go far up the river to get the required elevation. The reclamation service has always refused to irrigate by pumping, but insists that its schemes shall operate through gravity. It would not utilize one of the two ditches already constructed for this reason and the added reason that they are only small canals. It went farther back the river and tunneled the mountain that breaks off, into the palisades. Its tunnel will come out close to where the pumping stations are situated on the Grand, about 15 feet lower than the upper private ditch and the same elevation above the lower ditch. The plan is then to run the big ditch down the middle of the short slope of fruit land, cutting in two the small fruit lands. As land in this belt is worth from $500 to $3,500 an acre and as the big 80-foot government ditch would bisect all orchards and injure them as homes, the damage is considerable. Many orchard owners would have to go half a mile or a mile to cross from one side of their orchards to the other. Then the government wants a 120-foot right of way and this would cut out an acre or so from some of the orchards.

Would make Owners Pay Twice for land

The government plan of paying only for improvements, it is claimed, amounts to this--that it would be making the land owners pay over again for the strip taken from them after having spent years of labor in experimenting upon the best fruit that would grow on the short slope protected by the palisades, and a great deal of money in development. For the reclamation in fact acts only as a contractor. It constructs its canals, pays for them, operates them for 10 years. During that time the people through whose land the ditch is running pay for the construction in 10 annual installments, and then the ditch belongs to them and they operate it. But the orchard owners of Palisades declare that the men to get the benefit will be the ones at Grand Junction, while the Palisades orchard owners will have to pay their share of the installments to the government for 10 years, although their own ditch would have been sufficient to supply their wants.

This is the situation which Attorney Goudy and Mr. Potter will present to the secretary of the interior in their petition, but the decision will effect more than the Palisades orchard owners for the point to be settled in regard to the question of whether the reclamation service shall pay only for improvements and land in running a ditch over patented lands is one which has not as yet been settled.

SAYS ASSOCIATION IS IN FAVOR OF PROJECT

WASHINGTON, May.23.--(Special.)-F. C. Goudy of Denver and D. D. Potter of Palisade, representing the Palisade Protective Fruit Growers' association, are here in connection with the proposed construction of the government project to irrigate lands in Grand valley and through Senator Hughes have arranged to have an early hearing before the secretary of interior.

In explaining the object of their visit at this time Mr. Goudy said this afternoon that the association he represents was not opposing the government project but on the contrary would like to see it consummated as it would greatly benefit the Grand valley region and add materially to the general value of property in that part of Colorado.

The question he and Mr. Potter will take up with Secretary Ballinger is that of proper compensation for lands now owned by orchardists which will be crossed by the proposed ditch to be constructed by the government. As planned by the reclamation service the government ditch will extend for at least ten miles through lands which have been cultivated and which are worth from $200 to $3,500 per acre. The actual damage to these lands will be, the owners believe, upwards of $400,000. An agreement was reached in the matter of compensation between the Palisade Protective association arid the Grand Valley Water Users' association at which the damages were to be fixed at $165,000, but the reclamation service scaled this to $75,000. This sum, the Palisade owners deem inadequate. They will present their views on the subject to Secretary Ballinger and if he cannot allow them what they regard proper damages they will request him to set a time for a hearing at which they may present testimony to establish the justness of their request for higher compensation. They will urge also that the proposed government main canal be extended along higher levels than now proposed so that it will not be necessary to have it cross their cultivated lands and orchards. If this should be done it will remove the necessity for paying any compensatory damages.

Daily News, May 29, Denver Republican, June 3 1909

WILCOX KNOCKING GRAND VALLEY AGAIN

Editor Busily Engaged Fighting Reclamation Service

Lute M. Wilcox, one of our most persistent knockers on the Eastern Slope, is again tearing off large chunks of depression in regard to the outlook of the High Line in his "Field and Farm".

Lute seems to have a dyspeptic grouch on the Grand Valley. He jeered at the fire pot protection theory until it was conclusively shown to be absolutely reliable. He has fought the Reclamation Service at every turn of road and throws us a back handed compliment at every possible opportunity. We print his latest eruption on the High Line not as information but as indication of the amount of misinformation Lute can cram into a stickful of type when he sets about it seriously. If there is a statement in the following article that is correct The News will publicly print the name of the man that discovers it.

The government has made a nice mess of its High Line reclamation project in Mesa county. In the first place Uncle Sam had only $125,000 available, for a work that would cost over two million dollars and this very fact is the rat in the meal bag. The powers at Washington have also suddenly discovered that the government has no right to go into partnership with private citizens as was attempted when the reclamation service required the people of Grand Junction to put up $100,000 before the work could begin. The enabling act under which the service is working makes no provision for a bonus and it was a stick-up to ask the people for a cent. The main obstacle right now occurs in the bull made by the engineers in establishing the grade line so as to cut through the orchards of the Palisade people. While the federal status of 1890 provide that the government shall have the right to run a ditch across homestead or pre-emption lands and may have right of way for that purpose, the statute intended that the government should take the land only by right of eminent domain which under the government must pay the land owners for the value of the property taken and for all improvements upon it. The Reclamation Service, however, sneaked on the proposition and tried to confiscate the land by simply paying for the improvements thereon and of course this stirred up a great amount of dissatisfaction which can not be kissed off by the smooth ducks at Washington. The only thing to do now is for all about it. If it will relinquish to settlement the 35.000 acres withdrawn under the project we will guarantee to build a higher and better canal on the basis of $40 the acre under the district law of the state and will put up a bond to have the work completed within one year. This enterprise would cover 7,000 more acres than Uncle Sam could irrigate and the canal will cut through nobody's orchard.

Daily News, May 29, Denver Republican, June 3 1909

STATUS OF HIGH LINE CONTRACT IS STILL IN DOUBT

Attorney General Will Render Opinion of Legality Soon

BALLINGER WAITS REPLY

BY SHELDON CLINE.

WASHINGTON, D.C., May 28.--The present fate of the Grand Valley High Line project now rests in large degree upon the decision of the Attorney General concerning the legal points involved in the present contract. It was expected that as the matter had been submitted to him nearly a week ago that he would give out his decision today. In this all those who were working in behalf of the project here were disappointed and from present indications no decision will be rendered before the middle of next week. In the meantime, your committee have not relaxed their efforts to bring all possible pressure to bear upon Secretary Ballinger and the latter has definitely informed them that should an adverse decision be rendered by the Attorney General he will give them a formal hearing in the matter.

Government Can Not Co-operate With Outsiders

ATTORNEY GENERAL SAYS GARFIELD CONTRACT IN GRAND VALLEY WAS NOT CONTEMPLATED BY RECLAMATION LAW.

WASHINGTON June 2.--(Special.)--The attorney general has rendered an opinion to the secretary of the interior in relation to the validity of co-operative agreements between the government and individuals or associations under the reclamation act, holding that such agreements are not contemplated or authorized by the act in question.

This opinion, it is believed, will result in having the contract abrogated which was entered into between former Secretary Garfield and the Grand Valley Water Users' association for the joint construction of the Grand valley project.

Before taking final action, Secretary Ballinger will give Senators Guggenheim and Hughes and the representatives of the Water Users' association who are here a further opportunity to be heard in the matter. To Senator Hughes, who called upon him today, the secretary expressed a desire that the project should be undertaken and carried to completion by the government, He is having the question of funds examined into and if it is found that funds are available doubtless work will not be delayed by the finding that the co-operative plan is not in conformity with the reclamation act

Daily News, May 27 and 28 1909

BALLINGER DOES NOT APPROVE PRIVATE CONSTRUCTION

ASSURANCE THAT WORK ON CANAL MAY SOON BE RESUMED

Encouraging Telegrams From Congressman Martin

LEGAL PHASES NOW UP

Secretary Ballinger of the Department of the Interior does not favor private construction of the Grand Valley project.

Private capital has broached the subject to him but he does not regard it with favor.

This is the information brought to Grand Junction yesterday afternoon in a telegram to President Cherrington of the Grand Junction Chamber of Commerce from Congressman John A. Martin of the Colorado congressional delegation.

The telegram is as follows:

C. E. Cherrington,

Grand Junction, Colo.

I have just had an interview with Secretary Ballinger. Private construction of the Grand Valley project has been broached to him but he does not favor it. I emphasized the objections. The matter is now before the legal department and under investigation and it will be some days before a report is made. Also the Reclamation Service is being overhauled. The Secretary is entirely friendly to the project but is temporarily going slow.

I will write you in full.

JOHN A. MARTIN.

The above dispatch is cheering news to the people of the Grand Valley. It demonstrates that the prompt manner in which the people of the valley took up the question of the construction of the big ditch has had its effect in Washington. The sending of D. W. Aupperle and Senator Horace T. DeLong to Washington has been demonstrated as a wise move and the energetic manner in which the Chamber of Commerce, the Water Users' association and allied citizens Tuesday demonstrated by their strong telegrams the feeling of the valley in regard to private construction was most opportune. It left no room for doubt in official circles in Washington as to the feeling of the people.

Yesterday afternoon Secretary Mahony of the Chamber of Commerce and President Cherrington received a number of replies to their telegrams of the day before. They read as follows:

Daily News, May 27 and 28 1909

Washington. D. C., May 26.
T. F. Mahony,
Chamber of Commerce, Grand Junction. Colo..

Your telegram received, and I have this day presented the matter to the honorable secretary of the interior. There seems to be no purpose of other than government construction. The delay, I believe, is temporary, awaiting a legal opinion from the attorney general and investigation of funds available for project.

CHAS. J. HUGHES JR

Washington. D. C., May 26.
T. F. Mahony, Secretary,
Grand Junction. Colo.

Will act in conformity with your telegram of today, with Secretary Ballinger.

A.W. RUCKER

Washington, May 26.
T.F. Mahony,
Grand Junction, Colo.

I so informed the secretary a week ago; but will do so again, and see rest of Colorado delegation immediately.

EDWARD T TAYLOR.

Washington, D. C.. May 26.

Situation looks encouraging here. Bend your energies against Magenheimers direct. He promised us he would not interfere.

D. W. AUPPERLE.

TAFT WILL COME HERE IN AUGUST

Local Committee See Him and Receive Assurance of His Support and Promise to Come Here

----

BALLINGER TO GIVE COMMITTEE FORMAL HEARING

BY SHELDON CLINE.

WASHINGTON, D. C., May 27.—The Grand Junction committee, Secretary Aupperle and Senator DeLong and the Colorado congressional delegation called on President Taft this morning and had an extended interview regarding the High Line project. They were introduced to the president by Senator Guggenheim and cordially received. The present status of the project was outlined to the president and his interest enlisted in behalf of government construction. The committee also tendered him an invitation to visit the Grand Valley, see the canal for himself and familiarize himself with the details. The president expressed himself as very favorable to accepting the invitation and told the committee that if his proposed visit to Colorado in Au-

Daily News, May 27 and 28 1909

gust occurs according to present plans he would visit Grand Junction either immediately before or immediately after the Gunnison tunnel opening at Montrose. The arrangement for his visit are in the hands of Thomas Walsh.

Secretary Ballinger announced today that should the cooperative plan which is now under consideration by the judiciary department be declared illegal he would give the committee a full hearing on the subject. A definite announcement is expected next Monday or Tuesday and your committee stated tonight that they would probably remain here for another week

----

TO ENLIST INTEREST OF TAFT IN HIGH LINE

Committee and Congressional Delegation Lined up Solid For Government Construction to See President Today.

BY SHELDON CLINE.

WASHINGTON May 26.—The committee from Grand Junction here in the interests of' the Grand Valley project held an extended conference with the Colorado congressional delegation today and they are all lined up solid in favor of government construction. The telegrams and resolutions received from Grand Junction were read and every influence is being brought to bear upon Secretary Ballinger to have him issue an order for the resumption of the work. Senator Guggenheim is especially active and through his instrumentality a conference with President Taft has been arranged for tomorrow morning. It is expected that all the delegation and Senator DeLong and Secretary Aupperle will be present and seek to enlist his aid in securing government construction of the canal. An invitation will be extended to him to visit Grand Junction, to examine the orchards and canal and familiarize himself with the details of the project. It is expected that he will accept this invitation and go there during his western trip in August.

Secretary Ballinger has unofficially intimated that he does not favor private construction of the canal and today received the telegrams addressed to him from Grand Junction. He has not, however given out an official interview.

Daily Sentinel, May 27 1909

MAGENHEIMER SENDS EXONERATING LETTER

Interior Department Says No Cause for Blaming Him--Busy Day in Washington

----

No Local Fight Over Damages

This morning The Daily Sentinel received the following letter from C. C. Magenheimer, secretary and general manager of the Woodland company of Chicago, the company which purchased the bonds of the Orchard Mesa Irrigation District and which company is behind the construction of the big Orchard Mesa project

"Woodland Company, First National Bank Building,

"'Chicago, May 24, 1909.

"Mr. I. N. Bunting,

"The Daily Sentinel

Grand Junction, Colo.

"Dear Sir:--We enclose you letter received from the Department of the Interior Reclamation Service, in reply to a letter of mine informing the Department of the Interior that I had been accused of stopping the work of the Grand valley project.

"Yours very truly,

C. C. MAGENHEIMER.

The letter enclosed was as follows: Department of the Interior, United States Reclamation Service; Office of the Director

Washington. D. C., May 21, 1909.

Mr. C. C. Magenheimer,

Chicago, Ills.

Dear Sir: --Yours of: the 17th instant in regard to the present status of the Grand valley project, has been received.

I know of no basis for the claim that you had taken any steps to interfere with the operation of the Reclamation Service in regard to this project.

Very truly yours

MORRIS BIER.

Acting Director.

No Messages Today.

Secretary Mahoney, of the Chamber of Commerce has received no messages today from Washington and expects none before tomorrow. It is known that the day has been a busy one in Washington for the members of the Congressional delegation of Colorado and the Grand Junction representatives Messrs. DeLong and Aupperle. It was hoped to have a conference some time today with President Taft through the efforts of Senator Guggenheim. However, it is not ex-

Daily Sentinel, May 27 1909

pected that President Taft will take any personal action in the matter at least until after Secretary Ballinger had made a thorough investigation and decided on his policy concerning the project.

The editorial in the Denver News, reproduced herewith, is misleading in so far as there is no fight, locally, over the question of damages for the Palisade land owners as the Grand Valley Water Users' association has endorsed the official damage claims made by these fruit growers. There is no local fight that in any way is likely to interfere with government construction of the canal.

Senator Patterson's Advice.

Urging the people of the Grand valley, those of the Palisade section and of the Grand Junction section to get together on the High Line matter and cut out all local disputes over the amount of damage to be incurred by the passing of the High Line through a number of the upper valley orchards, the Denver News, of yesterday's date, contains the following splendid editorial:

The representatives of Colorado seem to be gradually unwinding the red tape which halts the building of the High Line canal. But there is something which the News believes can be done close at home to further the good cause: something in the power of those who most want the canal put through.

And that is merely to settle all the disputes which now divide the Grand valley against itself. The News has pointed out that the canal must pass through Palisade orchards. Let the people of the lower valley, who will benefit by the canal, stop all quarrels with the Palisade orchard owners who will be injured by the canal.

When the scheme was first broached the Palisade Protective association fixed the damages to the orchards of that district at about $147,000. The lower valley disputed the claim, declared it too high, and vowed it would never submit to the payment of such a price Now Palisade is claiming damages to the tune of half a million dollars--and the High Line halts.

Half a million dollars is, of course, far too much. But the News declines to believe that the Palisade people expect to collect such a sum. The original demand of $147,000 was, in the judgment of the News, little if any too high; and we believe a united, temperate effort of tile lower valley would have scaled down whatever extras were included in that demand. It is only a matter of two dollars and a half per acre on the land which the High Line will water; a mere trifle of expense compared to the value of the developed wealth.

The News urges the people of Palisade and the people of the lower valley to get together. Let Palisade make none but just and reasonable demands, and then let the lower valley refrain from random exclamations that those demands are too high. Palisade orchards are high-priced affairs---but they yield an income on the price; and that, after all, is the final test. According to the best information we can get, the men who made the original estimate of $147,00.0 damages based their figures on actual sales of similar land in similar locations. If that is truly the case then the people of the lower valley should not cavil. It isn't fair, and it doesn't pay.

Settle all these differences at home, and settle them quickly. Then, and we fear not till then, will it be possible to push through the canal against the sinister interests which seem now to be in its way.

Daily Sentinel, May 27 1909

Practically of the same import, and containing a similar announcement concerning Secretary Ballinger's attitude toward private capital building the High Line Canal, as the telegram from Senator Charles J. Hughes published in The Sentinel yesterday, a message was received late yesterday afternoon by President Cherrington, of the Chamber of Commerce, from Congressman John A. Martin, of this Congressional district. The message was as follows:

C. E. Cherrington,

Grand Junction, Colo.

I have just had an interview with Secretary Ballinger. Private construction of the Grand valley project has been broached to him but he does not favor it. I emphasized the objections. The matter is now before the legal department and under investigation and it will be some days before a report is made. Also the Reclamation service is being overhauled. The Secretary is entirely friendly to the project but is temporarily going slow. I will write you in full.

JOHN A. MARTIN.

At the meeting of the board of directors of the Chamber of Commerce last night the steps taken in the High Line canal matter and the manner in which the Colorado officials at Washington had rallied to the support of the local people's contention that only the government construction of the High Line canal is wanted was discussed.

PATTERSON AND NEWS THANKED

Water Users of Grand Valley and Chamber of Commerce Praise Work to Push High Line Canal.

URGE THAT GOVERNMENT COMPLETE THE PROJECT

Private Control Condemned in Resolutions; Protest Sent to Washington

----

[SPECIAL TO THE NEWS.]

GRAND JUNCTION, Colo., May 25.—Government construction of the High Line canal is the unanimous demand of the Grand valley, as expressed in mass meetings held today under the auspices of the chamber of commerce and the Water Users' association to consider what is believed to be a crisis in the affairs of the project.

Unalterable opposition to construction by private capital was expressed in the strongest terms in the resolutions adopted in telegrams sent to Secretary Ballinger, to the Colorado congressional delegation, to former Senator Thomas M. Patterson, Governor Shafroth and other men beseeching their aid in securing the continuance of the High Line project by the reclamation service.

Those friends of the valley who already have exerted powerful influence in its behalf received the warmest commendation and the first official action taken at the mass meeting was to enthusiastically adopt resolutions of thanks addressed to former Senator Thomas Patterson.

Senator Patterson Thanked by Water Users

The resolutions were introduced by Geo. A. Currie, a member of the Water Users' association and a large land owner. In a ringing speech, Currie thanked Senator Patterson in behalf of the Water Users' association for his stand in favor of the project which means so much for the. future of this valley. The resolutions were as follows:

"Resolved, That the Grand Junction chamber of commerce, on behalf of the entire people of Mesa county, express to the Honorable Thomas M. Patterson their appreciation of the interest he has shown in our High Line canal project and for all the valuable assistance he is rendering us through his newspapers, The Rocky Mountain News and The Denver Times, in our efforts to have the said project constructed through the United States reclamation service. We heartily indorse his editorials as ably presenting the views of our people and the reasons why the government should construct this canal.

"Resolved, That these resolutions be printed in our local newspapers and that, a copy be sent to the Honorable Thomas M. Patterson by the president of the chamber of commerce."

Private Control a Blight, Says Mass Meeting

The proposition made by C. C. Magenheimer of Chicago to Secretary Ballinger offering to build the canal himself on the district plan was vigorously condemned. It was declared by one of the speakers that as Magenheimer already controlled water rights that would revert to him if the government relinquished its claims, the Grand valley virtually would be in his power. The sentiment was freely expressed that no private enterprise could do the work so thoroughly and so economically as the government. Several speakers predicted that land values would fall 50 per cent if the government definitely abandons the enterprise.

Many speeches were made in a similar vein and excitement reached a high pitch, culminating in a set of resolutions drawn up by the chamber of commerce and addressed to Secretary Ballinger couched in the most emphatic language and asserting the most decided opposition to private construction of the ditch and asking the secretary to carry out the contract entered into by his predecessor, Garfield, with the water users of the Grand valley. The resolutions were as follows:

"Whereas, word has been received in the chamber of commerce of Grand Junction that the interior department has ordered the reclamation service to discontinue the work of construction on the Grand valley project, commonly known as the High Line canal; and

"Whereas, No reason is known and no reason has been given why the work of the construction of the canal should be discontinued at this time; and

"Whereas, The people of the Grand valley have complied with all the requirements of the interior department necessary to secure the aid of the reclamation service in building this canal by forming a Water Users' association according to the rules of the department; and

"Whereas the government through the interior department has entered into a contract with the Grand Valley Water Users' association to build this canal; and

"Whereas, the Grand Valley Water Users' association has complied with all the terms of this contract and all the requirements of the interior department and the reclamation service in order that the work of building this canal by the chamber of commerce we have been unable to learn of a single person of the Grand valley who is not heartily in favor of the construction of this canal by the government instead of by private capital.

"'Therefore, be it resolved, that we the board of directors of the Grand Junction chamber of commerce, are most heartily in favor of the building of this canal by the government in accordance with the terms of a contract with the Grand Valley Water Users' association and are absolutely opposed to its construction by private capital."

Several telegrams received by the Daily News correspondent concerning encouraging accounts of the interviews between Secretary Ballinger and the Grand Junction representatives, Senator H. T. DeLong and Secretary D. W. Aupperle, were read to the meeting and although the secretary still refuses to discuss the reasons for the suspension of work on the project, it was believed that the influences now being brought to bear on him would result in an order for an early resumption of the work. After the meeting Senator Patterson was called up on the long distance telephone and was apprised of the action of the citizens in thanking him for his splendid work in behalf of the canal and urging him to continue his efforts. Governor Shafroth also was communicated with and promised to help in every way in his power and to himself telegraph Secretary Ballinger.

Protest to Ballinger.

Telegrams to the congressional delegation in Washington were sent, as well as the following one to Secretary Ballinger: "Hon. R. A. Ballinger, Secretary of the Inferior, Washington, D. C.

"'The chamber of commerce of Grand Junction, at a meeting held today, again reaffirms all its previous action for the building of the High Line canal by the government. We will not entertain any proposition for private building of this canal, believing such a proposition to be a bad injury to the land owners, as well as the community at large. We urge you to go on with the contract as prepared by Secretary Garfield.

(Signed) "CHAMBER OF COMMERCE,

"T. F. MAHONEY, Secretary."

Following the mass meeting the Water Users' association held a special meeting and adopted a resolution declaring in favor of government construction of the canal first and last.

Declared unfriendly Act.

The offer of private capital to do the work was characterized as an "unfriendly act" against the best interests and the wishes of all the people of the valley and concluded by expressing a hope that the government would be able to complete the project without undue delay, but favoring government construction even though the canal might not be completed as rapidly as by private capital. The water users also sent a telegram to Ballinger similar to the one sent by the chamber of commerce.

Tonight, Carl A. Bickel, editor of a local paper, who was on the committee which drafted a telegram earlier in the day, received a reply from Magenheimer in Chicago in which he declares he did not desire to build the canal without the consent of the government, or against the wishes of the people of the Grand valley.

URGES THAT SHAFROTH ACT

Taylor Advises Governor Use Influence in Behalf of High Line.

[SPECIAL TO THE NEWS.]

WASHINGTON, May 2$.—

Congressman Taylor today sent messages to Colorado suggesting that Governor Shafroth interest himself in the High Line canal matter and lend his influence towards preventing the abandonment of this irrigation project by the government. In a dispatch to I. N. Bunting of Grand Junction, Taylor said that if Governor Shafroth would investigate the High Line project and either take his findings in person to Washington or communicate them to Secretary Ballinger, the government might find a way to take up the project at once and not postpone it on account of a lack of funds.

Taylor informed Bunting that every effort is being put forth by members of the house to have the department of the interior order work on the project commenced.

Daily News, June 3 and 5, Denver Post, June 6 1909

INTENTION IS STILL IN A SHROUD O F MYSTERY

Ballinger Continues to Hold Matter in Abeyance

COMMITTEE IS ACTIVE

BY SHELDON CLINE

"WASHINGTON, June 2.-- Secretary Ballinger did not give out an inkling today of the status of the Grand Valley High Line project. The Grand Junction committee called on him accompanied by Senator Guggenheim but he was uncommunicative and stated that he was not yet in a position to make any statement. He would not intimate the nature of the report made to him by the attorney general concerning the legality of the contract.

Both Secretary Aupperle and Senator DeLong have called daily at the offices of the secretary and everything is being done to bring the matter to a head. Senator Guggenheim has also been very active and has been taking exceptional interest in the negotiations and has exerted considerable influence to effect a speedy adjustment.

Hopes are entertained that Secretary Ballinger will make an announcement of some character tomorrow.

U.S. TO RESUME WORK IN GRAND VALLEY AT ONCE

None of Expenditures Will Come From Private Persons.

BALLINGER'S DECISION

Committee From Grand Junction Will Return From Washington.

Washington. D C.. June 5.~ Secretary Ballinger today announced his decision in the Grand Valley irrigation project case which has been pending in the interior department for the past month.

Reviewing the case the secretary stated that the agreement with the Grand Valley Users association provided the association was to co-operate with the government in construction of the project, furnish the money and labor while the government was to provide the cash for its share of the cost of the construction and that the association was to issue certificates in payment for work and labor done and material furnished which certificates were to be used to reduce the construction, charges against the lands within the project.

When the agreement was first submitted to Secretary Ballinger he ques-

Daily News, June 3 and 5, Denver Post, June 6 1909

tioned the validity of it and decided to submit the matter to the attorney general for an opinion. The attorney general held that the reclamation act does not authorize such an arrangement and that the secretary can not commence the construction of any portion of a project until the funds necessary are available in the reclamation fund.

There is no question of the feasibility of this project and the government is willing to enter upon the work when funds are available. An examination of the reclamation fund shows that the $175,000 is now available for this project during the years 1909 and 1910, and that thereafter funds will be available to continue the work in a satisfactory manner. The preliminary work which has been suspended will be resumed at the earliest possible date.

Messrs. DeLong and Aupperle of Grand Junction, who have been here representing the Grand Valley Water Users' association, will leave for home tomorrow.

IF THE High Line situation is "up in the air" as currently expressed in the realty offices and on the streets the Daily News suggests that the best thing that we can do, at the present juncture, is to wait until said situation comes down. It is imperative, however, that Secretary Ballinger give the people of Colorado some definite answer in regard to what the government will or will not do in regard to its future construction. Apparently he has seen fit to question the legality of the act of Secretary Garfield in signing our contract. On the other hand the people of the Grand Valley accepted the Garfield contract at its full value and complied in every respect with its provisions. There may be some technical faults in the contract. It would probably be impossible to draw up a human document that would not be fallible in the eyes of shrewd attorneys, but the fact that the Garfield contract was not just what Secretary Ballinger would like to have it does not remove the imminent need of the canal and its early construction. Secretary Garfield and the officials of the Reclamation Service in 1909 found that the High Line canal project was not only a feasible one but one that was very proper for the government to undertake. Simply because an administration has changed we cannot see how or why it should so affect the status of the undertaking. If the High Line canal was proper in the opinion of the Reclamation Service under Secretary Garfield why is it not equally so under the administration of Secretary Ballinger. The people of the Valley have spoken and declared in emphatic terms in favor of government construction. We have been tested and tried in that regard and have been loyal to Uncle Sam and our desire that his agents do the work. We bespeak equal loyalty from Uncle Sam. Technical difficulties are poor favors to present a people who have been led to expect great benefits. Unexplained delays are not the kind of sauce that should be served to a people who have met the government at every turn and complied with their slightest wish. Better fruits should be ours. Secretary Ballinger has been on the Western Slope and in the Grand Valley. He has some personal knowledge of our needs and of our prospects. The Daily News cannot believe that he would desert us now simply because of routine or temporary difficulties. The Reclamation Service may have been "hitting it. up" rather heavily of late but should the Grand Valley project suffer because of this? Colorado yearly pours into the treasury of the Reclamation Service some half million of dollars. The Gunnison project has been absorbing all of this to date but the Gunnison project will be completed within six weeks. This half

Daily News, June 3 and 5, Denver Post, June 6 1909

million should come back to Colorado and to the Grand Valley project. In justice to the people of the Grand Valley Secretary Ballinger should, make some statement and give us something definite to go on. We should not be kept "up in the air" any longer.

Denver News, Jun 8 1909

WATER USERS BALLINGER'S CRITICS

Nebraskans Score Refusal to Cooperate With Local Men in Case Similar to the High Line.

RESOLUTION DENOUNCES RED TAPE AND QUIBBLES

Novice Put in Authority and Work Tied Up for Years, is Complaint.

[SPECIAL TO THE NEWS.]

SCOTTSBLUFF, Neb., June 7.—Secretary of the Interior R. A. Ballinger has got into hot water over some of the reclamation service projects. In fact, the secretary has got into so much hot water that if he could only cool it enough for irrigating purposes there would be no need to worry about the snowfall in the mountains for many years to come. Ballinger has made a number of changes in the management of the reclamation service projects and none of these changes is relished by the people most interested, namely, the farmers who expect to use the water from the government projects.

One of these changes was to stop the work on the High Line canal in the Grand valley, Colorado, pending an "investigation." Another was to declare illegal and void the use by the service of the money subscribed by the locality interested to get things under way. This was ex-Secretary Garfield's way of dodging the stoppages otherwise caused by the perennial short appropriations. Ballinger also has stopped the so-called "cooperative plan" of construction, until now in use on the North Platte project in Nebraska. Under this plan, the Water Users' association would hire men to work on the government project, and pay these men in "scrip." The "scrip" thus issued was taken by the secretary of the interior in part payment for water rights under the project in question. In effect, it came to this, that the farmers paid for part of their water in work rather than in cash, and the work and the issuance of the scrip was at all times under the control of the government project engineer. Ballinger has declared the scrip plan illegal, and the North Platte water users are savage. About 25,000 acres which could be irrigated in this way must now be left barren until money is available from other sources.

The farmers' wrath was increased by a side issue which arose about the same time. It would seem from the reports that filter in from Washington and elsewhere that the new secretary of the interior is of a suspicious turn of mind, and inclined to look carefully into everything that his predecessors have done. It is further rumored that while he held at the very first that his new subordinates would bear watching, he was for a time undecided whom to put on the job. At last he selected one E. A. Keyes as a sort of traveling sleuth to uncover any lurking villainies in the reclamation projects.

Keyes is not hampered with age, being only 28 years old nor with experience, save such as he may have gained in the Philippines. But there is no doubt he had loads of good intentions. He arrived here inspecting the North Platte project about the time that Ballinger's order stopping the cooperative plan also floated in. As a result the following resolutions were prepared by the board of directors of the North Platte Water Uses' association:

"Whereas, It has come to the attention of the board of directors of the North Platte Valley Water Users' association, through the public press and through certain individuals, that the policy of the present secretary of the interior is to restrict the operations of the reclamation service, notably in the suspension of the so-called cooperative plan; and

"Whereas, The North Platte valley and the various commercial bodies therein have received visits from one E. A. Keyes, who presents himself as a personal representative of the secretary of the interior and who sets out the statement that the administration of President Roosevelt in general and of Secretary Garfield in particular were unstatesmanlike, and that President Taft and Secretary Ballinger propose to inaugurate reforms; and

"Whereas, the said Keyes under pretext of examining the engineering features of the North Platte project has given his attention mainly to conferences with R. G. Leavitt, whose plans have always looked to his own profit at the expense of the general welfare; now therefore, be it

"Resolved, that the North Platte Valley Water Users' association by its board of directors hereby expresses its emphatic disapproval of any action looking to the discomfiture of the reclamation service. By experience we have learned that satisfactory service can not be rendered by federal departments which are burdened by interference from Washington, and as an instance of this we cite the unsatisfactory, erratic and inefficient conduct of the land office.

"We regard the cancellation of the cooperative plan as an unwarranted interference and believe that it would have been an agency of great good. In the absence of any opposition to the plan we regard the suspension as an evidence of hostility to the general cause of reclamation.

"We regard the appointment of agents such as Mr. Keyes as useless and unnecessary expense. His activity in this valley has been such as to reveal his total ignorance of irrigation, engineering, law and Western conditions. We resent investigation by inexperienced men and we expect governmental agents to preserve a neutral attitude and not lend all their attention to those interests which have considered public welfare only from the standpoint of exploitation of the public.

"In a general way we regard the criticism of the Roosevelt administration the restoration of power sites, the published interviews with the secretary of the interior wherein he threatens in vague and uncertain terms the reclamation service and other public governmental activities, as uncalled for and unwarranted.

"We express confidence in the general policy laid down by ex-President Roosevelt and Secretary Garfield for the conduct of reclamation work. We believe the service to be generally free from serious errors and we regard the attitude of the present administration by Secretary Ballinger as a serious menace to the welfare of this community, particularly with reference to those communities which are yet unirrigated.

"We call upon all good citizens and particularly upon our senators, representatives and public officials to call the attention of President Taft to the mischievous interference with reclamation work and to demand of him that the well considered and time tried plans of the

Denver News, Jun 8 1909

reclamation service be given due consideration and that the people on the projects concerned be consulted before material changes are made at the dictation of promoters and others who probably have neither knowledge of the requirements of the work nor concern for the public welfare."

HENRY WILL ARGUE PLEA TO BUILD HIGH LINE

Proposes Ditch 35 Feet Above Present Survey; Valley Opposes Scheme.

WITHIN thirty days at the outside, and probably within half that time, T. C. Henry of this city will file a formal statement with the secretary of the interior, relative to the Grand Valley Highline canal. The statement in question will undertake to show that the Highline canal is a project which should be left to private enterprise. It will, in effect, be an elaboration of a letter written by Henry to F. H. Newell, chief of the reclamation Service, of which a portion is as follows:

"I will engage to build a gravity high line canal on a grade about 35 feet above the one adopted by the reclamation engineers, thereby avoiding all right of way across fruit orchards, etc. of a capacity to irrigate all the lands irrigable therefrom, for bonds of the existing Irrigation District enlarged to embrace the same, not exceeding $40 per acre on the basis of a cubic foot per second for each 100 acres to the satisfaction, and approval of the State Engineer of Colorado; complete it by May 1, 1910, without exacting land "units," bonuses, commissions or "subscriptions in work," provided that the government lands segregated shall be restored to entry, and all needed right of way over government lands be granted such irrigation district. All necessary bonuses and guarantees as the law requires shall be immediately provided."

The statement about to be filed by Henry will undertake to show that he has a better line for the canal than that proposed by the government engineers; and that it can be built by a private company for less money and in less time than by the reclamation service. Henry's plan is to take the water from the Grand river above the mouth of Plateau creek, some distance above the proposed intake of the government project. He plans to take the water out on the south side of the river, carry it along that bank until nearly opposite Palisade and then cross the river with an inverted siphon to the lands to be irrigated on the north bank. By this means, Henry claims he will avoid damaging the Palisade orchards, since it will bring the new ditch above the highest old ditch, and therefore on unused land. Owing to this line being higher up the slope, Henry counts that it will have about 7,000 more acres beneath it and subject to irrigation than would be the case with the government ditch.

Henry claims he has plenty of financial backing in this matter and that he will present the evidence of such backing to the secretary of the interior.

There is no question as to the attitude of the citizens of Grand Junction and the Grand valley in regard to the building of the High Line canal, however feasible the proposition may appear to Henry. They are a unit against the construction of the system by private capital.

When asked if the condition of the land titles, or lack of titles, in the land to be irrigated did not rule out private enterprise, Henry answered that it did not. He cited the Otero, Henrylyn and Riverside districts as three irrigation enterprises which were put through while half the land to be watered was held by the government, though filed upon by settlers. He declares that an irrigation district formed on this land under the state law would have no trouble in selling bonds for the project.

Henry adds that he is not in any way responsible for the stoppage of work on the government project and that he has made no attempt to influence Secretary Ballinger on the matter.

No engineers of the reclamation service could be seen last night. Other advocates of the government plan to build the canal claim that Henry's plans are not practicable. They claim that the government engineers discussed the plan to take the water out at the place indicated by Henry, but found that to do this and go above the Palisade orchards would add more than a million dollars to the cost of the project.

Daily News, Jun 5 1909

CONFERENCE TODAY MAY DECIDE FATE OF HIGH LINE

ANTICIPATED BALLINGER WILL BREAK HIS LONG SILENCE

WASHINGTON, June 4.-- Arrangements have been made for a conference which will be held tomorrow between Secretary Ballinger and your committee of Messrs. Aupperle and DeLong at which the Grand Valley proposition will be threshed out. It is expected as a result of this meeting that the matter will be satisfactorily settled or at least that Ballinger will make some definite statement of his own position. Although no information is obtainable as to the probable attitude which the secretary will take high hopes are entertained that the outcome will be favorable to the project. Your committee believe that Ballinger will issue an order for the resumption of the work.

RUMORS THAT CONTRACT IS HELD INVALID UNCONFIRMED

Sensational and unconfirmed rumors that the contract between Garfield and the Water Users' association had been declared invalid by Attorney General Wickersham are so far without the slightest foundation in fact. The Denver papers printed such a story several days ago and last evening under flaring headlines those dispatches were reproduced in the whimper. At best such a report is only a guess and how good a one only subsequent events will reveal

The following telegrams were yesterday received from Washington by The News:

"Decision of Attorney General regarding contract not yet announced." — Aupperle.

"Nothing definite regarding opinion of the Attorney General. Expect that tomorrow it will be known."--S. Guggenheim.

"Ballinger still declines to make any statement regarding purport of decision rendered last Saturday by Wickersham." — Cline.

President Delaplain of the Water Users Association has received neither confirmation nor denial of the truth of the dispatches from Washington published in Denver papers and reprinted in the evening whimper which states that the decision of the Attorney General on the legality of the High Line Contract finds it void. "The Attorney Generals report has not yet been made public," said Mr. Delaplain, "and of course we can at best only surmise its contents. Secretary Aupperle will certainly wire me as soon as anything definite occurs and as yet I have received no message. It will be a great disappointment if the contract is declared invalid, however, it ought not to prove an insurmountable obstacle in the way of the Government speedily undertaking the work."

The Reclamation office had not received any word from Washington regarding the matter.

Daily News, Jun 6 1909

BALLINGER PLEDGES GOVERNMENT CONSTRUCTION OF HIGH LINE DITCH

WORK WILL BE RESUMED WITH PRESENT OLD ALLOTMENT OF FUNDS—MORE MONEY AS FAST AS AVAILABLE—COOPERATIVE PLAN ABANDONED

Active Efforts of Guggenheim, DeLong, Aupperle and other Grand Valley Boosters Successful. in Keeping Uncle Sam on Project—Yesterday's Interview Has Joyful Outcome

BY SHELDON CLINE.

WASHINGTON, JUNE 5.--THE GRAND VALLEY PROJECT WILL BE CONSTRUCTED BY UNCLE SAM. THIS WAS THE GLAD NEWS BROUGHT OUT FROM THE CONFERENCE ROOMS LATE THIS AFTERNOON BY SECRETARY AUPPERLE OF THE GRAND VALLEY WATER USERS' ASSOCIATION, HORACE T. DE LONG AND SENATOR GUGGENHEIM. THE BROAD SMILES ON THE FACES OF THE TRIO INDICATED THAT THEY HAD BEEN SUCCESSFUL IN THEIR WORK. SECRETARY BALLINGER, AFTER HAVING READ THE OPINION OF ATTORNEY GENERAL WICKERSHAM OH THE CONTRACT AND LISTENED TO THE ARGUMENTS OF THE COMMITTEE AND PERUSED THE LETTER$ AND TELEGRAMS SENT TO HIM FROM THE GRAND VALLEY AND FROM GOVERNOR SHAFROTH, EX-SENATOR T. M. PATTERSON AND MANY OTHER PROMINENT MEN STATED THAT HE WOULD GUARANTEE THAT THE GOVERNMENT WOULD IMMEDIATELY PROCEED WITH THE WORK OF CONSTRUCTION ON THE BIG RECLAMATION DITCH.

Work, however, will not be rushed this year. The Grand Valley project will be given all available funds, which will probably include the $125,000 which was allotted the project under the Garfield contract. While there will not be as great an amount of work done on the canal this season as was first anticipated considerable excavation will be accomplished and work will go forward steadily on the tunnel above Palisade. As the funds of the Reclamation Service are increased in 1910 and 1911 the work will be sent ahead with.

Senator DeLong will leave Washington Sunday night for a short trip in the east and middle west before returning to Grand Junction. He expects to reach Grand Junction on June 14.

WASHINGTON, June 5. At a conference between Secretary Ballinger, Senator Guggenheim, Senator DeLong and D. W. Aupperle an agreement was reached today recommitting the government to the Grand Valley project. Some few details remain to be worked out before work is resumed. The secretary will not allot more for the work in '09 and '10 than actually available but it is stated that the work will be vigorously pushed with funds at hand.

Besides the dispatches from The Daily News correspondent, Mr. Sheldon Cline, Postmaster Edwin Price received the following dispatch from Senator Horace T. DeLong:

I leave Saturday night and will arrive Monday, Jun 11 We were successful in keeping the government interested. While work may be slower at first than first hoped for it will be vigorously pushed later Senator Guggenheim was a wonderful help to us.

Whoop---hurrah!

HORACE T. DE LONG.

There is a dispatch in the telegraph office for President M. O. Delaplaln of the Water Users association, but as that gentleman could not be located last evening its contents are not known. It was learned, however, that it was from Washington.

Project Engineer Sands was greatly pleased when told by the Daily News of the result of the conference at Washington.

"Secretary Ballinger's action means that the government will certainly build the canal," he stated, "and I am very glad that the work is to go on. I have no official information from Washington at all and do not know just what will be done in regard to immediate construction."

The Daily News flashed the story of the government's decision in regard to the High Line from one end of the valley to the other. At Fruita and Loma the news was received with rejoicing. In Grand Junction everyone who heard the news expressed the deepest satisfaction. A great burden had been lifted from the shoulders of the people of the valley by the secretary's action. The long period of suspense was over as Ballinger's promise in regard to the completion of the work of construction by the government is final.

"It's splendid news," declared Mr. Orson Adams when phoned to by the Daily News and told of the news from Washington. "I haven't heard of anything that pleased me more in a long time."

"Great news," declared President Cherrington of the Chamber of Commerce. "'It means that the era of the valley's greatest prosperity is surely at hand." "Fine work," stated County Commissioner Tom Dwyer, who just arrived from the Plateau Valley. "The boost that the Grand Valley will receive by the construction of the canal by the government will mean much for the

Daily News, Jun 6 1909

whole county and will be a big boost for the Plateau Valley too." "I am certainly pleased to hear it." said V. C. Talbert. "Now that it is finally settled for good we can all draw a breath of relief. It's splendid news." Business men stopped work during the height of the Saturday rush to give out expressions of the satisfaction at the favorable results of the efforts of Messrs. DeLong, Aupperle and Guggenheim at Washington in securing from Secretary Ballinger his promise that the work would be done by the government.

By Associated Press.

WASHINGTON. June 5.--The reclamation act contains no provisions for the agreement between the government and the Grand Valley Water Users' association at Grand Junction, Colo., looking to a construction of the Grand Valley project and the agreement in question cannot be carried out, according to an opinion of Attorney General Wickersham.

The opinion was today sent to the association at Grand Junction accompanied by a letter from Secretary Ballinger. It held the secretary cannot commence the construction of any portion of the project unless the necessary funds are available in the Reclamation funds.

It was stated possibly $170,000 may be allotted to this project in 1909 and 1910. Preliminary work will be resumed as soon as possible.

Daily News, Jun 10 1909

T. C. HENRY is loose again.

In the Grand Valley we understand T.C.

He is an accomplished builder of irrigation ditches on paper.

He is said to have one of the finest collection of blue print irrigation systems on record in Colorado. Now, despite the understood restrictions to his activity in the Grand Valley due to his agreement with the people of this valley three years ago and his recent contract with the Magenheimers, Henry is out on the open range flaunting a blue print in each hand and spitting a continuous series of sentences from his mouth concerning his gravity plan for a High Line in the Grand Valley.

He declares, states the Rocky Mountain News, that he will file a formal statement with the Secretary of the Interior sometime within the next 30 days undertaking to show that the High Line should be built by private capital and per se by T. C. Henry.

Henry has all the rising powers of yeast or Banquo's ghost without the saving essential quality.

One would believe after the long series of rebuffs Henry has received in the Grand Valley that he would not again attempt to inject himself in the Grand Valley affairs. He has been shown on a number of times that interference from his office in regard to €he construction of the High Line was about as welcome as an invasion of the smallpox and that if given an opportunity the people of the Grand Valley would cheerfully undertake to demonstrate this fact to him personally.

There is no danger that his "formal statement" will receive any particular amount of consideration from Secretary Ballinger. Within the last ten days Secretary Ballinger has publicly stated that he did not favor private construction of the Grand Valley project and that he did favor government construction of the same. On what grounds Henry believes that he will receive a hearing from that Secretary is not clearly understood. Apparently the secretary's declaration is so explicit as to satisfy even one of the persistency of Henry that private construction is not to be entertained in regard to the High Line. The only theory advanced that has any particular weight is that T. C has read recently several press rumors to the effect that Director Newell of the Reclamation Service was to be removed and bases a hope that in case a new director assumes charge of the bureau, his wild-eyed scheme will secure a hearing it would not otherwise receive from an experienced man.

The people of the valley, however, have taken steps to remove the shadow of Henry from across the Grand Valley. Any irrigation scheme presented by him would not receive enough support to justify the formation of a district.

Daily News, Jun 8 and 10, 1909

ANGRY OVER REJECTION OF CO-OPERATION PLAN

North Platte Men Hit Hard--T. C. Henry Breaks Out of Captivity

----

In the Rocky Mountain News of Tuesday were two articles of great interest to the citizens of the Grand Valley. One was a dispatch from Scottsbluff, Nebraska, relating the anger of the Water Users of the North Platte project in Nebraska over the declaration of Secretary Ballinger that the cooperative plan of constructing ditches was illegal and would not be continued under his administration. The North Platte project has been partially constructed under the cooperative plan and the action of Secretary Ballinger in stopping the work on the Grand Valley project because of our plans to adopt the cooperative hit the North Platte men in the same way and with more disastrous effects. About 25,000 acres are involved under the North Platte project. The North Platte men are also aroused over the action of the department of the interior in sending out E. A. Keyes, a special investigating agent of the department to investigate, the Reclamation Service. Keyes has been reported due in Grand Valley shortly. As a result the farmers of the North Platte country adopted a series of resolutions which fairly sizzled with wrath.

The second story was to the effect that the ubiquitous T. C. Henry of Denver, a gentleman who is generally credited with being one of the Grand Valley's greatest obstacles to progress, has announced that within the next 30 days he will file a statement with Secretary Ballinger declaring his wish to build the High Line on the gravity plan on a grade of about 35 feet above the present government survey. He states he will do this at a cost not to exceed $40 per acre. He plans to build his intake above Plateau creek and avoid the Palisade orchard district.

The people of the Grand Valley have on one specified occasion declared by an overwhelming vote that they did not want Henry to have anything to do with the canal under any circumstances and on a more recent occasion have declared that they did not want the canal built by any capitalist or combination of private capitalists to construct the ditch. Secretary Ballinger is in full sympathy with the people of the Grand Valley in their desire to have the Canal built by the Reclamation Service and has declared within the last two weeks that he did not favor the construction of the ditch by private capital. Henry however, is apparently made of the kind of stuff that can't be repulsed with anything short of an application of the "big stick" and from expressions heard in Grand Junction yesterday it is evident that there are men in the Valley who would pay a premium to have a chance at being at the power end of the stick. Henry's plans have been demonstrated time and again as not feasible and impractical.

In Tuesday's Republican is printed a screed advocating private construction of the High Line canal in the Grand Valley. It is signed by an "'Outsider" but from what literary investigators would call "'internal evidence" it can be safely accredited to the inspiration at least of Henry. Henry has evidently broken loose from the reservation in which it was supposed that the Magenheimer boys had corralled him last March. His efforts in regard to disorganizing the sentiment of the people of this valley, however, will be fruitless.

THE HIGH LINE CANAL AND BALLINGER'S DECISION AS DENVER PAPERS VIEW THEM

It is interesting to note the widespread comment that the action of Secretary Ballinger regarding the High Line canal is causing. Perhaps the best editorial comment yet made on this action is by the Denver News, while the Denver Republican goes out of its way to criticise the present line of survey or right of way for the canal.

The following is the editorial that appeared in Monday's Denver News:

Secretary Ballinger announces that the reclamation service will build the High Line canal, and will resume the work as soon as possible. Just when that will be, Secretary Ballinger does not know. The reclamation service is rather short of funds; and Mr. Ballinger does not believe that the service can legally accept the money advanced by private citizens and by communities to hurry the work. It will therefore be necessary to check up accounts pretty carefully before fixing a date at which the work will be renewed

So far, so good. We could wish the style of the reply were a bit less oracular and that the secretary had been able to promise that work would begin at once, and continue without halting until finished. But he is doubtless doing the best for us that can be done and we must look to our senators and representatives for the rest. It is something to know that canal will be put through, even if one can't be sure of the date. The red tape which is tying Secretary Ballinger's hands in this matter is the same that delays and makes costly almost all government work; and must be reckoned with in any plans based on governmental procedure. But as the News pointed out before, this is a case where the canal must be built by the United States government or not built at all.

The High Line when completed will water nearly 60,000 acres of land whose

Daily News, Jun 8 and 10, 1909

soil is a mineralized silt that even Egypt cannot equal. An extension of the High Line---an extension sure to be made---will bring another 40,000 acres of the same sort of land under ditch. The hundred thousand acres thus to be reclaimed are worth more, agriculturally speaking, than any million acres of land in New England or New York. With a stake like this in sight, Colorado's congressmen cannot afford to be baffled. They must put the work through.

---------f

The Denver Republican on Monday contained the following editorial: Although the co-operative agreement entered into with the Water Users' association by Secretary of the Interior Garfield to construct the High Line ditch in the Grand valley has been pronounced illegal by the attorney general, Secretary Ballinger has determined to undertake construction of the ditch by the Reclamation Service without the co-operative feature as soon as the funds of the service make it practicable.

This seems to insure construction by the federal government, and since it is what the people of the valley want, no outsider has any right to object, whatever opinion may be entertained respecting the practicability of completing the work at an earlier date by private enterprise.

There is one thing, however, which common sense suggests ought to be taken into consideration, and that is the possibility of securing a better line than the one chosen by the Reclamation Service engineers.

The selected line runs through some of the most valuable orchard land in the state, and the destruction of these orchards should if possible be avoided. It is not alone a question of the additional cost which, in the payment of damages, this destruction would involve, but also the permanent loss to the valley of that much productive area. The owners of the land to be reclaimed by the High Line ditch might be willing to pay this damage, but the net result would be a permanent loss to the valley as a whole, which needs as large a productive area as it may be possible-to secure.

Consideration of this phase of the problem is suggested by the report that a much better line can be secured than the one located by the engineers of the Reclamation Service. According to what is said respecting its location, this new line would miss the Palisade orchards entirely, thus saving them to the valley and avoiding the payment of heavy damages to the owners.

We are confident that Secretary Ballinger is too prudent a man to order construction on the line chosen by the Reclamation Service engineers without first making an investigation respecting the practicability of securing a better line.

Daily Sentinel, Jun 8, Daily News, Jun 15 1909

NEWELL MAY RETIRE AND MR. BALLINGER MAY MAKE OTHER CHANGES IN SERVICE

It will be of great interest to scores of local people to know that, unless there is a change of program, Frank H. Newell will retire from the office of director of the United States reclamation service some time during the coming fall, probably to be succeeded by some man not now in the government service. Who that man will be has not yet been determined, but the indications are that he will have a comprehensive knowledge of the law in addition to an engineering education.

Mr. Newell, unless he so desires, will not sever his connection with the reclamation service, but will be retained as an engineer, probably a general consulting engineer.

Mr. Newell is well known in this city, having made several visits to Grand Junction in past years in connection with the High Line canal project and other irrigation projects. Some time ago The Sentinel was informed that there was a probability of Mr. Newell being retired from the office of director of the service, but no definite announcement was made.

The statements made above are contained in an extensive dispatch sent yesterday from Washington to the Salt Lake Herald.

Again there come reports to the effect that entire harmony does not prevail between the secretary of the interior and the reclamation service.

Following announcements embodied in the above paragraphs the dispatch says in part:

About the time of the change of administration it was learned that certain western senators were anxious to secure the removal of Mr. Newell from office. That effort failed. In fact, those who talked most about securing the removal of Director Newell were the last to make any serious effort to accomplish their purpose. The agitation they started was of short duration. In time it completely died out.

But it develops that Secretary Ballinger is not altogether satisfied with the way the reclamation service has been conducted. Although he finds no evidence of corruption or crookedness, he finds more or less dissatisfaction, some little friction, and what he is inclined to regard as looseness in the conduct of that bureau of his department.

The situation, in brief, is this: Although Secretary Ballinger finds little fault with the engineering work of Mr. Newell, he inclines strongly to the opinion that he can find a more satisfactory administrative officer to take general supervision of the reclamation service to become its business manager, so to speak. He recognizes that Mr. Newell's abilities lie along the lines of engineering rather than administration, and it is that fact which inclines him to the opinion that a change should be made. By appointing a new director he believes he can overcome those unsatisfactory features of administration which do not meet with his approval.

Secretary Ballinger is not ready to recommend a change at the present time,

Daily Sentinel, Jun 8, Daily News, Jun 15 1909

as he desires to get more detailed information before taking such a radical step. During the summer the secretary will personally visit several government irrigation projects, especially those that have given rise to complaint and to trouble. He will find out for himself on the ground what has led to unsatisfactory conditions and who is to blame.

If his investigation fails to bear out his present impressions, and if he finds that no fault attaches to Mr.

(Remainder missing0

LAUD GUGGENHEIM FOR EFFORTS IN BEHALF VALLEY

Aupperle and DeLong Back From Washington -Much Pleased With Outcome

The Grand Valley's two emissaries to Washington in behalf of the High Line canal reached Grand Junction Sunday and yesterday. Secretary D.W. Aupperle arrived Sunday and Senator DeLong in on No. 3 Monday. Both men were fatigued by their long trip, which lasted exactly 30 days and both expressed themselves as glad to reach home.

"We have nothing particularly new to report," declared Senator DeLong. "The Daily News correspondent, Mr. Cline reported everything as fast as it happened and kept the valley informed of everything that transpired. Secretary Aupperle and I will write a report for the consideration of the Water Users which will be read to them Tuesday afternoon. Both Secretary Aupperle and Senator DeLong are loud in their praises of the efficient work performed for the people of the Grand Valley by Senator Guggenheim.

"Senator Guggenheim worked very hard for us, declared Senator DeLong. "We would have been badly off if it had not been for his valuable assistance. Senator Guggenheim is a model senator in the way that he performs his work for the people of Colorado. During the day he can always be found at his desk in the Senate watching out for the interests of the state. In the tariff discussion he has proven himself a consistent protectionist and Republican. He has voted to protect Colorado productions at every opportunity, fought to have them kept on the protected list when threatened and at the same time watched after the routine affairs of the people of Colorado at Washington. Senator Guggenheim stands very high in Washington and has great influence."

"We hope that the word will soon come from Washington ordering the resumption of the work," stated Secretary Aupperle. "When we left everything looked favorable for the early issuance of such an order."

Before returning home Senator DeLong visited Buffalo, New York and the Niagara Falls. Secretary Aupperle came direct to Wisconsin, where he visited with relatives for two days and then came on to Grand Junction.

Daily News, Jun 16 and Aug 4 1909

WATER USERS ENDORSE BALLINGER'S PLAN

Secretary Aupperle and Senator DeLong, the committee representing the Water Users' association, who have spent the last four weeks in Washington in the interests of the High Line project yesterday presented to the meeting of the association a formal report of the incidents of their trip and its results. The details of what had taken place were already familiar, having been fully narrated in the splendid Washington reports by The Daily News special correspondent, Mr. Sheldon Cline, which were received every day during the progress of the negotiations and published exclusively in The News.

Accompanying the report of the committee was a letter from Secretary Ballinger in which the latter took occasion to repeat his oft repeated sentiments of friendship for the valley and the project and the assurance that the government will build the ditch as rapidly as funds are available. It did not contain anything definite as to when actual resumption of work will occur.

In addition to the many informal congratulations which Secretary Aupperle and Senator DeLong received from the others present for their splendid work in behalf of the project they also received a formal vote of thanks which was tendered them and the Colorado congressional delegation who has assisted in the work.

Secretary Ballinger's position was endorsed by a unanimous vote.

His letter is as follows:

Mrs. Haywood Brown of Mack is spending a few days with friends in the city. (sic!)

Department of the Interior,
Washington, June 5, 1909.

Mr. D. W. Aupperle
Secretary Grand Junction Water
Users' Association
Grand Junction, Colo.

SIR:

The importance of constructing government works for reclamation and irrigation of land in the Grand Valley Colorado, especially to the north of Grand river, has been urged upon this department by the body you represent, and also by many individuals interested therein. Realizing the importance of this work and that large interests are involved I have given the matter careful consideration. I am desirous of doing all that may legally be done to secure reclamation of these lands through the agency of a government irrigation project.

It is unnecessary to recite in detail the correspondence between various parties interested in this project and this department leading up to the execution, under date of February 20, 1909, of an agreement between the United States and the Grand Valley Water Users' association, under which the association was to make available for use by the United States in the construction of the project, certain subscriptions theretofore obtained, in money and labor, to the amount of $125,000. This contract involved also resort to the so-called cooperative plan, including the issuance of certificates by the Water Users' association in payment for work and labor and materials furnished which certificates were to be used to reduce construction charges against land within the project.

Doubting the legality of such an arrangement, I concluded it advisable and to the ultimate best interests of all parties

Daily News, Jun 16 and Aug 4 1909

concerned, to submit the matter to the attorney general for his opinion. He holds that the reclamation act of June 17, 1902, (32 Stat, 388) which is the source of authority in the Secretary of the Interior in respect of the construction of reclamation works,

(lines missing)

Secretary of the Interior "can not commence any construction of any portion of the work unless the necessary funds are available in the 'reclamation fund', and that the agreement can not be carried out.

This opinion but confirms my own views that I would not be justified in proceeding further with this project along the lines contemplated by that agreement, nor at all, unless money for the work may be set apart out of the reclamation fund, and for this reason work on this project was by me temporarily suspended.

It remains, therefore, to determine whether existing conditions of government reclamation work in the state of Colorado and throughout the county and the amount of money now in the reclamation fund, or to become available therein, will justify the undertaking of this project at this time or in the near future.

The preliminary examinations and surveys have satisfactorily demonstrated this to be a feasible project, and the department is willing to enter upon, the work when funds are available therefor and certain legal investigations now under consideration have been completed, and to prosecute it with all diligence.

The director of the reclamation service has considered the financial aspect of the matter to some extent and if an opinion that there is money in the fund to the amount of, possibly $170,000 which may be appropriated for expenditure in connection with this project during the years 1909 and 1910, and that very probably funds will thereafter be available to continue the work in a satisfactory manner. If it be found that conditions permit, the preliminary work, which has been suspended, will be resumed and prosecuted with reasonable dispatch,

Very respectfully,
R. A. BALLINGER, Secretary.

BALLINGER WILL NEVER GIVE UP GRAND VALLEY PROJECT

Promises All Projects Now Under Way Will be Finished

CONFERENCE OVER

Our Allotments Are Still Secret-Montrose Fares the Best

BY DON STEFFA.

Special to the News.

PORTLAND, Ore., Aug. 3.--The reclamation conference between Secretary Ballinger and the engineer of the service concluded today and the secretary returned to Seattle. The allotments for the various projects now under construction are under consideration and will not be known definitely for several weeks.

Asked to comment on the report that the Grand Valley project would be relinquished by the government and turned over to private capital, Secretary Ballinger declared enigmatically that "all projects now under way will he completed as fast as funds permit but no new ones will be undertaken."

This is taken to mean that the Grand Valley project will share in the present allotment and that work will be pushed up to the limit of the money available.

The sentiments expressed unofficially by some of the engineers present at the conference tends to strengthen the opinion that the project will be awarded the $175,000 tentatively assured them by the secretary last June.

It was brought out at the conference that the service needs at least twenty million and that it has only eight. The Gunnison tunnel project will get a great part of that. Director Newell declared tonight: "The Gunnison tunnel project will be completed during the next twelve months, indeed my department is going to make a special effort to finish all the larger projects next year including the Pathfinder Dam in Wyoming and the dam at Bellfourche, S.D."

None of those who participated in the conference would make any statement of the exact amount of money any project is to receive and this leads to the belief that that is still to be decided by Ballinger.

By Associated Press.

PORTLAND, Ore., Aug. ~--- Secretary of the Interior Ballinger tonight said the announcement of the appropriations to reclamation projects would be made October 1. Director Newell said after Ballinger's conference today that different irrigating schemes as prescribed by the supervising engineers was simply overpowering. Ballinger in an informal statement said that the present work would be completed before any new projects were undertaken, Eleven million dol-

Daily News, Jun 16 and Aug 4 1909

lars will be available he said while the new projects would take fifteen million.

Private advices received yesterday from Portland would indicate that a change for the better has taken place in Ballinger's attitude toward government construction for the big ditch here and that the original allotment will be made and that all together the affairs of the project are assuming a rose hue.

Daily News, Jun 15 and Aug 12 1909

JUBILATION FEAST IN HONOR OF BALLINGER

Thank Secretary Amid Rejoicing Over His Pledge

A jubilation feast by the Directors of the Water Users association and the Chamber of Commerce over the successful outcome of the negotiations with Secretary Ballinger regarding the Grand Valley project was held at the La Court last evening. The dinner was in the nature of a tribute to Secretary Ballinger. He came in for high praise at the hands of the various speakers for his promise that the Reclamation Service will build the canal rapidly as funds are available. Secretary Aupperle and Senator DeLong for their important share in securing this promise from the secretary were also the recipients of the highest praise.

Senator DeLong acted as toastmaster and introduced the speakers with appropriate and witty remarks at the same time reflecting the general tone of the dinner, one of rejoicing over the stand taken by Secretary Ballinger toward the project upon which the ultimate destiny of the Grand Valley, rests.

E. A. Keyes, the personal representative of the secretary who is on a tour of the west making a report to the department independent of reports furnished by the Reclamation service and ex-Senator Henry M. Teller and Project Engineer Sands were the only guests present.

Secretary D. W. Aupperle the first speaker on the program narrated an account of the Washington trip, the details of which are already familiar. Engineer Sands spoke on the project from engineering standpoint. Secretary Mahoney of the Chamber of Commerce responded to the toast "The Unity of the Community" and emphasized what indeed was touched on by all the speakers the perfect unity of the entire valley in favor of a government project.

Senator Teller spoke on the history of the Reclamation act and its importance. He also took occasion to add his word to the general rejoicing and thanks to Secretary Ballinger and for his decision to carry out the project. He reiterated his entire sympathy with the project and assurances of hearty support of the plan of government construction.

Mr. Keyes in a non committal speech briefly tendered his thanks for his cordial reception here and declared that with both people and city he had been most favorably impressed.

SPOKANE, Aug. ll.--Secretary Ballinger and ex-Governor Pardee had a final debate tonight when Ballinger took the platform to answer Pardee's charges in regard to the taking up of government land upon which valuable water power sites were located. Immediately afterward Ballinger left for Billings, Montana, where he will meet with congressional investigation committee.

"I have records from Washington to prove that not a single water power site has been taken by private individuals since the millions of acres were thrown open last April." he declared. He declared that today there were more water

Daily News, Jun 15 and Aug 12 1909

sites withdrawn from entry than at any previous time. He also stated that his department, while not advertising, was making substantial progress along the line of Roosevelt policies. Gov. Pardee later furnished technical descriptions of dam sites with maps of surveys, which he stated, proved conclusively that dam sites had been taken up under the Ballinger administration and that these could not have been taken up under the Garfield administration.

Pueblo will be the next convention city without doubt and B. A. Bowler of Phoenix, Arizona will be the next president of the congress. Arthur Hooker of Spokane will be the permanent secretary. These selections will be confirmed by the congress

---

BALLINGER SCORED FOR POLICY IN DEPARTMENT

"We Want Men Who do Things," Declares Pardee—Pinchot Smiles

SPOKANE, Wash., Aug. 11.-- By Associated Press.

WITH the stage carefully set, the actors picked before an overwhelming audience the present policy of the administration of the Department of the Interior in relation to the Reclamation Service received a terrific scoring today at the hands of ex-Governor Pardee of Oakland, California. Secretary Ballinger sat in the auditorium and heard his policies criticised and his methods arraigned without any visible sign of emotion. Gifford Plnchot sat in another part of the building and smiled. Before the debate closed Ballinger left the building but Pinchot remained until its close. Senator George Turner of Washington, a resident of Spokane and an old political friend of Ballinger, took up the argument in defense of the secretary.

The battle royal between the followers of Pinchot and the Roosevelt policies and the Ballinger men will break with full fury upon the congress tomorrow. All that has been expected will happen. After the attacks of today Ballinger will almost certainly take the stage and make a personal defense of his administration. When Ballinger entered the auditorium today he was received with cheers. His remarks were listened to and he was immediately followed by ex-Governor Pardee. The strenuous former executive of California discarded his written notes, which he held in his hand and opened the attack upon the secretary. He opened by stating that he stood for Roosevelt and Roosevelt policies.

"Roosevelt was a president who did things first and talked about them afterward," he said, "and that's the kind and character of men we want in public office now." He then praised the policy of former Secretary of the Interior Garfield declaring that under Garfield public lands had been withdrawn from entry with the belief that these withdrawals would save this territory from corporate exploitation. "Now," he declared, "Ballinger has again put these lands up for entry and each tract has in its boundaries water power sites. I do not oppose development by private capital but I do oppose giving away immense rights to private corporations which in a few years will enable their owners to hold the same political control over the country as do street railway and other public utility companies now control the cities. We know of the corruption that has resulted from railroad control. Shall we now hand out a new form of corporate power over our institutions?"

Senator Turner followed Pardee and stated he believed that it was in bad taste to make remarks concerning Ballinger in the way that Pardee had done. He declared Ballinger had done only what he should do as an official. Pardee's address evidently struck a popular tone with the delegates as they cheered the speaker long and loudly

In speaking to the delegates Ballinger said that it was not the intention of the government to oppose private construction of Reclamation projects or those projects which were available to private capital. He stated that the government's desire was to take on projects of great magnitude covering larger areas of land. He stated that the great danger in private construction was that private capital frequently took up small projects and thus destroyed opportunities for the watering of larger tracts by the government. He went into detail concerning the law governing the Reclamation Service appropriation.

Daily News, Jun 6, Jun 11, Aug 11, Aug 27 and Sep 4, 1909

BELIEVES GOV'T WILL BUILD CANAL

Secretary Aupperle issues Statement to Local Press

PATIENCE IS NEEDED

Secretary Aupperle of the Water Users' association last evening issued the following statement to the press of Grand Junction, detailing the present High Line canal situation to date as it is understood by the Water Users. The statement reads as follows:

To The Press of Grand Junction; There seems to be some uneasiness locally because of a financial uncertainty as to the condition our High Line project occupies at the present time in the government list.

This feeling of uncertainty has been caused by the press reports of Secretary Ballinger's statement at Portland recently that "The projects now under way must be finished before new work can be taken up". The question which is disturbing the minds of some of our people is whether our project is "now under way" or whether it is classed as "new work".

When we take into account all the facts connected with our case we can come to no other conclusion than that we belong to the list of projects classed as "now under way".

In November 1907 the Secretary of the Interior submitted an alternative proposition to our people, proposing that the government should withdraw from the project and release the public lands from the restrictions placed on them by the withdrawal under the Reclamation act, leaving the situation open for private enterprises, or the government would allow $50,000 to be used in taking up immediately the work of making permanent surveys preparatory to undertaking the construction of the system as early as funds are available. The proposition was fully discussed at public meetings and at meeting of organized bodies, and the later proposition, that favorable to government construction was heartily endorsed, not a single vote was cast again it and not a voice was heard opposed to it at any of the meetings. A report was made to the Department and the allotment of $50,000 was immediately made and work on the permanent survey was begun. Since that time, more than a year and a half the "'Grand Valley Project", the official name given it by the Department, has appeared in all its alphabetical positions in every publish list of the government projects.

At the annual meeting of the engineers held in July 1908 an allotment of another $50,000 for completing the permanent survey, purchase or rights of way, etc.

In February 1909 a cooperative contract was entered into between the Department and our Water Users' association. This contract provided that our association should provide $125,000 of work on the system and that the government should expend an equal amount on construction work on the system during the same time, in addition to the $50,000 previously allotted. Later the Attorney General held the cooperative feature of the contract to be illegal, but Secretary Ballinger made a substitute proposition on June 5th 1909, which was duly accepted by the association. This proposition has been published but I will quote a part of it which bears directly on the secretary's position in the matter. "The preliminary examination and surveys have satisfactorily demonstrated this to be a feasible project, and the Department is willing to enter upon the work when funds are available therefor and certain legal investigations now under consideration have been completed and prosecute it with all diligence.

The director of the Reclamation Service has considered the financial aspect of the matter to some extent and is of the opinion that there is money in the fund to the amount of $170,000 which may be apportioned for expenditure in connection with this project during the years 1909 and 1910, and that very probably funds will then be available to continue the work in a satisfactory manner."

Some of our people look upon the apportionment of only $170,000 for 1909 and 1910 as an indicator that the government will extend the construction period over many years. There is .no foundation for this fear. When Senator DeLong and I arrived in Washington in May we learned that the Department found itself confronted by larger demands for 1909 and 1910 than had been anticipated, and in order that the large projects may be completed and the lands placed upon a producing basis. The men in charge have a very creditable ambition to complete these larger projects at the earliest possible date. When we discovered that they were needing the money for these projects in order to complete them in 1910 we voluntarily proposed that a limited amount be given us for that time, with the understanding that our project shall be pushed to completion when these are out of the way. The secretary's position in this matter is expressed in that part of the statement quoted above which says. "And to prosecute it with all diligence" also, "and that very probably funds will thereafter be available to continue the work in a satisfactory manner."

In the light of the facts as I believe I understand them, I do not hesitate to predict that work will be resumed soon and continued at a moderate rate during the balance of 1909 and 1910, and that the work will be pushed to completion during 19ll and 1912.

The work in this, the 5th district, is well advanced. This district consists of Colorado, part of Wyoming, South Dakota, Nebraska, Kansas and Oklahoma. The larger projects in this territory are the Uncompahgre, the North Platt and the Belle Fourche. Director Newell is quoted as having recently said that these will be completed in 1910. This will clear the slate in this district of all heavy work and the most expensive projects in the other districts will also be practically completed, while the receipts from the lands supplied with water under the completed projects will constantly increase.

There is also a mistaken idea that the Uncompahgre project has cost more than has been contributed by this state to the Reclamation fund, consequently the state has no claim for more work. The accounts at Washington showed that on April 3 last the state had contributed $4,315,027.44. The total expended up to that date was $3,549,767.68, leaving a balance of $765,259.76. It was estimated that expenditures fort the following two months, to the end of the fiscal year, June 30th, would not exceed $65,000

Daily News, Jun 6, Jun 11, Aug 11, Aug 27 and Sep 4, 1909

with a strong probability that they would run less. The receipts were running an average of $50,000 per month, so the state would have had on June 30th fully $800,000 to its credit, with the receipts running an average of $50,000 per month.

When we left Washington in June it was confidently expected that work would be resumed in a very short time, the order to take up the work being withheld to get a little legal question adjusted. There was an unforeseen delay in this and Secretary Ballinger started on his western trip before the matter was arranged and it may remain open until the secretary returns to Washington. It is not a question having serious bearing on the situation but the secretary desired to have it disposed of.

Simmering the points bearing on the case as it now stands we find that,

The government has on several occasions committed itself definitely to our project.

That this has been published for more than a year and a half among the accepted and active projects.

That the permanent surveys have been made, even to cross sectioning a portion of the canal line.

That construction work was actually begun under direction of the Service when more than five thousand dollars worth of material, supplies, and machinery were purchased and the building of a permanent "camp" was begun at the tunnel site.

That the government has made a definite contract with the railroad for right of way.

That Secretary Ballinger in his document of June 5th stated that work was "temporarily suspended" and that he stated in the same document that the Department is willing "to prosecute it with all diligence".

All these points go to show that Secretary Ballinger's reference to "new work" can have had no bearing on our project. To refuse to go on with the work on this project would be an abrogation of an agreement made in good faith in a straightforward manner between the government and the Grand Valley Water Users' association. I have always had implicit faith in the integrity of the government of the United States and I do not believe for one minute that this administration will violate an agreement made with us.

Delays are annoying, that is true, but others, even those who deal with private enterprise must bear with them. I am fully convinced that just a little more good natured patience will bring a resumption of work in the very near future, which will result in the government building in a reasonable time an irrigation system which will be a monument to the Reclamation service and a benediction to the Grand Valley for centuries to come.

D. W. AUPPERLE.

Daily News, Jun 6, Jun 11, Aug 11, Aug 27 and Sep 4, 1909

THE rumors that have been circulating in the papers of the West for the last few weeks to the effect that Secretary Ballinger was considering the removal of Director F. H. Newell of the Reclamation Service may be but rumors and it is to be hoped that they have no greater heft. The Reclamation Service should be operated by men who understand irrigation. And there is no man in the United States who knows more about irrigation and the United States Reclamation Service than Newell. When Senator Newlands of Nevada first opened his campaign for the passage of the Reclamation Act his aide and strongest ally was Newell. Since that time Newell has become the recognized leader in irrigation matters in America. He is not an engineer and not a lawyer but he is par excellence an irrigationist. He can trim a promoter with a shady scheme with greater quickness and dispatch than any other known bureaucrat in Washington. It is the earnest hope of every man in the West interested in the development and the success of' the Reclamation Service that Newell be retained. His retention will do much to insure a continuance of confidence in the effectiveness of the Reclamation Service that might otherwise be lost.

BALLINGER WILL KILL SOME 'SNAKES'

Refuses to Explain His Allusion

By Associated Press.

WASHINGTON, Sept 3.--"Incidentally," said Secretary Ballinger, at the close of his first day in his office here after a hurried trip from the Pacific coast, "I intend to kill some snakes." The remark was made in response to questions relative to the Ballinger-Pinchot controversy over water power sites, the Cunningham Alaskan coal entries and other conservation matters. The illustration was drawn from an incident in the secretary's life when, 20 years ago, as a cowboy at Medicine Lodge, he found the camp infested with rattlers, and in one night killed twelve.

Secretary Ballinger refused to explain the purport of his allusion, and declined further to discuss his controversy with the forest bureau until the heads of the various divisions of his department have prepared their reports. These he will send to President Taft.

After a visit to the president at Beverly Secretary Ballinger will return west to continue his inspection of reclamation projects.

Daily News, Jun 6, Jun 11, Aug 11, Aug 27 and Sep 4, 1909

BALLINGER KEEPS MUM AND SAWS WOOD

BY SHELDON CLINE.

WASHINGTON, June 12.--Secretary Ballinger stated this evening that he could not tell definitely just when work on the Grand Valley project will be resumed.

He stated that there were some questions which the department is now investigating in connection with this work and until they have been disposed of, work cannot be resumed. The secretary declared that those who are interested in the project are being kept informed as to the investigations being conducted by the department but when these will be completed cannot be stated now.

Ballinger Stays Away

National Conservation Congress Does Not Get Expected Excitement

----

By Associated Press.

SEATTLE, Wash., Aug. 26.--The national conservation congress, which it was predicted would witness a personal controversy between the chief forester and the secretary of the interior opened in the auditorium at the exposition today with Gifford Pinchot on the platform applauding the peace declaration and Secretary Ballinger, a hundred and five miles away studying an irrigation project, when E. H. Libby, president of the Washington conservation association called the congress to order he said it would benefit all people not to exploit the policies of any set of Washington officials.

He said the object of the efficiency meeting would be if politics were injected. Pinchot sat on the platform and led the applause that greeted the declaration for peace.

Daily News, Sep 14 and 17, 1909

The High Line Situation

SUNDAY morning the Denver morning dailies published a dispatch from Washington which stated that Secretary Ballinger of the Department of the Interior had declared that he was unfavorable to the commencement of work on the Grand Valley project this year.

Yesterday Secretary Aupperle of the Water Users' Association received a letter from the Secretary dated last Friday in which he states that he will visit the project with President Taft on September 23rd and that he will make no definite decision on the matter of the High Line until he reaches here.

Apparently the Washington correspondents for the Denver papers based their declarations that Ballinger did not favor the resumption of work on the Grand Valley project to the fact that he had recently reiterated his disapprobation of the "co-operative plan" which was inaugurated by the Reclamation Service under the friendly regime of Secretary Rudolph Garfield and under a misapprehension as to the true status of the Grand Valley project.

The Grand Valley project is not a tentative proposition under the consideration of the Reclamation service. It is an approved project upon which work has been started. As understood by the Grand Junction Water Users' Association, and their understanding is based on their contracts with the government, the Grand Valley project is an "old" project, not a "new" project. A sum amounting to $170,000 has already been appropriated for work on the ditch by the government.

Still The Daily News in common with every other citizen of the Grand Valley will await the coming of Secretary Ballinger -with considerable eagerness.

The News believes that Secretary Ballinger should no longer continue the policy of silence and inactivity in regard to the project. Secretary Ballinger's attitude towards the Grand Valley project since his assumption of office has been most unfortunate for the people of the Valley. The News believes that they deserve better treatment than they have received in the past. Under Secretary Ballinger's predecessor a contract that would have been most binding between private parties was entered into and duly signed. This contract, which provided for the cooperative plan of work, was disapproved of by the present secretary.

Another contract was then entered into last May but no work has been accomplished under it. When Secretary Ballinger arrives in Grand Junction next week he should be prepared to make a clear and convincing statement to the people of his intentions towards the High Line canal. Work should be started and started at once. A contract with the government should have the binding force of a contract between private concerns and the faith of the people in the government should not be shaken.

Daily News, Sep 14 and 17, 1909

BALLINGER POSITIVELY COMING; TELEGRAPH TO NEWS SO STATES

Declines to Make Statement in High Line Canal Matter Till Visit is Made

----

SECRETARY BALLINGER will be in Grand Junction with President Taft. And he may remain in Grand Junction for several days for the purpose of investigating the Grand valley thoroughly.

Doubts that Ballinger would be with the president were expressed yesterday by some of the officials of the Water Users' association. Dispatches from the East indicating the make-up of the Taft party had failed to state that the secretary of the interior would be present. This fear is allayed by a personal message from the secretary to The Daily News.

After Ballinger has thoroughly investigated the Grand valley project he will make a statement in regard to the intentions of the department respecting the prospects of immediate work. Due to many conflicting reports concerning the High Line and the attitude of the secretary of the interior in regard to it the Daily News yesterday telegraphed Secretary Ballinger and asked him for a statement concerning the project. The following reply was received by the Daily News last evening:

"K.A. Bickel, Daily News, Grand Junction, Colo.:

"I expect to be in Grand Junction with the president. I cannot make a definite statement respecting the project until after a further investigation.

"'R. A. BALLINGER, Secretary."

Every effort should be made on the part of the people to see that Secretary Ballinger has full opportunity to make the investigation he refers to. This investigation cannot be made within the short limits of the hour that has been assigned Grand Junction for a visit of President Taft, and it is to be hoped that Secretary Ballinger may be induced to remain in the Grand valley for several days and examine the project from the ground.

Daily News, Sep 24 1909

EARLY COMPLETION OF HIGH LINE CANAL PROJECT IS FAVORED

President Taft and Secretary Ballinger Impressed with Need of Government Aid in Development of Irrigation Projects Under Way--Would Recommend Further Undertakings

----

Brought face to face with the tremendous possibilities under irrigation and having their mental and optical vision focused on the actual centers of reclamation activity, President Taft and Secretary of the Interior Ballinger have declared themselves in favor of pushing to completion the projects NOW UNDER WAY, and since the High Line canal has been considered by the government as being no longer a NEW project but one UNDER WAY, it is safe to assume that the government is possessed of the intention to go ahead with the work.

In the absence of any specific declaration in this regard, reliance must be had on the composite utterances of these two men while on the Western Slope. President Taft acknowledges that he has been most forcibly impressed with the wonderful results of irrigation as applied to the arid and semi- arid regions and has extolled the products of this part of the country to the skies. He has also expressed the wish that even more extensive development be achieved throughout the great fruit belt of the west.

Add to this the declaration of Secretary Ballinger at Montrose yesterday to the effect that it was his opinion that projects now under way should be completed and new ones undertaken as soon and as rapidly as possible, it would seem that while no official statement has been forwarded to the people of the Grand Valley, it is but reasonable to presume that there is no intention on the part of the government to let work on our local project suffer from delay or Inaction.

It is hoped and expected that a definite announcement is now in the forming and will be promulgated in the very near future. The fact that a special conference was called for yesterday afternoon en route to Montrose for the special purpose of going deeply into the situation and the subsequent statements made in the latter city are cumulative evidence which can not be disregarded in the absence of more positive proof of the government's plans.

All has been done that can be done, this valley has strong friends at court and never was a better case presented than has been laid before the federal powers. Therefore let us take courage and comfort from the facts at hand.

Daily News, Sep 24 1909

MONTROSE, Sept. 23, (Special) ——

From words that have been dropped it is believed that government irrigation work is only in its infancy. In his speech here today President Taft said that whatever constitutional features were invoked could be overcome like many other similar matters were overcome for the "general good."

MONTROSE, Sept. 23, (Special) "My present tour which has brought me into intimate connection with perhaps fifteen of the largest irrigation projects which are under way has strengthened my opinion that it should be the policy of the National government to push the completion of the projects under way or contemplated, however, the reclamation service must act within the bounds of the conditions imposed upon it by circumstances."

This was the utterance of Secretary Ballinger in making an address in Elks park here this afternoon, in discussing irrigation projects. This remark would tend to encourage the belief that the high line canal would be pushed to completion if it were not for the temporary hindrances.

In a talk made by Senator Horace T. Delong of this city at the Gunnison tunnel opening yesterday, he touched on irrigating projects of the Western Slope and in a way referred to the Highline canal, which brought smiles from both President Taft and Secretary Ballinger, significant of approval of its completion.

Senator DeLong stated in his speech that he hoped both President Taft and Secretary Ballinger would issue the orders that to bring about the early construction of the high line in which our people are interested.

MONTROSE, Sept. 23--(Special).-- At 5:18 this afternoon President Taft opened the headgates which allowed the waters of the Gunnison tunnel to flow in their way across the plain, thus completing the first irrigation project started by the United States government, and also the largest irrigation enterprise in the west. He accomplished the act by placing a pure gold bell upon a silver plate, thus completing an electrical connection which opened the headgates. A moment later the water appeared at the mouth of the tunnel, and a great shout went up, which echoed and re-echoed around the surrounding hills. Simultaneously with the opening of the gates electric signals sounded in several places in Montrose and Delta counties, thus announcing the completion of the project. The president's trip through the state today was one continuous ovation from Glenwood Springs, where he was called out at 6 a. m., until 10 o'clock tonight, when he left Montrose. His day was filled with crowds, receptions, speeches, county fairs and general western hospitality. Grand Junction and Montrose vied with each other in crowds novel features and entertainments. As he was leaving tonight, he called the Colorado newspaper men to him and said:

"I want you to convey to the people of Colorado my heartfelt wishes for their continual prosperity. I was particularly impressed with your agricultural products. You have a great state; you have given me a good time, I wish you well."

The people cheered as the muddy wall of water rushed forth. In all his speeches today President Taft touched briefly on the subject of irrigation.

Just after the tunnel opening one of the thirty Colorado sheriffs who accompanied Taft across the states knocked down H. L. Daniels, superintendent of the tunnel, who was trying to shake hands with the president. The sheriff drew a large revolver. The President heard the quarreling and stopped for a moment on his way to the train. He was told that nothing serious had happened and proceeded.

Ballinger made a short speech here tonight referring to the reclamation service in a general way.

In speeches Taft praised Colorado climate and expressed wonder at the results of irrigation.

MONTROSE, Sept. 23.---President Taft spent the day on the Western Slope, amid a succession of scenes typical of the great western country. In many respects the day was the most interesting since leaving Boston.

Before traveling to the west portal of the tunnel on a narrow gauge train, Taft visited the Montrose county fair, and after some. formal speech-making was given a touch of real western life---a relay race by cowgirls. Miss Bertha and Elise Hull, 16 years old, riding in relay, three horses, won two half mile races by a quarter of a mile from two older sisters.

The President today had ample opportunity to study the effect of irrigation. For a long time his train passed through country where only sage brush grew. Then suddenly, it burst upon veritable oasis, where waving green fields of alfalfa and miles of orchards told of the miracle wrought by the touch of water.

Almost every stop fruit was presented the president. One little box a yard long, just held ten peaches in a row.

The scene of the opening of the tunnel was picturesque. Thousands of people lined the ditch a hundred and fifty feet deep, leading to the mouth of the tunnel.

The President stood on a little stand at the edge of the cut. He made electric connections which caused a squad of laborers several miles way to knock out the bulkheads holding back the Gunnison river.

Smiles Indicate Canal Action

At 2 o'clock this morning those from this city who accompanied the Presidential party to Montrose returned and while no definite statement could be made by any of them they had broad smiles on their faces and intimated that there was a reason for it..

Senator DeLong said: "We have carefully watched and heeded every word that has been said by either Mr. Taft or Mr. Ballinger on the trip from this city to Montrose and during the time they were there and while we cannot say positively that the high line canal will be pushed to completion we have heard enough during the past twenty-four hours to cause us to look for something favorable at an early date.

Turning to the reception given the Presidential party at Montrose Mr. DeLong said: "I certainly must confess that Montrose carried out a program for the entertainment of the Presidential party that was a credit to the state of Colorado. There was not a hitch in the entire afternoon's reception and everything was so pleasing that even members of the Presidential party could not help but be favorably impressed.

Daily News, Oct 1 and Nov 2 1909

PALISADE MAY ACCEPT GOVERNMENT'S OFFER

Would Settle Mooted Right-of-Way Question and assure Immediate Work.

IF APPROVED PALISADE WILL HOLD ELECTION

Local Reclamation Officials Would Have $175,000 to Expend During Fiscal Year on Project

----

News Special Service.

PALISADE, Sept. 30.---At a meeting tonight of the executive committee of the Mesa County Irrigation District Land Owners Protective league it was decided to call a general meeting of the land owners in the Stub and Price Irrigation districts to consider the proposition of accepting a tentative offer made by the government reclamation project of constructing the Grand Valley project main canal, the High Line, along the right-of-way of the stub ditch. The water users under the Price and Stub ditches were to receive their water from the government canal for about $40 per acre. The water users under the High Line not in the irrigation districts will receive their water at about $70 per acre. The proposition involves the turning over to the government the water power rites of the two ditch plants which are estimated to be worth about $300,000, and will settle to the complete satisfaction of the majority of the people in the Palisade district the mooted right-of-way question and this (illegible) High Line above the orchards. Although the matter has not yet been formally brought to the attention of the land owners under the ditches in an informal way, a great number of them have expressed themselves as favorable to the plan.

It is stated here on excellent authority that this change in the right-of-way will remove the last objection or obstacle in the way of issuance of an order from Secretary Ballinger to immediately resume work on the construction of the ditch and will put into the hands of the local reclamation officials $175,000 to be expended on the ditch during the present fiscal year.

Project Engineer Sands has been in Palisade today generally inspecting the camp and conferring with a number of local men, presumably in regard to the proposed change in the right-of-way of the canal. If the land owners meet the proposal with favor an election will be called in the districts and the proposition voted upon. This election will probably take place in the next four weeks. Right-of-way Agent J. A. Alexander is expected in Palisade within the present week.

Daily News, Oct 1 and Nov 2 1909

SENATORS TO VISIT HIGH LINE TODAY

SENATORIAL PARTY

Senator T. H. Carter, Montana.
Senator Charles J. Hughes, Colorado.
Senator Wesley L. Jones, Washington.
Congressman Taylor, Colorado.
Congressman Martin, Colorado.
Director F. H. Newell of the Reclamation Service.
C. B. Eaton, official reporter.
S. J. Murphy, sargeant at arms for party.
George S. Scharching, land editor Chicago-Record Herald
A. C. Johnson, Washington correspondent Denver News.

The United States senate irrigation investigation committee will be in the city today and investigate the high line project. The list of senators in the party today is not large but more will join the delegation. The senators are accompanied by two of Colorado's congressmen, E. T. Taylor, congressman-at-large, and John A, Martin, of the second congressional district. There are three newspaper men accompanying the party.

The senate committee originally was to consist of more than twenty but several of them could not make the trip and the party which was to have gathered at Denver ready for the start Sunday eyeing was very small, consisting of Chairman Carter, Jones of Washington and Hughes of Colorado, besides Mr. Newell, chief of the reclamation service.

Railroad officials will join the special train from time to time along the trip and yesterday J. C. Dailey, superintendent of the Utah lines of' the D. & R. G. arrived in the city from Utah in his private car and joined the party at Montrose yesterday afternoon. Frank A. Wadleigh, assistant general passenger agent of the D. & R. G., is also with the senate committee.

The special train which is in the service of the senatorial party, consists of a baggage car, day coach, diner and sleeper, but private cars will be attached on the different divisions of the various railroads.

Today the Water Users association will assist the senators in their work of investigating the high line irrigation project which the government has under way. The special train will be run to near Palisade where they will be allowed to view the starting point of the canal and in the afternoon automobiles will carry them to the other parts of the proposed canal and the lands which are to be watered by it. The entire day will be devoted to investigating the project. The committee is out for business and they will not be banqueted as is oft times the case when a distinguished party is on a tour of the country. The word came to the Water Users association that it was the desire that all formalities be dispensed with as the party was out for business. There will however be a public meeting in the evening at which time matters of business pertaining to the completion of the high line project will be. taken up.

MONTROSE, Nov. 1.--The visit of the senatorial committee on irrigation tonight brought little consolation to the homesteaders and renters under the

Daily News, Oct 1 and Nov 2 1909

Gunnison tunnel project who will bring raw land under cultivation when water is turned on in the Gunnison tunnel.

The matter of payment for water by the homesteaders was one of the main points presented to the committee in the resolutions which were adopted by the board of directors of the Water Users' association. It was pointed out in the resolutions that it would be impossible for any one bringing raw land under cultivation to pay one tenth of the cost of the water the first year. Senator Carter, chairman of the committee, in response to the resolution, said that the Reclamation law allowed the Secretary of the Interior to use his own discretion in regard to the payments for water and that the Secretary could and would be governed he was sure, by the exigencies of individual cases. Senator Carter and Senator Jones, after the meet stated to The News correspondent that personally they were in favor of any legal sort of cooperative plan which would facilitate the completion of projects. It was their opinion that the majority of their committee were in favor of this and that the committee would work towards that end. Senator Carter said that the script plan, so called, or any plan which comprehended the issuing of evidences of indebtedness which bore no interest and yet circulated in a way as money would not be either legal or practical. It was his opinion that this script bearing no interest, would become subject to speculation and fluctuate in value, thus defeating the ends for which it was issued. He thought that some scheme could be and would be devised whereby the water users could perform work on canals or other parts of a project and be credited therefor on their water contracts. Both of the senators expressed their approval of the proposition in favor of the national bond issue. For various reasons, however. Senator Carter said that a straight bond issue would not be possible owing to the fact that the Reclamation fund was a separate entity from other of the government financial operations and that evidences of indebtedness on the part of this department, if in the form of bonds, would undoubtedly become mixed with other government bonds, and would interfere with the self replenishing feature of the reclamation fund. He favored, however, the, issuing of some form of warrants, which should bear interest but could be separate from the bond issues of the government. It was his opinion that the committee on irrigation favored this plan and would work for it. The committee did not favor any amendments to the existing reclamation law. The senators also stated that the delay of existing projects meant actual financial loss to the government and for this reason projects like the Grand Valley project and others ought to be hurried to completion and the committee favored this.

Daily News, Nov 3 and 12 1909

SENATORS GIVE VIEWS ON HIGH LINE

As a finale in the entertainment of the Senate Irrigation committee a mass meeting was held in their honor at the Margery Hall last night when the different members of the committee and others of note gave talks. The meeting was presided over by Judge Sprlgg Shackleford. The first speaker introduced was Senator Horace T. DeLong who urged immediate construction of the High Line Canal and stated very clearly his reasons for so urging. Following Mr. DeLong came Director Newell of the Reclamation Service.

"The committee here tonight," said Mr. Newell, "is not to carry out old laws but to recommend the enaction of new legislation which will advance the reclamation work." Mr. Newell at this place gave a short sketch of the cooperative plan of the reclamation work bringing out many strong points in favor of this system. "However," said he, "the vital point in your case is,---Shall the Secretary of the Interior build your canal or not? And this injustice to you should be settled soon. Long ago when the government construction of the Grand Valley project was first proposed to me, I weighed it against the claims of the Gunnison Tunnel Project and this proposition had perhaps the best of it. However, I was besieged by men of private capital until it was impossible to proceed with this project and the Gunnison Tunnel was started. At present I cannot see the way clear to the finishing of this project at once or as quickly as you would like to have it done." Here Mr. Newell gave a long insight into the workings of the reclamation service. Going on he added, "I am very anxious to get this work here completed. The reclamation people can do this without the backing of the people interested in the work. In the past the service has been too ready to go ahead and push projects without the direct cooperation of the people, but hereafter the reclamation people will remain more in the background. The high line canal is essential and will be built sooner or later. I would advise you at present to send a committee of one or two men to Washington and dog the trail of the officials there until a definite understanding is reached. In regard to the cooperative plan which was to be adopted I will say that I believe it to be a very good thing. The people have the satisfaction of paying for the water as they go along and it takes no funds from the government supply."

At the conclusion of Mr. Newell's remarks Senator Carter of Montana chairman of the Irrigation Committee, was introduced and made a very interesting talk. He said in part, "This committee has been visiting the various government projects, finished and uncompleted, for two reasons: first, to report on the physical conditions and second to report on the attitude of the people. We stopped here not under instructions but to visit work that is only in prospect and I will say that we have profited by our sojourn here. We have visited about fifteen projects since we started and at every place it has been the custom to say that their country is the finest on earth without exception. But here I am not governed by custom. From my own personal observations I can truthfully say that this valley is surpassed by none. Your fruit, apples, peaches and other products, is the best I ever saw any place

Daily News, Nov 3 and 12 1909

and you should be proud of the land where they are raised.

"This is the kind of project which the congress had in view when it passed the reclamation act. It is our business to suggest new laws and additions to legislation by which more of this reclamation work can be carried on. To do this we must have funds. The amount allotted this work is small in comparison to the labor which should be done. From the sale of government lands and the receipts of the Water Users' association in the country we will obtain about $15,000,000 per annum for the next five years. This is not enough to half complete the work now underway and in my mind it would be unwise to issue bonds because the reclamation fund has been dedicated to this work. I do, however, feel certain that congress will enact legislation allowing the Secretary of the Interior and the Treasurer to file and register bonds for this work which should bear a small rate of interest and be payable in a few years after the water has been put on the land. Anyone who sees this land here cannot say that it would be unwise for the government to invest $3,000,000 in this valley for a few years."

Senator Warren said: "Under our commission we are here to hear what you have to say yet to be very careful what we say ourselves. I have seen rudiments for irrigation and the results as found in this valley and I am very, very much pleased with them. You should be proud of your past. The money ought to be on hand to finish all uncompleted and committed work, and unless we can secure the necessary legislation to bring this about I shall insist that the government take its hands off this project and give those people a chance who can get the work done."

Senator Jones gave a little insight into the workings of the department of the interior. "When the reclamation act was first passed," said he, "each senator had a pet scheme which he wanted to be put through and for which he brought great pressure to bear" on the secretary of that department. In this way so many projects were started that there was not enough money to complete them all and as a result many have been delayed for some time. It ought not to be this way but it is. If Secretary Ballinger could go down into his pocket and take out the money for the completion of the unfinished work he would have all the projects done up in short order, be he cannot do this because he hasn't got the cash."

So far as I have any capacity for accosmplishing anything for your cause I will be most glad to do so," said Senator Hughes of this state. "I shall help you people in every way possible. If I can accomplish any good by hanging around Washington officials. I shall do so, and I pledge myself to help enact any laws which shall advance this great work.

"'The east with her great majorities in the house and her influence in the Senate is enabled to secure any appropriation they want but when the West wishes a small sum it is almost impossible to get it. After the census, and a new grouping of congressmen is in the West we shall have a more nearly equal chance with the easterners."

At the close of the meeting Congressman-at-Large Ed. T. Taylor of Glenwood Springs made a few statements as to the stand of the house of representatives. His remarks were loudly cheered by the audience.

When the session had come to a close many Grand Junction citizens went up and met the distinguished visitors and a large delegation accompanied them to their train which left at 10:00 p. m. for Thistle Junction, Utah.

The senators as a whole expressed their deep appreciation of the treatment accorded them while in this city and all expressed the hope that they should again be able to visit Grand Junction in the future. The men were presented with great loads of the best fruit which could be obtained and it was greatly appreciated.

The senators and the senatorial party arrived in the city early yesterday morning and spent the forenoon at Palisade investigating the high line canal at that point and were driven over some of the land to be watered by the canal. They left late last night for the west and will visit projects in several of the western states.

The senate committee was entertained at the La Court hotel by the Water Users' association at lunch in a very informal way. There was no speechmaking and they truly enjoyed the occasion. The tables were nicely decorated with apples raised in the Grand Valley and the ice cream and several other dishes were served in delicate dishes made of apples carved in bowl shapes.

HIGH LINE SOON TO BE BUILT IS A STATEMENT FROM DENVER

News Special Service.

DENVER, Colo., Nov. 11.---The Times says tonight that the high line canal in the Grand river valley is to be completed by the government and that work will be commenced within the next few months. The cost will be from three to four millions of dollars and the other ditches in the vicinity are to be taken over and controlled by the government. The canal is to be built along the lines outlined by Garfield with but a few changes.

This information comes from a man closely associated with President Taft and who is in Denver. He would not allow his name to be made public, but states that at a conference recently with Taft and Ballinger this was the decision reached.

Daily News, Nov 19 and 28, Daily Sentinel, Nov 27 1909

BALLINGER WIRES HE'S READY TO BUILD THE CANAL

A MESSAGE RECEIVED IN THIS CITY THIS AFTERNOON FROM RICHARD A. BALLINGER, SECRETARY OF THE DEPARTMENT OF THE INTERIOR, BRINGS THE LONG HOPED FOR AND MOST IMPORTANT ANNOUNCEMENT THAT SECRETARY BALLINGER WILL ORDER CONSTRUCTION WORK TO BEGIN IN EARNEST UPON THE GRAND VALLEY HIGH LINE CANAL JUST AS SOON AS THE LINE OF SURVEY IS DEFINITELY SETTLED AND DETERMINED.

THIS LINE OF SURVEY, WHICH IN ALL ITS PARTS HAS NOT BEEN FULLY SETTLED UPON, WILL BE DETERMINED IN ITS ENTIRETY WITHIN A PERIOD OF FIFTEEN DAYS. THEREFORE, THE CONSTRUCTION WORK ON THE GREAT PROJECT THAT MEANS SO MUCH TO THIS CITY AND VALLEY WILL BE UNDER WAY WITHIN THE PERIOD OF A MONTH.

THE MESSAGE FROM BALLINGER CAME TO OFFICIALS OF THE WATER USERS' ASSOCIATION. ITS CONTENTS WILL BE HAILED WITH JOY THROUGHOUT THE CITY AND VALLEY.

U.P. TO INVADE W. SLOPE?

News Special Service.

GUNNISON, Nov. 18m.—A corps of Union Pacific surveyors, direct from Cheyenne, is known to be occupied at Kobler Pass, above Crested Butte and on the North Fork side of the divide. It is not believed that the U. P. has any intentions of building in this direction, but rather portends railroad activity by the Colorado & Southern in building down from Gunnison and Baldwin to Paonia, Delta and Grand Junction.. Twenty years ago the U. P. owned the old South Park road, later the Denver, Leadville and Gunnison and now the Gunnison branch of the C. & S. It operated coal mines at Baldwin and still owns such coal property in that vicinity and on the Anthracite branch of the North Fork. As near as can be determined it is these coal properties that the surveyors are determining.

Daily News, Nov 19 and 28, Daily Sentinel, Nov 27 1909

BALLINGER'S ASSURANCE

Secretary Aupperle of Water Users Receives Encouraging Message Announcing Intentions

Secretary Aupperle of the Water Users association received a telegram late yesterday afternoon from Secretary Ballinger of the interior department which indicates clearly that the government intends to "proceed in good faith" with the High Line work as soon as location of the line for the canal at Palisade is definitely determined.

The original survey ran about midway through the Stub District which would cause heavy damages to the land owners. Several months ago a proposition was made to the two Palisade districts which, if accepted, will raise the government canal twenty-five feet higher than the present survey and place it on the Stub Ditch line. If this is done gravity water will be supplied to both districts from the government system. Negotiations have been under way for some time between the districts and the government with every prospect of an agreement being reached very soon.

If the present plans are carried out the land owners in the Stub district will be protected from loss by damage to their orchards, and it will also be a great advantage to both districts in providing gravity water at a very reasonable cost. It is understood that a majority of the landowners of both districts are favorable to the agreement now under consideration.

By Associated Press

WASHINGTON, Nov. 27.--Secretary Ballinger today issued an order granting the request of the property owners along the banks of the Grand River for an irrigation project in that section of the country. The proposed ditch will irrigate 50,000 acres of land.

Daily News, Nov 19, 20, 30 and Dec 8 1909

BALLINGER MAKES A STRONG DENIAL

By Associated Press.

WASHINGTON, Nov. 19. —In replying to the charge made in a magazine, Secretary Ballinger today denied the reports concerning his conduct in his department. He emphatically denied that he had ever "Advised or ordered or let his support to any effort to perpetuate fraud upon the government." He reviews the history of the Alaskan coal lands and declares no coal claims are pending and says an investigation of all claims with a view to determining their validity will be vigorously prosecuted. He denounced it as "Pure fabrication" the report that he is interested in shares of the Alaskan Petroleum company, that in his private practice after his resignation as Commissioner of the General Land Office he had been consulted about the preparation of articles of incorporation under the law of 1908, which were approved by the president, permitting the entry of 2,500 acres of coal lands, but that it is not true that he had given advice looking a fraudulent act against the government.

NEW IRRIGATION DISTRICT IN THE VALLEY

Will Organize in Garfield County Along the Grand

ELECTION DECEMBER 16

News Special service.

GLENWOOD, Nov. 17.--The county commissioners of Garfield county issued a call today for an election in three precincts of the Grand valley in this county to vote on the question of organizing what will be known as the Grand Valley Irrigation District. The election will be held on December 16.

The proposed ditch will run along the foothills of the Bookcliff range from a point about two and half miles west of Rifle and thence down the river to the Mesa county line. There is a large ditch over part of this land put in by the Havermeyers who own several hundred acres of land in the vicinity of the proposed ditch. This is known as the Wilcox canal and waters a part of the land. The farmers generally favor the new ditch as it will assure better water advantages and water holdings.

Daily News, Nov 19, 20, 30 and Dec 8 1909

TWO DISTRICTS TO VOTE ON BOND ISSUE

Elections are Called for December 16 in Upper Valley

$176,000 AMOUNT ASKED

At a joint meeting of the board of directors of the Mesa County Irrigation district and the Palisade Irrigation district held in this city yesterday it was decided to call elections to vote bonds for a dam for the two districts. The election will be held on December 16 in each of the districts.

It was agreed at this meeting that bonds to the amount of $176,000 should be put to a vote of the districts, each district to vote on the issue of $88,000. A member of the board stated last night that the directors did not know just how the election might terminate as the question was one that had two impartial sides to it, but the matter will be thoroughly gone over and all points discussed by the time of the election so that every voter in the two districts may know just how he stands by the time election day arrives. In the face of the proposal from the government to take over the ditches of the two districts when the high line is completed makes it a difficult question to decide, but it is the belief of most every one that the government project is so far off that it becomes necessary to vote bonds and put in a dam.

The directors accepted the bid of J. J. Lumsden of $151,571 for the bonds. The issue includes a sufficient amount for preliminary work to the construction of the dam, which will be near the present intake above Palisade.

NEW IRRIGATION DISTRICT TO BE ORGANIZED

The board of county commissioners met yesterday morning to act on an application for a new irrigation district which is proposed for a tract of land near Palisade. The district is to be known as South Palisade Heights Irrigation district and will water 540 acres of land and fifteen property owners have signed the petition which includes the entire list of holders of the number of acres involved. The district is just south of East Palisade Irrigation district. The commissioners will take final action on this matter tomorrow, but will be in session today

Daily News, Nov 19, 20, 30 and Dec 8 1909

OFFICIAL HAS MANY FRIENDS IN GRAND JUNCTION

Morris Bean is Removed From Charge of Legal Department.

Morris Bean, formerly chief legal advisor of the Reclamation service and well known in Grand Junction which he has visited a number of times on matters in connection with the High Line, was deposed Saturday by Secretary Ballinger, and Anthony Campbell, attorney general of New Mexico, installed in his position. Bean's last visit to Grand Junction was at the time when the cooperative script plan was put into effect and he presided at the meeting at the court house when the subject was discussed. He has many friends in the Grand Valley. It is reported that he will remain in the Interior department in a minor capacity. F. H. Newell, director of the Bureau, will hold his position it is reported, despite the statements that it was contemplated to remove him and install City Engineer Thompson of Seattle.

Daily News, Nov 30 1909

BALLINGER ISSUES ANNUAL REPORT

WASHINGTON, Nov. 29.--Secretary of the Interior Ballinger has just issued his report for the past fiscal year. He makes some interesting recommendations. One of the most important is that regarding the use of public lands. The secretary says: "Specific legislative authority should be given the department to classify and segregate our public lands into well-defined divisions according to their greatest apparent use, with the continuing power of reclamation to meet changing conditions and the increased knowledge of their character which may necessitate the transfer of lands from time to time from one class to the other. Full legal effect should be given such classification when made, so as to prevent entries, under laws applicable to one class, of land belonging to another class, except after application for a review of the classification; in other wards, if lands are classified as coal land, they should be enterable only as coal lands, and in case it be shown that they are improperly classified as coal lands and are, in fact, agricultural they should then be enterable only as agricultural lands. This to my mind, is the only way the government can scientifically and effectively proceed to conserve the natural resources of the public domain. To leave to the self-interest of applicants the option to claim a particular tract of land as more valuable for mineral, agriculture, timber, phosphate, oil, or gas, or for power sites, is to invite confusion, fraud and contention, and this has been the source of a major portion of the difficulties of the Interior Department in administering the public domain.

"As regards new legislation, the present coal-land laws respecting the states and territories, as well as Alaska, should be supplanted by an act fully meeting existing as well as future conditions. The inducements for much of the crime and fraud, both constructive and actual, committed under the present system can be prevented by separating the right to mine from the title to the soil. The surface would thereby be open to entry under other laws according to its character and subject to the right to extract the coal. The object to be attained in any such legislation is to conserve the coal deposits as a public utility and to prevent monopoly or extortation in their disposition. This may be accomplished either through a leasing system, by which the title would remain in the government, under proper regulation and supervision by the secretary of the interior, or through the sale of the deposits, and in neither case with restrictions on their mining and use which would con-

Daily News, Nov 30 1909

trol the minimum output and conserve the deposits as a public utility. I believe the most advantageous method will be found in a measure authorizing the lease or sale of the coal deposits in the lands, subject to forfeiture for failure to exercise the rights granted, under such reasonable regulations as may be imposed. An exploration period of at least one year upon a merit basis, at a nominal charge would insure to the applicant the necessary preliminary knowledge upon which to make the lease or purchase of the coal deposits and venture the necessary investment for operation. The maximum unit authorized for this use could safely be made form three to five sections, provided no greater surface right be granted than will give proper facilities to reach and extract the coal deposits. In case of the failure of the lessees or grantees to open and operate the coal deposits under reasonable limitations and to maintain an output reasonably suited to the deposits and in case of combinations as to price or limitation of output, title should be forfeited by proceedings in court for that purpose. Government mine supervision would doubtless be necessary to enforce the conditions and limitations under the grant."

Daily News, Nov 30, Dec 4 and 8 1909

THOUSANDS OF ACRES OF SEEP LAND ARE TO BE RECLAIMED

Lost Territory May be Saved at a Very Light Expense

Almost 1,000,000 acres of fine irrigated desert land, once in cultivation, has been lost during the past five years, due to too copious irrigation and the consequent encroachment of seep. The situation is regarded as grave by the Department of Agriculture and a portion of Secretary James Wilson's report is devoted to an examination of the subject. Colorado has less seep land than any irrigated state in the West save Idaho and Montana. Colorado is reported to have 75,000 acres of seep land, Idaho 40,000, California 100,000, Utah 150,000, Nevada 250,000, Montana 60,000.

Every encouragement is given to the irrigation farmer, however, by the department who declare that seep can be prevented without much trouble and that seep lands, if properly attacked, can be reclaimed. The department has issued rules in regard to the reclaiming of seep land which are as follows:

Drains should be located in accordance with the behavior of the water and the following facts should be determined:

The source of the surplus water is the leakage from canals, but more especially the waste water from irrigated lands which occupy the higher levels.

Soil water is frequently deflected in its course, as it percolates through the soil, by layers of hardpan. Water which finally reaches the lower lands is arrested in its downward percolation by comparatively impervious earth, and then rises year by year until the surface is saturated.

Water dissolves a considerable quantity of the salts which are present in the soil. When evaporation takes place at the surface the salts, many of them being injurious to plants, accumulate rapidly.

One drain should then be placed along the upper edge of the wet land approximately across the surface slope and deep enough to intercept the flow from the higher land. Frequently the depth must be from 5 to 7 feet. A few drains then should be placed along the lower end of the tract to remove surplus water supplied directly by irrigation or rain. These should be located in depressions and not constructed until after the intercepting drains have cut off the supply from outside sources. Drainage wells sunk from 8 to 10 feet are effective on lands bordering on gravel formations. These wells collect free water until the well is full. The well can then be tapped by drains at convenient depths and discharged in a proper outlet. The cost of draining is put at $15 to $20 per acre and the land should be worth at least $100 per acre in old districts.

Daily News, Nov 30, Dec 4 and 8 1909

We're For Ballinger

GOOD news from Secretary Ballinger,--the news we've been waiting for, ever since he issued his order suspending the work that had been begun under Secretary Garfield. Work on the high line, says Secretary Ballinger, will be resumed as soon as an agreement can be reached regarding the right of way. This, surely, ought to require no great time. The government, we understand, has offered the Price and Stub people water at a reasonable figure, in return for which, if those water users make a contract with the government, the High Line will be constructed entirely above the Stub right of way.

This proposition looks good. We see no reason why the contract should not be entered into, and the work commenced at once. Then, in a few years, the High Line will be an actuality; our years of waiting will be at an end; the entire Grand Valley, for the first time, will be under irrigation.

If this contract is signed for the right of way, and work begun within the next month, we don't care what Collier's says about Secretary Ballinger; we're for him. We're for any man who builds that canal.

NEW DIRECTORS ARE ELECTED IN DISTRICTS

Three Irrigation Systems Select Men For Positions on Board

MAY BE SOME CHANGES

Three irrigation district elections were held yesterday for the purpose of electing a director to the board in the Price, the Stub and the Orchard Mesa districts.

In all of the districts rival candidates contended for the position and in some the race was close although the vote was not heavy in many.

In the Price district Harris defeated Johnson by a majority of 149. The vote by district was as follows:

District No 1	Harris	214
"	Johnson	15
District No 2	Harris	19
"	Johnson	64
District No 3	Harris	9
"	Johnson	23

In the Stub district Bowman defeated Thompson by a substantial majority. The vote in this district was very light.

District No 1	Barber	22
"	James	8
District No 2	James	11
"	Barber	9
District No 3	James	3
"	Barber	2

As a result of the election in the Price and Stub ditch it is reported that a number of changes will take place in the personnel o£ the executive management of the ditch.

Daily News, Dec 8, 17 and 19 1909

RECLAMATION TO BE GIVEN A SPECIAL MESSAGE BY TAFT

Advises Against Congressional Investigation of Sugar Frauds

Reports Big Cuts in Army and Navy Estimates for 1910

MESSAGE IN BRIEF.

Scores Zelaya Government in Nicaragua.

Reports all other foreign relations pleasant.

Asks for assistance to improve consular service.

Approves of civil pensions for government employes.

Believes Congressional investigation into Sugar frauds would hamper Department of Justice at this time.

If tariff commission reports excessive duties in new tariff law will recommend revision.

Cuts down army expenditures $45,000,000.

Cuts down naval expenditures $38,000,000.

Recommends coast defenses at Chesapeake bay.

Asks for legislation looking towards simplification of federal judicial procedure.

Recommends higher postal charges on magazines.

Recommends postal savings bank.

Recommends ship subsidy bill.

Asks for statehood for New Mexico and Arizona.

Promises special message on reclamation and conservation.

Asks for $50,000 to fight the white slave traffic.

Recommends observation of Semicentennial of Negro freedom

Believes nation just entering into period of great prosperity.

By Associated Press

WASHINGTON, Dec. 7.—Promising a special message in the near future in which he will consider the topics of reclamation and conservation, President Taft in his first message to Congress dismissed the subject with but a small paragraph in this 12000 word document.

The message is a compressed review of public needs and while comprehensive in many departments leaves others with but scant mention. The tariff is largely dealt with. The president asserts that he does not be-

Daily News, Dec 8, 17 and 19 1909

lieve that the tariff is accountable for the present high cost of living which he declares is a world wide movement and not affected by any single tariff.

Foreign affairs, declares the president are in a most excellent condition. The joint fisheries convention with Great Britain has completed its regulations which will soon be reported to Congress. He hopes for the early appointment of a commission from the United States, Japan and Russia to prepare enactments for the regulation of the fur seal fishing in the North Pacific. The results of the Second Peace Conference, he finds, were generally satisfactory. The questions arising out of the misgovernment of the province of Congo under private ownership are being satisfactorily settled by Belgium and the president believes that better things can now be hoped for, The report of the Liberian government commission, sent to Monrovia last year to consider the needs of this little nation, is referred to Congress. South American affairs are in excellent condition.. The affairs of the nation with Japan are in a pleasant state and the same condition exists in China. In Central America the only discordant note is found in the conditions in Nicaragua of which the president states:

"It would be no less unnecessary to rehearse here the sad tale of unspeakable barbarities and oppression alleged to have been committed by the Zelaya Government. Recently two Americans were put to death by order of President Zelaya himself. They were reported to have been regularly commissioned officers in the organized forces of a revolution which had continued many weeks and was proceeding in an orderly fashion in control of about half of the Republic, and as such, according, to the modern enlightened practices of civilized nations, they would be entitled to be dealt with as prisoners of war.

"At the date when this message is printed this Government is proceeding with deliberate circumspection to determine the exact truth in relation to these reports and upon the course in the premises most consistent with its dignity, its duty to American interests and its moral obligations to Central America and to civilization." On the subject of domestic affairs the president shows the urgent necessity for economy and in order to avoid a deficit for the incoming year the president states he has directed the Heads of Departments to prepare estimates and to make them as low as possible. The result will be a saving of $55,663,000 over the outlay of the present year. A small surplus of $712,000 is expected at the end of the next fiscal year. He recommends the adoption of the old age pension system for government employes although he states he recognizes that there is a strong feeling against its establishment.

Frauds in the custom departments in New York are referred to, the president stating that the government was taking every step to protect the interests of the people and prosecute the wrong doers. He stated that he did not believe that a Congressional investigation of the evils at this time would result in any good and would hamper the work of the Department of Justice. Referring to the tariff the president states that the "Maximum and Minimum" clause gives the executive great powers or retaliation against nations entering into a tariff war with this country but that he knew of no country unduly discriminator. The tariff commission was at work investigating the schedules and states the president, "if the rates of duties imposed by the present law are excessive--I shall not hesitate to call Congress's attention to this fact and indicate the necessity for action predicated thereon."

In the war and navy departments great cuts in the estimates are noted due to the necessity for economy. A cut of $45,000,000 is noted in the war department and $38,000,000 in the navy. The president, however, recommends the building of adequate coast defense at Chesapeake bay but recommends the stoppage of work at Pearl Harbor in Hawaii.

The expedition of legal procedure in the Federal courts is touched upon, the president believing that reforms should be instituted looking towards greater speed and less expense. A special message is promised on the "Inter State Commerce Act." Higher rates for magazines are suggested as one step whereby the postal deficit can be overcome and the president states he can seen no danger to establish financial institutions from the establishment of postal saving banks. Advising that the merits of any constitution be mixed up with the question of statehood for New Mexico and Arizona, the President recommends that both states be admitted.

Under the subhead of "Conservation of Natural Resources" the president states he will soon prepare a special message on this question.

The White Slave traffic is dealt with and the president asks for an appropriation of $50,000 to pay for the activities of agents to punish those involved in the detestable traffic. The semicentennial of negro freedom occurs in 1913 and the president recommends proper celebration. He closes stating:

"I have thus in a message compressed as much as the subjects will permit, referred to many of the legislative needs of the country, with the exceptions already noted. Speaking generally, the country is in a high state of prosperity. There is every reason to believe that we are on the eve of a substantial business expansion and we have just garnered a harvest unexampled in the market value of our agricultural products. The high prices which such products bring mean great prosperity for the farming community, but on the other hand they mean a very considerably increased burden upon those classes in the community whose yearly compensation does not expand with the improvement in business and the general prosperity. Various reasons are- given for the high prices. The proportionate increase in the output of gold, which today is the chief medium of exchange and is in some respects a measure of value, furnishes a substantial explanation of at least part of the increase in prices. The increase in population, and the more expensive mode of living of the people, which have not been accompanied by a proportionate increase in acreage production, may furnish a further reason. It is well to note that the increase in the cost of living is not confined to this country but prevails the world over, and that those who would charge increases in prices to the existing protective tariff must meet the fact that the rise in prices has taken place almost wholly in those products of the factory and farm in respect to which there has been either no increase in the tariff or in many instances a very considerable reduction.

W.H. Taft

The White House. Dec. 7 1909

Our readers and subscribers will notice in this issue, announcement of The Park Hill Heights Realty Company, Denver, Colorado, that on the 15th of December. they will advance the price of

Daily News, Dec 8, 17 and 19 1909

their unsold lots. We, personally, have seen this addition ourself, and will say to our readers, that it is one well worthy of consideration. Any of our readers intending to invest some of their spare money in real estate, cannot do better than to buy at least a pair of these lots, thereby laying a foundation for a home as well as an investment.

NEW IRRIGATION DISTRICT CANAL SOON TO BE BUILT

GRAND VALLEY, Dec. 23.--A corps of engineers will be put to work in a few days to survey the line of the canal which will water 15,000 acres of land in the district which was organized by an election held on December 16. The election held in the different precincts of this district resulted in not a dissenting vote.

After the survey has been completed arrangements will be made to start work constructing the canal at a very early date and as several steam shovels will be used the work will no doubt progress very satisfactorily, The bonds have been contracted for and there will be no delay when the bond issue is once voted.

Federal Building Bill Has Been Introduced

If congress passes the bills introduced by Senator Guggenheim Grand Junction will soon have a new federal building for the senator has introduced a bill among others for the appropriation of $100,000 with which to erect a handsome new government office building in this city. The bill will be pushed through by Senator Guggenheim if it is possible for him to do so as it will be remembered the senator wired the News recently to the effect that he would not only introduce the bill but would do all in his power to have it passed.

Should the appropriation be made work will be commenced on the erection of the building in the near future as the site is already selected and it is known that the federal architect has been working on the plans. There is every reason to believe that the appropriation will be made at this session.

Daily News, Dec 8, 17 and 19 1909

TWO IMPROVEMENT BOND ELECTIONS TO BE HELD TODAY

An election will be held today in the Palisade and Mesa county irrigation districts for the purpose of voting on the question of a bond issue for joint improvements in the two districts. It is proposed that each district vote bonds separately but the money is to be used in making such improvements as will benefit the two districts, the rebuilding of the dam above Palisade being the principal work in view. The bonds, if the issue carries, have been contracted for and work will be commenced in the very near future if the money is voted.

Daily News, Dec 21 and 24 1909

BALLINGER'S CAREER IS TO BE INVESTIGATED

ULTIMATUM IS DELIVERED TO THE PRESIDENT

WASHINGTON, Dec. 20.--- Ballinger this afternoon served upon President Taft virtually an ultimatum to the effect that an investigation is the price of his remaining in the cabinet. It is said Taft reluctantly admitted a disappointment of his hope that the country would accept as final his vindication of Ballinger in the dismissal of the charges brought by L. R. Glavls, former special agent of the office. Ballinger's attitude is supported by leading Republicans of both houses, who feel the festering sore is of such a character as to poison the whole system of the party and that the time is ripe to lance it.

Attorney General Wickersham and Postmaster General Hitchcock accompanied Ballinger carrying the ultimatum to the White House. It is known tonight that other cabinet members sympathize with Ballinger.

Tomorrow is cabinet day and there is little doubt that it will be the important question. Among the things contributing to the demand was a speech by Representative Hitchcock, Democratic, a few days ago demanding action upon a resolution providing an investigation of the land offices and also advice received by Ballinger from a personal friend, recently of the federal department of justice as investigator, urging such an ultimatum. An important aspect, as viewed by the party leaders, is the effect upon the future of the party and Taft's administration should the attacks upon the integrity of a member of the cabinet be made with impunity and allowed to pass unanswered and even unchallenged. There is a possibility of the controversy endangering legislation by bringing to a practical standstill the whole policy of conservation of the natural resources during the conflict between those in whose hands lie the administration of that policy. There is a desire to do justice towards cabinet officers whose personal and official integrity have been assailed.

Daily News, Dec 21 and 24 1909

Ballinger's Wish

SECRETARY BALLINGER'S demand for an investigation should be met with general approval. Ballinger has been attacked both officially and personally in many of our largest most reliable and influential magazines of the country.

He demands that the charges be investigated. He ought to get what he asks for. The attacks upon Ballinger are a source of weakness to the administration. They embarrass the president and are injurious to the whole workings of the Department of the Interior, including all departments of the work of Reclamation and Conservation. If Mr. Ballinger is unfit for the position he now holds, due to corporation leanings, antagonism to our national policies, or because he is by profession a land fraud lawyer, and the last is a very remunerative profession in the northwest, the people and the president should know it and Ballinger should go. If he is none of these things, but has simply crossed wires with some of the more enthusiastic and overzealous officers of his bureau the people and the president should know that also and be able to judge the importance of the attacks made upon him accordingly. Let us have the cards laid on the table so that all may see. The time has come for a show down in the Interior Department.

Henry's Organ

THROUGH the columns of the Whimp we are being entertained with a serial from the fluent pen of T. C. Henry. T.C. is attempting to re-create a feeling in favor of his plans for a High Line canal as against the government project. The government project was never too warmly endorsed by the Whimp, who took occasion last winter to enter into a malignant attack upon Project Engineer Sands, which resulted however, in merely discrediting the Whimp and making the Project Engineer more popular. The people of the Grand Valley are not inclined to turn to Mr. Henry for canal advice nor to his journal for inspiration.

At a meeting of the Water Users' association held yesterday morning, it was decided that Secretary D, W. Aupperle of the association should be sent to Washington in the interest of the High Line canal proposition. Mr. Aupperle will leave for Washington tomorrow.

Daily News, Dec 28 1909

GLAVIS CHARGES CAMPAIGN GRAFT

Says Ballinger Asked to Have Him 'Hold Up' On Alaskan Cases Until After Election of 1908 as Funds Were Coming in Slow

----

INVESTIGATION TAKING ON A VERY LIVELY ASPECT

WASHINGTON, Jan. 28.—The proceedings of the Ballinger-Pinchot inquiry took on a livelier aspect this afternoon when Louis R. Glavis, continuing his testimony against Secretary Ballinger told of various interviews he had had with the latter while he was in and out of the government service. Glavis declared that in one of these interviews in October, 1908, Ballinger told him that he was having a hard time trying to collect campaign contributions, and that two men involved in the Cunningham claims, who had been liberal contributors in the past, declined to contribute because he was angry at not being granted patents for Alaska coal lands. Glavis said Ballinger asked him to hold up on the Alaska cases until after the election, he agreed to do this because he had his hands full in another case. One of the interesting developments of the day was the distribution of a quantity of press matter which purported to interpret and point out the significance of the testimony. It came in an envelope of the American Conservation association of which Gifford Pinchot was recently chosen president.

Reading from Attorney General Wickersham's report to the president on the Glavis charges, Attorney Brandeis, for Glavis, quoted this sentence: "He (Glavis) might have added that he had never taken any action whatever to bring these criminal prosecutions which he advised the land office must be brought before October, 1908, to escape the bar of statute of limitations."

Glavis declared this was not true; that he had prepared the report for Commissioner Dennett in June, 1908, but learning that he was to be in Oregon soon, did not send it. He talked the entire situation over with Dennett, who said he did not think there should be any criminal prosecution; that he thought it was sufficient if the claims be cancelled.
Glavis said, "If I had not been taken off the work in May, 1908, I would have had my final reports in the land office in the fall of that year."

Daily News, Dec 28 1909

AUPPERLE LEAVES IN INTEREST OF HIGH LINE PROJECT

D. W. Aupperle left yesterday afternoon for Washington, D. C., in the interests of the High Line canal. He will also join with the representatives of other irrigation projects in lobbying on the passing of the $30,000,000 bill for the benefit of all the irrigation lands of the West. The bill will come up soon before congress. Mr. Aupperle took with him several boxes of the finest apples in the valley to be given out to the government official and members of the National Press association as examples of Grand Valley's products. He is loaded up with letters of instruction to many of the leaders of both houses of congress as well as other government officials. (Illegible) shows the many advantages of location which this city possesses.

Charley Wallis, the deputy sheriff, went to Pinon Mesa yesterday morning to summon witnesses for the Fruita case,

----

H. J. Nelson was up from Fruita yesterday on business.

T. C. Henry has secured a hearing in another magazine, the Field and Farm, and is a column and a half knocking the High Line and the valley very strenuously. The Field and Farm and a local evening paper are the only publications in the state that will print T. C.'s stuff even at the paid rates T. C. gives to sheets that will take the matter.

A letter from D. W. Aupperle states that he saw Orson Adams on the streets in Washington, D. C. last Saturday.

FENCING OF PUBLIC LAND IS GOVERNMENT CHARGE

SEIBER CATTLE COMPANY SUED

Federal Authority Seeks Destruction of Fences Placed on Lands Unlawfully

----

DENVER, Dec. 27.---The government today filed suits to secure the destruction of fences which have been illegally placed around public lands in Colorado. Along with the suits were those against the Seiber Cattle company of Mesa county, Lafe Young and B. B. Brown of the Montrose district and Ora Haley of Wyoming. A large tract of the land claimed to be illegally fenced in is in Mesa and Montrose counties.

Daily News, Dec 29 1909

LAYTON ENGINEERS RUNNING LINE IN THE MOUNTAIN SNOW

Irrigation Tangle in Norwood Country is Being Cleared Up and Action Promised

News Special Service.

NORWOOD, Dec. 29.--The element of uncertainty has not passed entirely from the Paradox railroad proposition, although there are some favorable signs of its construction, states the Naturita Valley Record. Notwithstanding the fact that the weather is adverse to the promotion of outdoor projects, Engineer Donovan, at Norwood, has engaged the services of three Redlands young men who worked for Engineer Meeker on the big ditch survey. This trio consists of Hugh Jacobs, Jerry Foster and Victor Sharp.

It was planned to enter the field Thursday morning. The Cone region, for some reason unknown to a vast majority of the people who are taking an interest in the enterprise, was the objective point of the force that had made ready to get busy. With the additional snow that has piled up within the last few days it would appear to the novice at least that running lines and setting stakes at this time in the Cone country will have to be abandoned until conditions are more favorable, regardless of the object of the survey.

The irrigation tide of events is a long ways from having entered into one straight and narrow way, under one management, a proposition to be devoutly wished for, if it shall have the effect to do away with the ever-recurring adjudication of water rights, and if the one-management plan shall give to every man his due.

Two new investors in water ways have entered the field, and quickly signalized their presence by the purchase of the Redlands ditch and reservoir. The new investors are Dr. Harvey and son of Trinidad, known as blue ribbon business men of that town. The sale calls for work to begin in not to exceed thirty days from the date of the sale or forfeit rights. One of the middlemen in this transfer is an old-timer here and was very familiar with the pat statement that "the Redlands people have no water," is now in possession of information that has set him right on this point. Inspection of supply, work done and records has led this man to say that the Redlands Canal company had a first-class water proposition.

The Harveys have made an offer for another ditch and have given the owners thereof until the first of the year to make answer.

The stench and the shades of the Gurleys still hover near, if the report be true that they are now out with their grappling iron trying to draw into their net men and money to assist them in developing enterprises in these parts that lag for lack of means.

Daily News, Dec 29 1909

PARADOX VALLEY THE RICHEST UNDEVELOPED SPOT IN STATE

---I .W. McCONNELL

I. W. McConnell, of the government reclamation service, who was in charge of the Montrose tunnel, while there, visited Paradox valley, and on his return expressed the opinion that the "Paradox valley is the richest undeveloped district in the United States.--Montrose Enterprise.

The above statement comes from a man who is familiar with conditions and who has no special interest in the Paradox country. If the capital which is proposing a railroad into that undeveloped country can be made to see this there is no doubt but what the road will be built in the very near future.

The News correspondent at Montrose had an interview yesterday with Civil Engineer Johnson who is now at work on the proposed Paradox railroad. Mr. Johnson states frankly that he has been instructed by his company to get all of the data necessary of the west end of Montrose county. It will probably take six months to do this and do it thoroughly. The building of the road, says Mr. Johnson, will depend upon his report. If the company is satisfied that the country justifies the building of a railroad the road will be built. "It is not a question of money," he stated, "but a question as to whether the railroad through that country will pay. It is a cold blooded business proportion. If the men back of the project believe that it will be a paying proposition it will be built,"

The engineer reported that as far as he had investigated the proposition looks good to him--much better than he had expected.

Daily News, Dec 30 1909

SITE FOR INTERURBAN TERMINAL PURCHASED

Line to Fruita to be Completed by June 1, is Statement of General Manager Sunderlin---Bad Weather is Cause of Delay in Construction

"ALL ABOARD for Fruita, June 1st.'" This was the date set for the opening of traffic on the Grand Junction and Grand River railway between Grand Junction and Fruita by General Manager E.A. Sunderlin yesterday.

President Thomas Curtis and General Manager Sunderlin were in Grand Junction yesterday after a two day inspection trip of their properties about the Grand Valley.

"Work would have been started on the line by this time," declared Mr. Sunderlin, "but for the bad weather. At present we are glad we did not start construction because the cold weather would have delayed us and caused considerable expense.

Superintendent Brinkerhoff expects to have everything ready to begin moving dirt as soon as the cold weather gives way. "I don't expect heavy work to begin until about March 1st. If we do not begin work until March 1st we will have everything ready to open our line for traffic by June 1st and we. confidently plan to open the line at that time."

Word was given out yesterday at the Ramey-Udlock Co., general agents for the Town and Development company, that the street railway had purchased 54 lots in block 154 and 156 on South street along the right of way of the Rio Grande railroad. This sale was confirmed by General Manager Sunderlin.

"We have purchased these lots and expect to add another piece of property to this and on this tract will build a new terminal barn and our car shops.

We are expecting the structural steel for the shops to get here at any time and we will start the work of construction on the buildings as soon as the weather permits. The car barns will be about five times the size of the present barns, permitting the use of six tracks under the roof. We expect to use the balance of the tract for switching purposes. Plans are now being prepared for the barns and shops."

"Machinery is now being shipped into Grand Junction for the power house which, when installed, will increase the power producing capacity of the plant by three times."

Messrs. Curtin and Sunderlin have just returned from a trip to New York City and made this visit for the purpose of inspecting the work on the line and looking over their general interests. They spent two days at Silt where they have the Antlers Orchard Development project which is destined to become one of the big orchard districts of the upper Grand Valley.

The Antlers district will have an exhibit of considerable size at the Denver National Apple Show. Messrs. Sunderlin and Curtin will leave for Colorado Springs today. The recent cold snap has prevented Supt. Brinkerhoff and Engineer Mosley from doing any field work, The work of laying out the permanent line, however, has progressed well towards Fruita. Hundreds of car loads of material are now piled up in the yards of the company and with the opening of construction work the road bed can be laid very rapidly. Superintendent Brinkerhoff estimates that the line can easily be built within three months from the date of the opening of work.

BALLINGER=PINCHOT DISPUTE WILL NOT BLOCK CONSERVATION LAWS

By Associated Press

WASHINGTON, Dec. 29.---"Acts rather than declarations," the president reported in a terse way as he described the program for conservation of natural resources. It is said that the president declared when the present session of Congress ends that he will be able to point to the first practical conservation statutes enacted as an answer to the criticism directed toward the White House. Taft is represented as holding nothing in the Ballinger-Pinchot dispute that has a bearing on his own legislative program. He will proceed regardless of it. He will soon begin the preparation of a special message on conservation and has received the assurance of both houses that while some trouble is expected in the lower house his recommendations will be enacted in law before adjournment.

In his declaration of the message the president will have the cooperation of Ballinger and Senator Nelson, chairman of the committee on public lands. It is asserted that in the construction work of legislation that the president had to depend on Ballinger to draft the proposed statutes to meet the intricacies of practical conservation.

New laws are ready for submission to both houses with the special message.

NEW YORK, Dec. 29.—Pinchot today found support in the controversy with Ballinger from an unexpected quarter. Addressing the Economic and Historic association, Sir Horace Plunkett of Dublin said that he know more about America's resources than 99 per cent of its own citizens and said "I regret Pinchot's business keeps him away. I must decline to enter the controversy but will say in regard to Pinchot that he is a remarkable public servant. He is not only doing the work with great affection for it but is trying to broaden the administration position." Plunkett declared the wastefulness of the American farmers must be changed or lands will be bankrupt before the end of the century.

Daily News, Dec 30 1909

Morgan Given His Preliminary Yesterday

Russel Morgan, charged with the robbery of A. L. Dunham of Montrose one night last week, was given his preliminary hearing yesterday before Justice Sweeney. The case was taken under advisement until Friday, and Morgan was placed under bond but for failure to give bond he was remanded to the county jail.

Banquet at Palisade Tonight

News Special Service.

PALISADE, Dec. 29.~Tonight at the home of Mr. R. H. Bancroft will occur the annual banquet given by Mr. Bancroft to his Sunday school class of the Baptist church.

Estrada Will Continue Revolutionary Campaign

By Associated Press

WASHINGTON, Dec. 29 — Doubting the sincerity of President Madriz in his effort to secure a conference with the revolutionists looking to the establishment of peace, General Estrada has declined to suspend hostilities, according to a cable received tonight by Dr. Castrillo, representative of the provisional government here.

Prominent Pueblo Man Dead

By Associated Press

PUEBLO, Dec. 28.~Henry Herman, aged 65, a brother-in-law of DaniSel Guggenheim, head of the American Smelting & Refining Co., died today as a result of a long illness.

Mercer, PA, Dispatch and Republican, Daily News, Jan 25 1910

BACK TO THE LAND.

Development of West Through Irrigation Is Attracting Many City People to the Farm.

The great development now going on all over the West, opening up to settlement extensive areas of fertile land by the construction of large irrigation systems, is slowly swinging the pendulum of population from the city to the farm. An interesting feature of this shifting of population, and one of profound economic importance to the country, is the large number of citybred people who are seeking homes in the open country. This is accounted for in some degree by the fact that country life in the West has grown more and more attractive as population increases. The farms are small, agriculture is intensive, communities are compact and neighborly, rural delivery is established, the country telephone is in operation everywhere, schools are centralized and social conditions are actually more suburban than rural in character. With eight or more farm houses located along each mile of the highway, there is no isolation or loneliness. With cheap water power from the mountain streams and from huge dams trolley lines are being extended through the valleys. All these factors are bringing the farmer into the stimulating current of the world's daily thought and action, quickening his ideas and broadening his viewpoint. Cooperation in the farm work, in producing and marketing crops, in social and educational activities, have increased financial returns and have made home life attractive and pleasant.

The daily mail of the statistician of the United States Reclamation Service, at Washington, D. C., contains many inquiries from town people who, having grown tired of the never ending grind of commercial life, are seeking the independence and freedom of the country. The successes of those who have ventured into the New West are sufficiently numerous to predicate a continuance of the movement to the farm and should encourage the city man who has laid by a little bank account to establish his family in their own home, where the landlord and the grocer will no longer dominate. A few simple statements from some of these new settlers in the West might be cited as indicating the measure of reward which follows industry and perseverance when applied with moderate capital by men whose experience in farming is limited.

On the Huntley project, in Montana, 350 farm families have located on the 40-acre farms since May 21, 1907. These settlers came from all parts of the United States, although the majority probably are from the Mississippi Valley States. The lands were absolutely virgin, covered with grass and sage brush, and no crops were possible without irrigation. Fully 80 per cent of the newcomers were ignorant of irrigation and an unusually large number had no experience in farming. Notwithstanding these drawbacks the crops of 1909 were as a rule satisfactory, and only a few partial failures were recorded.

W. V. Waller, formerly in the grain commission business in Omaha, Neb., writes: "I like this country and think I have one of the finest farms in Montana. Don't want to sell; expect to live here the rest of my life. I like the climate and never saw things grow as they do here when irrigated. Notwithstanding that I didn't get my ditches in proper shape, my gross return from a 37 ½ acre farm was $809.50. I also farmed another 40-acre tract for my mother-in-law, largely

Mercer, PA, Dispatch and Republican, Daily News, Jan 25 1910

by my own labor, and cropped $864 worth. My oats yielded sixty bushels per acre; from six acres of millet I sold $300 worth of seed."

J. H. Patterson, formerly a candymaker in Omaha, with the labor of his son <?> years old, cultivated 30 acres and sold products valued at $1,583.

J. H. VanderBaan, formerly a furniture maker, from Grand Rapids, Mich., with the assistance of a grown son, cropped 23 acres and received $899.60 for his products. Mr. VanderBaan had a total capital of $1,000 to start with. He has built himself a comfortable farm home, a good stable and a root cellar.

A. Oberhauser, who formerly worked a smelter in Anaconda, Mont., writes that he would not sell his land for $100 per acre. It will cost him $34 per acre, with years to pay for it. He hired labor for forty days. From 32 ½ acres he took crops worth $949 the first year from sod land.

Henry Bliss, an experienced farmer, harvested in 1909 his second crop from 28 ½ acres. From five acres in sugar beets he sold $355 worth; from four acres in potatoes $600 worth; other crops, hogs, poultry and produce, $622.75 worth. Total receipts, $1,575.75. He operated this farm without employing outside labor, his two boys, 11 and 12 years of age, assisting him.

Walter E. Bartlett, formerly from Illinois, located on an 80-acre farm on the Shoshone project, in Northern Wyoming, March 1, 1909. From 12 acres of oats he cropped 1,250 bushels, which he sold at two cents a pound; 15 acres of wheat produced 315 bushels, which brought $1.35 per cwt. on the farm; five acres in potatoes yielded 42,000 pounds which sold at $1.10 per cwt. The balance of the land is in alfalfa and furnished excellent pasture this fall.

Philip Scheve gave up tailoring in Denver and moved to Palisades, Colo., in 1900, where he purchased 60 acres of raw land for $495, all of which he borrowed. In 1901 he set out 16 acres in peach trees. For four years in the summer he worked out for his neighbors to earn enough to live on while his orchard was in growing. In the winter he went back to Denver to engage at his trade. In 1909 he sold seven acres of his orchard for $10,000. He values the remaining nine acres at $22,500. He has erected a residence costing $6,000. From five acres of bearing orchard he harvested this year a crop worth $3,000. Starting with a capital of actually only $25 in cash and borrowed money to the amount of $495, Mr. Scheve has accumulated property in nine years worth not less than $20,000.

L. I. Motz, a day laborer from Guthrie Center, Iowa, came to Grand Valley in 1904, with a family of six and a capital of $60. He borrowed $500 to make a first payment on 20 acres of raw land, and in the next year borrowed the same amount for the second year. In the third year he sold 10 acres of three-year old peach trees for $10,000, and the next year sold the balance for a like amount, making a total return in three years of $20,600 from an investment of $1,060, of which $1,000 was borrowed.

G. W. Bowman, formerly in the coal and lumber business in Leadville, on a capital of $932.50, of which $400 was borrowed, started an orchard in the Grand Valley, Colo., in 1894. In 1900 his crop receipts enabled him to repay his indebtedness and to place to his credit $1,100 besides. From 1904 to 1909 his gross receipts have been $55,000, or more than $9,160 annually, an average of $305 per acre per annum.

Mercer, PA, Dispatch and Republican, Daily News, Jan 25 1910

Approximately 1,000 farms on Government projects, varying from 40 to 80 acres each, are now ready for settlers. The settler goes upon the land, selects his farm and makes his homestead filing with the necessary annual payments for water right and maintenance charges. These vary from $3 to $6 per acre. Some capital, of course, is required, the amount depending upon the experience of the settler and the character of farming he engages in. Probably not less than $1,000 is desirable, and double that amount would be better. A large influx of settlers is expected this spring to take up these farms.

Settlers should not forget that the country is new and being a desert there is a lack of trees and green verdure. The winds in the spring of the year in many sections of the Western country are frequently disagreeable, although severe storms are practically unknown. The soil responds quickly to irrigation and all vegetation grows rapidly so that windbreaks can be provided in a few seasons. Sunny days abound in the arid country, the atmosphere is clear and full of ozone, and the climate is healthful and invigorating.

FAKE SCHEMES IN IRRIGATION INJURES STATE

“Fake irrigation schemes based on Colorado’s land are being foisted on Eastern investors, and bonds are being sold here that may eventually injure Colorado tremendously” said C.C. Magenheimer, secretary and general manager of the Woodland Company of Chicago, in an interview to the Daily Republican of Thursday.

“Most of these schemes are never heard of in Colorado, but projects of meritorious irrigation enterprises are reminded of them in all cities of the East. Your state engineer should be vested with authority to commend or condemn all irrigation projects as soon as they are broached, and a commission should be appointed by the governor to assist the state engineer in this work.

"A number of very bad irrigation schemes have been financed of late by the sale of bonds in the East, and the success of the meritorious projects now being completed in the state allows the unscrupulous promoters to get in their deadly work in all parts of the country. The newspapers, engineers, attorneys and banks of Colorado should get busy at once and see to it that the bad ones are run down. The state should take hold of the matter officially at once.

"Papers like the Wall Street Journal and the Chicago Tribune have contained articles recently bearing on the danger of the irrigation situation in Colorado, and have thrown out warnings to the people

Mercer, PA, Dispatch and Republican, Daily News, Jan 25 1910

of the state to guard their water rights more carefully, and to see to it that Eastern investors are not 'stung' in the manner in which they were by many Colorado mining promoters during the several, booms that have prevailed in the state."

In commenting on the Orchard Mesa project the Republican says:

The Orchard Mesa project, extending from Palisade to Grand .Junction, has been financed and built by the Magenheimers, G. J. Magenheimer being here at this time with his brother. This project is 75 per cent complete and will furnish water during the season of 1910 to 12,000 acres of the richest fruit land in the world.

The power house and the main canal of the Orchard Mesa project are completed, and only two laterals remain to be built. It has been an expensive project to construct, the total cost amounting to $100 per acre. A much larger project is in contemplation by the Magenheimers in Routt county, water rights for its completion having been secured.

PAGE PHOTO: *Mercer, PA Dispatch and Republican, Daily News, Jan. 25, 1910*

Daily News, Jan 15 and 20 1910

GRAVES SUCCEEDS PINCHOT

WASHINGTON, Jan. 12.—President Taft this afternoon appointed Henry S. Graves, director of the Yale forest school, as chief forester of the United States to succeed Gifford Pinchot. Also appointed Albert F. Potter, at present acting forester, as associate forester. Both are known as Pinchot men and have served under Pinchot and both are in sympathy with the policy of his administration.

By Associated Press

WASHINGTON, Jan. 12.--The house did not agree on the senate amendment to the Ballinger-Pinchot inquiry resolution and sent the measure to conference. The Bennett "White Slave" bill passed in spite of objections by Mann, father of a rival bill to sidetrack it. In the senate the greater part of the session was taken-up with the introduction of new bills.

STATEHOOD IS PREDICTED FOR TWO TERRITORIES

By Associated Press

WASHING'TON, Jan. 12.--The bill providing separate statehood for New Mexico and Arizona was considered by the house committee on territories today and will probably be reported to the house tomorrow. Chairman Hamilton said he expected the bill would pass the house but could not predict what the senate would do with it. He had no doubt but what it would become a law at the present session, however.

Daily News, Jan 15 and 20 1910

GLAVIS IS FIRM IN QUIZ

Cross Examination Causes a Stir in Ballinger Case But Develops No Relief From Charges.

By Associated Press

WASHINGTON, Feb, 14.~The Ballinger-Pinchot Congressional inquiry began to bristle with interest today when John J. Vertrees, counsel for Ballinger, undertook the cross examination of Glavis. The climax was reached when Vertrees announced that a box belonging to Glavis and left in the grand jury room in Seattle had been broken open a few days ago and a number of letters were missing from the files of the land office in Seattle, copies of which secondly had been published, were found therein. Glavis angrily declared if any letters were found in his belongings "a frame-up" had been prepared against him. His denunciation of the federal officials who stoop to such a trick to win favor of their superiors called out a demonstration if applause from the spectators. Throughout the hearing the sympathies of the spectators, a majority of whom were women, clearly has been with Glavis. Vertrees announced that C.A. Christianson, Glavis' successor, would be called to testify to the finding of letters. He offered as evidence a letter from Christianson giving the details of the alleged discovery and the committee got into a long wrangle over the admissibility of the matter, which was put over to tomorrow.

In consideration of the executive session Vertrees drew from Glavis the fact and made no charges against anyone in the land office or in the department of the interior. Glavis declared if he had found evidence of corruption he would have taken the matter to the grand jury instead of the president. He said he thought the facts warranted an opinion. Ballinger and Dennett had acted improperly and he charged them with official misconduct.

Daily News, Jan 15 and 20 1910

Committee is Chosen for the Investigation

GOVERNORS TALK ON CONSERVATION OF OUR RESOURCES

By Associated Press

WASHINGTON, Jan. 18---A joint caucus to select the congressional committee which will investigate the Ballinger-Pinchot charges, tonight chose McCall of Massachusetts, Olmsted of Pennsylvania, Denby o£ Michigan, Madison of Kansas, Republicans; James of Kentucky and Lloyd of Missouri, Democrats; Cooper of Wisconsin, led the bold disgruntled Republicans.

Rainey of Illinois who, with James, was named by the Democratic caucus was rejected and Lloyd named in his stead. There was objection to both James and Rainey because they are alleged as being extreme partisanship objectors, claiming to express the view of Taft. Five other insurgents bolted with Cooper, Lenroot, Neuson and Carey of Wisconsin and Davis and Lindberg of Minnesota. In the house tomorrow it is expected the Democrats will vote to repudiate tonight's action.

By Associated Press

WASHINGTON, Jan. 19.---While the house was engaged today in an extended debate of the Mann "White Slave" bill, a consideration of which will be resumed tomorrow. The senate disposing of the district of Columbia appropriation solution calling on the secretary of the navy to report whether, in the present fiscal year, any powder was bought from the trust.

By Associated Press

WASHINGTON, Jan. 19.--"States rights" was the burden of most of the addresses at the conference of governors today with particular reference to conservation of natural resources and the regulation of public service corporations. Governor Shafroth of Colorado read statistic to show the magnitude and value of the water power of the country and the great possibilities of its development. He contended the ownership of land by the federal government in a state was proprietary right and not political or governmental, that the United States held lands the same as an individual and the government land rests with the state.

GOVERNMENT INVESTIGATING STATUS OF HIGH LINE LAND HOMESTEADERS

Two Special Agents Spent Week Here in the Detailed Examination of Claims

MANY SIGN THE FORMAL STATEMENTS

CLAIMING that they are agents of the Department of the Interior two men, J. P. Golden and H. F. Henrich, whose Denver address is said to be at 714 Ernest and Cramer Building, have been busy in the Grand Valley during the last week obtaining detailed information in regard to the exact standing of every local homesteader on High Line government land.

They have also been investigating the coal land filings in the Grand Valley.

The two men work together. They have a complete list of all of the homesteaders in the valley. Each homesteader is approached and asked a series of questions in regard to the date of his or her filing, the time that the filing party has actually spent on the claim, the amount of work done on the land and various other questions.

After the cross examination is complete the questioners' requested the homesteaders to sign the document and swear that it was a truthful declaration. Some of the homesteaders did this and others refused.

Just what the object of the agents are is not known in Grand Junction.

It is stated here that several months ago a certain individual interested in the private construction of the High Line, raised the question of the legality of many of the filings and declared that it was his opinion that the majority of the homesteaders could be ousted off their claims and the land subjected to the Carey Act. It may be that a formal complaint has been made to Secretary Ballinger and that the Investigation is the result.

In an editorial in the Liquor Organ last evening the silly insinuation was made that Special Agent E. A. Keyes of the Department of the Interior, who visited Grand Junction last summer investigating the Grand Valley project as well as all other western reclamation projects, was in reality Jack Ballinger, the son of the secretary, and that Keyes or "Ballinger" knew about as much as a kid on his first tour on the Choo choo cars and the work did was nil." The attack on Keyes is in line with the Evening Organ's well known opposition to government construction of the High Line, which has been demonstrated in the past in vulgar personal attacks upon the person o£ Engineer Sands and the recent publication of a series of letters from T. C. Henry advocating private construction of the ditch.

The facts of the case are that Mr. Keyes is a well known engineer, a man who was an official on the first governmental survey for a ship canal across Nicaragua, later an official in the government service in the Philippines, a Mason in high standing and now holding a position in the Department of the Interior as an inspector of Carey Act propositions.

While here Mr. Keyes, in company with Engineer Sands, made a careful and painstaking examination into the Grand Valley project, occupying several days in field and office work and his report was both comprehensive and favorable to the project .The assault upon Keyes was most unjustified and uncalled for except as an attempt to place further obstacles in the path o£ the Water Users' association in speedily lining up matters so that construction work might be started.

DITCH TROUBLE IN MONTROSE MAY SOON END

By Associated Press

MONTROSE, Colo., Jan. 20.—Appraisers of the Water users association today reported that valuation of the Ironstone ditch at $1,000. This report makes it possible for the unofficiation of private ditches in the Gunnison tunnel district so that the government requirement that all private ditches be under the control of the water users association may be met. All other private ditch companies have been waiting for the Ironstone report.

AUPPERLE CONFERS WITH BALLINGER

==============

Arranges for a Meeting Which May Develop In Something Satisfactory for High Line

==============

BY GUY LA COSTE

WASHINGTON, Feb. 2.--D. W. Aupperle had a conference this morning with Secretary, Ballinger and Attorney Campbell, attorney for the reclamation service, starting negotiations on the matter of contracts which will probably result in a settlement of the difficulties to the building of the High Line ditch in the Grand Valley. Another meeting will be held Friday morning with Attorney Campbell, Bean and Director Newell.

Mr. Aupperle feels that an adjustment will be reached in the near future providing that the government shall accept bonds in payment.

ADMINISTRATION PROGRAM IS REVISED—RECLAMATION BILL SHOVED DOWN ON LIST

WASHINGTON, Feb. 2.--Approximately 769,167 acres of land was designated today by Secretary Ballinger as open under the enlarged homestead act. Of this about 672,840 acres are in Wyoming.

By Associated Press

WASHINGTON, Feb. 2.—The administration program as revised today by leaders in the senate and house and given the stamp of White House approval, comprises the following legislation:

Statehood for Arizona and New Mexico,

Postal savings banks.

To give the president the authority to withdraw from entry public lands.

Federal incorporation.

The creation of a court of commerce.

The creation of a legislative council for Alaska.

Later in the session there may be additional measures for the ship subsidy and providing for certificates of indebtedness of $30,000,000 to raise money for reclamation work. It was stated that the president had the situation well in hand in the house and that Republican "insurgents" and regulars alike, have agreed to put their shoulder to the whole and roll the Taft policies through that body

----

WASHINGTON, Feb. 2:--The senate committee on public lands today reported a substitute for the two administration bills authorizing the president for conservation purposes, to withdraw public lands from entry. A decision to report such a measure was reached after several hearings in which the necessity for such legislation was attacked and defended in a spirited controversy.

A three cornered fight was carried on in the committee. The factions were made up of senators who believe that the president, under the existing law, has the authority to withdraw lands from public entry and senators who have been in favor of the government giving him such power, but who contend that numerous acts of former President Roosevelt were illegal, and senators who have taken a decided stand against any such authority being conferred on the executive.

G. W. Bowman, formerly in the coal and lumber business in Leadville, on a capital of $982.50, of which $400 was borrowed, started an orchard in the Grand Valley, Colo., in 1894. In 1900 his crop receipts enabled him to repay his indebtedness and to place to his credit $1,100 besides. From 1904 to 1909 his gross receipts have been $55,000, or more than $9,160 annually, an average of $305 per acre per annum.

MONEY FOR IRRIGATION PROJECTS FORTHCOMING

WASHINGTON, Jan. 22.--Certificates of indebtedness or bonds to the sum of thirty million to complete and extend the reclamation projects will be authorized in a bill to be reported from the senate committee Monday. Most of the committee voted lot certificates of indebtedness as Cannon says there is little chance of getting a bond issue through the house. Borah of Idaho, framed the bill after a conference with the president. The securities will pay three per cent interest and be guaranteed by the government out of the reclamation funds

----

HIGH LINE CONTRACT ARRIVED YESTERDAY

Document is to be Presented to Irrigation Districts Today for Consideration

----

MAY SOLVE PROPERTY DIFFICULTIES

Final instructions and documents were received yesterday by Project Engineer Sands of the Reclamation Bureau from the Department of the Interior in regard to the proposed contract between the government and the Mesa County and Palisade Irrigation districts by which the government hopes to obtain the right of way now used by the Stub ditch for the High Line while passing through the Palisade orchard district.

These papers will be handed over to Attorney Straud M. Logan, attorney for the district today and a conference will be held between the Reclamation bureau men, the Water Users,' and Attorney Logan. Upon the decision of Attorney Logan the fate of the proposed combination largely rests.

The principal point at issue is the legality of the proposed merger in the light of the state irrigation district law. The first proposal was considered faulty by some in this regard and it is hoped that the new papers will clear all of the disputed points up and permit the proposition to be brought before the stockholders in the district.

Daily News, Jan 30 1910

"I BECAME CONVINCED IN 1909 THAT DENNETT WAS CROOKED", SAYS GLAVIS SAYS BALLINGER PROTESTED AGAINST TOO MUCH INVESTIGATION

WYOMING COAL LAND CASES ARE TO BE PROBED

By Associated Press

WASHINGTON, Jan. 29,—Louis R. Glavis concluded his testimony before the Ballinger-Pinchot investigation committee late today. He, will be recalled Monday afternoon for cross examination by the person against whom he has made his charges.

The testimony today developed that animosity long existed between the land office and the forestry bureau and the resentment of the land office people had started when Glavis called it "the forestry" as they termed it. The witness also declared representatives Mclachlan of California and Kinkaid of Nebraska were interested in Alaskan claims and that Ballinger, after being commissioner of the land office acted as attorney for Kinkaid.

Glavis asserted Ballinger suggested to him not to pursue an investigation against Mclachlan, saying there had been too much of that sort o£ thing in the past. When Glavis was asked the direct question if he thought Ballinger and Commissioner Dennett were in 1eague to do wrong in the Alaskan cases he replied, "Well I thought the cases would be better protected with them out of the way." He said he became convinced in the summer of 1909 that Dennett was "crooked" and took steps to secure carbon copies of Dennett's writings to Washington. Several of these letters were introduced as evidence.

By Associated Press

WASHINGTON. Jan. 29.--Official disclosures made today are that more than a million dollars is the estimated of the coal lands in Lander, Wo., a district which the interior department is trying to have cancelled ground as it alleges, the lands were entered pursuant to a conspiracy whereby the title was to go to an incorporation. All entrymen in the case reside in New York while the principal movers alleged a conspiracy had been indicated by the grand jury in Wyoming.

MA.IOR RICHARDSON IS TO BE RETAINED

By Associated Press

WASHINGTON, Jan. 29.—Secretary Dickinson today informed Alaskan Delegate Wickersham that he could not accede to his request that Major W. P. Richardson of the army be relieved from duty at Washington, as his services were required.

Wickersham based his request on the ground that the major was lobbying, "for his special work in Alaska" and

Daily News, Jan 30 1910

against the people. The major threatened Wickersham when he heard of the charges, Wickersham stated.

SOUTHERN MEMBERS WANT BOLL WEVIL APPROPRIATION

By Associated Press

WASHINGTON, Jan. 29.~Efforts by the Southern members to increase the appropriation carried in the agricultural bill for the study and demonstration of methods for controlling the Boll Weevil furnished the chief feature of interest in the house today. Sime of Tennessee, referring to newspaper criticism of Congressmen on account of the alleged misuse of the franking privilege, defended his colleagues against the accusation.

The senate was not in session.

ALASKA IS GEM SAYS BALLINGER

PROVINCE, R. I., Jan. 29.--"Alaska is a prize package, full of value which no man can yet estimate," said Secretary Ballinger before the conservation club of Rhode Island and the Economic club of Providence, here tonight. The secretary said the homestead statute signed by Lincoln was one of the most beneficent ever enacted for the upbuilding of a certain section of the country but an absurdity in forest lands of Alaska.

Daily News, Feb 4, 8 and 15 1910

SPECIAL ACT OF CONGRESS MAY PAVE WAY FOR COMPLETION OF GOVERNMENT CANAL PROJECT

Secretary of Interior, Colorado Senators and the Local Secretary of Water Users Hold Conference in Washington and Discuss Plan

----

CONSTRUCTION WOULD BEGIN EARLY IN SPRING

BY GUY LA COSTE

Special Correspondent in Washington for The Daily News.

WASHINGTON, Feb. 3.--If Congress passes a special act authorizing the Department of the Interior to accept the bonds of the Palisade Irrigation district as payment for the use of water from the High Line canal within the next thirty days, as is expected here, construction work on the Grand Valley project will begin about April 1.

As the result of a conference held in Washington today between Senator Guggenheim, Senator Hughes, D. W. Aupperle and Secretary Ballinger it was determined to take one of two courses satisfactory to the people in the Grand Valley most interested. One is to have the special act passed and the other is to accept the bonds of the Palisade Irrigation district as security on the contract for payment. Either of these propositions is satisfactory to Ballinger. In regard to the latter proposition it is stated that the Colorado irrigation district law prevented a district from putting up its bonds as security and that it is probable the first plan will have to be adhered to. If so the bill will be introduced in both houses at once and all of the influence that Senator Guggenheim and Senator Hughes can put behind it will be exerted to secure its passage.

Daily News, Feb 4, 8 and 15 1910

The following interesting bit of news is taken literally from the Denver News of Thursday and bears a Grand Junction date line:

"Telegrams received here this morning indicate that immediate work will be started on the High Line canal. Advices from Washington are to the effect that D. W. Aupperle, representing the Water Users' association, today, reached an agreement with the government for the taking over of the Palisade and Mesa district ditches and that the contract will be approved in Palisade. Secretary Ballinger has prepared a bill by which the government will be permitted to take over the bonds of the irrigation ditches. This was the only obstacle in the way of a settlement of the difficulties, and work may start this spring."

From this it would appear that our old college chum Aupperle has been fixing things up for the Palisade people so that all is left for them to do will be to obey orders and look cheerful. It reminds one of the old song which runs something like this'. "It may be so, it may be so, but it sounds to me like a--" etc. Let's wait a bit, D. W., and see how Palisade ranchers take to it. —Palisade Tribune.

For the second time since the establishment of the local weather station eighteen years ago, a faint auroral light was visible in the north west Sunday and yesterday morning

SECRETARY TO BE GIVEN RIGHT

MONDELL BILL CONTAINS CLAUSE PERMITTING IRRIGATION DISTRICTS TO PAY IN BONDS REPORTED OUT SOON

By Guy La Coste.

WASHINGTON, Feb. 15.--The public lands committee has heard the report of its special sub-committee and agreed to include the Mondell bill a clause giving the secretary of the interior authority to accept bonds in payment for irrigation districts, This is the clause that Secretary D. W. Aupperle is anxious to have become a law in a hope that this will solve the right of way difficulties for the High Line canal in the Palisades districts. The Mondell bill will be reported out of committee shortly.

Daily News, Jan 11 1910

Taft and Conservation

LATE Associated Press dispatches from Washington, D. C. announce that President Taft is hard at work upon his conservation message which he promised in his annual message several weeks ago. The purport of the advance rumors concerning the message are of a very encouraging nature to the men of the west. When President Taft was in the Grand Valley last fall he expressed himself as marvelously well pleased with the exhibition of the results of irrigation and reclamation His first opportunity to view the fruits of irrigation was at the Mesa County Fair and his admiration was unfeigned. Later on his way to the Northwest and again in the south he had many similar chances to observe the tremendous value of reclamation and conservation. Before the President completed his wonderful 13,000 mile swing about the continent he became an enthusiastic advocate of the work.

At the time of the penning of the annual message it was the hope of President Taft as well as his closest advisers that the Ballinger-Pinchot row would come to a head and pass over before the first of the year. This, unfortunately, is not a fact, and the president has determined to prepare his message on conservation without any reference to the coming investigation into the Department of the Interior and the Forestry Bureau.

That the president will recommend at least a $10,000,000 appropriation, bond or warrant issue, for the purpose of rushing to completion all government works in the Reclamation Service is assured. He promised us this and the president has shown a splendid capacity for remembering and holding to his promises. Likewise he will probably urge the formation of a definite Waterway policy in regard to the deepening of navigation on the Mississippi and Missouri rivers The reopening of navigation on the Missouri is of importance to the men of the Grand Valley as water transportation from Chicago via the Chicago-St. Louis canal and from St. Louis to Kansas City should mean a great lowering of freight rates into this territory as it will reduce the freight from the Chicago common points to the Missouri common points many dollars on the carload.

The Republican party has long supported the policy of Conservation and Reclamation. In its last national platform, the platform of 1908, it specifically endorsed the work of reclamation. The Democratic platform took an antagonistic position. Indeed, the fundamental theory of the Democratic party in regard to the activities of the federal government is opposed to federal aid in reclamation and conservation. The wildest eyed opponents of Gifford Pinchot and the forestry bureau in the west are the leading representatives of the Democratic press.

The fact that Colorado sent to Congress in 1908 a Democratic Congressional delegation has had a most unfortunate effect in Washington upon the government's policy of Reclamation and Conservation in this state. The leading exponents of anti-Pinchotism and anti-Reclamation are residents of Colorado. Senator Ammons and his anti-Pinchot campaign and T. C. Henry, whose politics extend no further than his own personality and a consistent opposition to the present administration, have given Colorado a reputation of not wishing the federal government to do any work in this state. The Rocky Mountain News has spread this idea in its policy of detraction of the Forestry Bureau. Having no Republican representatives in Congress to fight this impression the state has been helpless. Indeed if it were not for the energy and hard work of Senator Guggenheim, who is doing five men's work for Colorado at Washington, Colorado would be in a bad way indeed. The Grand Valley is looked upon as one of the spots in the state consistently friend-

Daily News, Jan 11 1910

ly to government construction yet the recent publication of a series of letters opposing the government construction of the High Line canal in the local Democratic organ has operated to spread a feeling that even the Grand Valley was changing in its opinion, which needless to say, is just the effect that T. C. Henry, the author of the articles, and his publisher wished to convey.

The President, however, was given ample assurance of the desire of the Grand Valley to see the government build the High Line and the Senatorial Committee were given another dose of the same doctrine. Whatever President Taft has to say in regard to Reclamation will be of deep personal interest to us in the Western Slope as it will bear directly upon our fortunes. The message will be awaited with eagerness.

DAMAGES FIXED AT SUM OF $174,042

City Officials to Discuss Probable Steps at Meeting to be Held Today

----

EXCEPTIONS TO BE FILED BY FEB. I

THE question of what action the city will take on the report of the Mountain Water referees will be taken up at the meeting of the Commissioners this afternoon. Of course the report of the Commissioners was much higher than any one of us had estimated it would be— higher by $100,000 than many of us believed possible. You can rest assured however, that the Commissioners will not be deterred in their aim to give Grand Junction a Mountain Water system by the report. The report of the referees place the valuation so high that it is doubtful if the city could float the bonds for the system as the initial expenditure for the rights will raise the original estimate so much higher than ever contemplated. "Yet the people of Grand Junction want Mountain Water and they will get it."—Statement of Water Commissioner Vedder.

Judge Sprigg Shackleford met with the Commissioners in the Kannah Creek water case yesterday morning and received their report and while the announcement of the exact amount was way beyond what it had been expected it would be it occasioned no great surprise yesterday for The News Saturday had published an exclusive story to the effect that it would be in the neighborhood of $170,000. When the announcement was made yesterday that the damages had been placed by the commission at $174,042 those who are vitally interested and that means most every resident in the city, were prepared for it. After The Daily News report of Saturday had been read the Kannah Creek case was a much discussed subject and many have since wondered what action the city officials would take in the matter.

The attorneys for the city have been given until February 1 to file their exceptions and the case will come up for a hearing at the opening of the March term of the district court.

The valuation is regarded in Grand Junction as very high. At the same time the testimony given in the district court before the Commissioners will give the County Assessor an excellent basis for a re-valuation of Kannah Creek property and the resultant taxes will enrich the county treasury by a nice sum.

Discussion of the report will be thoroughly gone in to at the Commissioners meeting this afternoon and a line o£ action outlined by the officials.

A Test

SECRETARY BALLINGER, in his annual report, made public Monday , morning, comes out in favor of the issuance of $30,000,000 in bonds for the carrying on of the work of the reclamation service. As the secretary points out, the bonds could be issued against the various projects and so be amply secured. They would in no way injure the government's credit, being a lien against the lands actually reclaimed, and payable by the owners of those lands.

It is difficult to see how even Cannon reactionism can find objection to this plan, and yet we hear that the czar of Danville has already announced that he will not permit the passage.

IRRIGATION BONDS ARE SOON TO BE ISSUED

Such is the Statement of Private Messages Received in Denver Yesterday--Taft Said to be Pushing $30,000,000 Fund for Projects

----

News Special Service.

DENVER, Feb. 10.---It was rumored here today that private messages were received in Denver to the effect that the $30,000,000 bond issue for irrigation projects in thee West would be taken up and acted upon favorably at this session of congress without a doubt.

It is said that President Taft is strongly urging that this matter be taken up and acted upon favorably, which comes as a result of his trip through the West and especially through Colorado last September. While the messages are said to be private ones there seems to be credit placed in them to the extent that the news was joyously received here. The report is also to the effect that the Grand Valley project in Mesa county is to get $750,000 annually for the completion of its High Line project.

Daily News, Jan 7 1910

CLARKE WOULD AID GRAND JUNCTION IN DEVELOPMENT

D. & R. G. OFFICIAL REITERATES POSITION AS TO LOCAL AFFAIRS

States He Will Make City His Temporary Headquarters at Fruit Shipping Season

----

MANY IMPROVEMENTS OUTLINED FOR SYSTEM

Vice President Horace W. Clarke of the Denver and Rio Grande railroad arrived in Grand Junction yesterday and went on record to The Daily News as being very much in earnest last Tuesday when he wired The Daily News that he was heartily in sympathy with any movement looking towards securing wholesale establishments or factories for Grand Junction.

"I believe Grand Junction is one of the booming towns of the state," he said, "and the Rio Grande wil1 go to any reasonable length to assist in its development. I will say that no one has ever came to me from Grand Junction and ever suggested any reductions in rates or anything else that the road might do to help put jobbing houses in this city. I have little to do with rates directly but I will say that I will guarantee to lease locations for factories or wholesale houses in the railroad property in Grand Junction along the tracks for $1.00 per year for a term of five years or even longer."

"We are planning to make the Rio Grande system one of the best in the country The road has had trouble of its own of late but we are changing all of that and improving the service wherever we can. If you've got any kick coming send it over to me. I want to hear about it.

"Next year I'm coming to Grand Junction about the middle of August when the fruit shipments begin to get heavy and live in the city, live right here in the yards on my car, and personally see that the shipments are moved out of here as they should be with no trouble about cars. I will make Grand Junction my headquarters during that period. We intend to see that in the future every car of fruit leaving the Western Slope gets out when it should and in the right way."

"I want to express my sorrow that the Grand Valley Special, engaged by the Mesa County Business association had the misfortune to be delayed en route to Denver. Accidents such as befell it cannot be avoided on any system but we worked hard to get your people the proper kind of a train and were disappointed that they did not get over as we planned they would."

The Clarke party will leave Grand Junction early this morning for Salt Lake.

PINCHOT STIRS TAFT TO ANGER

By Associated Press

WASHINGTON Jan. 6.--The Ballinger-Pinchot controversy was intensified today by the reading in the senate of a letter addressed by Pinchot to Senator Dolliver in which the course adopted by L. G. Glavis with the assistance of Price and Show of the bureau of forestry was warmly approved. Pincher suggested that the president was mistaken in facts when he removed Glavis whom he describes as a "most vigorous defender of the people's interests," despite the fact that the president declares that Glavis was unfit to remain in the public service.

Hale rebuked the chief forester severely for having ignored the recent order of the president because no subordinate officer should give information concerning affairs of the government except to superior officers.

By Associated Press.

WASHINGTON, Jan. 6.--After reading the newspaper reports of the Pinchot letters to Dolliver late today, Taft sent a hurry call to the cabinet members. Knox, McVeag, Wickersham. Wilson responded and said the letter was freely discussed though none of the cabinet officers would discuss the matter afterwards. It is reported that the letters aroused Taft to keen resentment and some action of the incident is not unexpected.

By Associated Press.

WASHINGTON, Jan. 6.--Insurgent Republicans of the house are not intimidated by the action of the administration in withholding patronage from them and are not scared into silence by any apparent combination of President Taft and Speaker Cannon.

This was the declaration Representative Norris of Nebraska one of the leading insurgents. Norris resented the stories printed to the effect that the insurgents are "lying down" under .the edict of the. administration "How can the president say we opposed his policies and declare that we are to be punished for it," he continued, "when we had not opposed his policies anyway. We do not yet know what his policies are to be. We have not even received his message on several important subjects." The fight of the insurgents in the house will continue

Daily News, Jan 7 1910

against Speaker Cannon and the house rule. We will not, however, oppose any policy of the president so long as it is Republican doctrine."

ALASKAN TRIP STILL IN VIEW

WASHINGTON, Jan. 6.--President Taft is still looking forward to a trip to Alaska in the spring. Today he declined to accept an invitation from a committee of New York congressmen to visit New York May 30, saying that he might be on his way to Alaska.

May you be enabled to continue throughout the year, the good resolutions with which you began it. We have pledged ourselves anew, to a better service for our customers and a greater business institution for the Valley.

There are many changes in view for us this year.

Methods the being perfected and equipments and working force increased, until with your help, we trust to enlarge our business, until we have far more than a local renown.

You have your part to do in helping us attain this ambition.

A determination to give us your trade, this year, will be just the uplift we need.

With the best wishes for a happy New Year, we are,

very truly yours,

THE C. D. SMITH DRUG CO.

Daily News, Feb 8 1910

KIEFER WINS REDLANDS CASE

Final Court of Appeal Reverse All Other Adverse Decisions

CLAIMANTS WIN ALL

Early Construction of Main Street Bridge Now Made Possible by Secretary's Action.

THE REDLANDS contest case has been won by the original claimants.

The news of the decision was received in Grand Junction yesterday in a dispatch from Attorney John S. Lacey at Washington to B. W. Kiefer, vice president of the Redlands Irrigation and Power company, and was immediately made public by The Daily News who bulletined it on the streets. Until The News bulletins were flashed no other news agency in the city had any information in regard to the final settlement of this famous case.

The decision gives back to the original claimants all their contested rights to the land. Seven of these claimants filed final proof last July and others will file final proof next summer. The decision is considered a most important victory and one in which every good citizen in Grand Junction cannot help but rejoice. The settlement of the suit will mean the quick development of the Redlands district which comprises about 6,000 acres of fine land, the early construction of the Main street bridge and a great deal of development and improvement work on the Redlands side of the Grand river.

The suit against the Redlands company was first announced late in August, 1907, when a group of individuals set up claim to the lands on the Red Mesa declaring that the filing of the first claimants were faulty. Many sensational allegations followed. The first hearing of the suit took place before Register Blair in Montrose in the following September and resulted in a decision against the company. The case was then taken to the Land Office where another unfavorable decision was found against the company. Then Redlands people persevered, confident that they were in the right and in the final court of appeal before Acting Secretary Pierce of the Department of the Interior the decisions of the lower courts were reversed and titles given back to the first claimants.

The victory of the company is only saddened by the fact that Frank Kiefer the original promoter of the project and its driving force from the date of its inception, died last fall, Speaking of the decision, Vice President B. W. Kiefer stated last evening:

"The decision was expected. We knew that right would prevail in the end. The result of this suit means a great deal for the future of the Redlands. One thing it means is the early construction of the Main street bridge. Many other matters will also be taken up and completed which have been delayed by this legal tangle. Within a few days we expect to make a formal statement to the people

Daily News, Feb 8 1910

setting out in full the company's position and its plans for the future in order that no further misrepresentation may do us any harm"

President Carl Osborne of Fruita was immensely pleased over the verdict.

"It pleases me more than I can tell," he said. "It has been a long and terribly hard pull but we won out as I always felt that we would."

REDLANDS HISTORY

The story of the Redlands case is a story of Western development in all of its phases.

A fertile mesa lay neglected and despised for twenty years. It was called the Red Mesa and many had declared that it was doomed to be desert for ever because of the high cost of placing water on it.

A daring Grand Valley pioneer, Frank Kiefer, interested a number of friends in a project to redeem it. Eastern capital was brought in and the great ditches built, the largest power plant on the Western Slope constructed, and hundreds of thousands of dollars expended in an effort to make this rich district one of the fairest gardens in the Grand Valley.

Thus the profit prospering, certain individuals believing that they saw a chance to obtain this land, in the western vernacular "pumped" the claims after planning one of the most remarkable raids in the history of the West.

Testimony at the first hearing brought out the fact that a deliberate effort was made by these men to muzzle the press by offering the various editors forty acre sections in case the raid was a success. It is on record in the courts that the three newspaper editors of Grand Junction were so approached and to the eternal credit of the then editor of The Daily News, Mr. E. A. Sherman, he refused the bribe and later on the witness stand exposed the whole nasty mess.

The original promoters in the Redlands Land and Irrigation company were Frank D. Kiefer, D. T. Stone, B. F. Kiefer, Robert Orr, A. J. McCune, William Osborn and others. About April, 1906, A. A. Piatt of Lansing, Mich., was interested and the company formally organized. Work went ahead in splendid fashion.

Then came the blow. On August 9, 1907, the announcement that the government was going to attack the Redlands claimants on the ground of conspiracy to defraud was flashed. It was even hinted that certain ones would be compelled to go to jail. Every effort was made by certain ones to frighten and intimidate people who were among the claimants. The sentiment of the general public were so pronounced against the efforts to paralyze the operations of the Redlands company that the editor of The Herald, who had considered the forty acre offer withdrew from the deal. One or two others also quit. But the case went on to a hearing before Register B. P. Blair at Montrose. The decision of the register was against the Redlands company. Then the case was heard by the Land Office and again, but on other grounds, the case was declared against the Redlands company. The Redlands took it up to the Secretary of the Interior.

Acting Secretary Pierce heard the case last winter. Governor Shafroth, Attorney John S. Lacey of Iowa and Attorney J. C. Carnahan of Grand Junction prepared the Redlands case for the final test. The result has been what has always been confidently expected by the Redlands people and their friends--that

Daily News, Feb 8 1910

the government would see that justice was done.

The reversal of the decisions of the lower officials means much for Grand Junction. It means the quick and fast development of the fertile Redlands with its 6,000 splendid acres. It means the quick building of the Main street bridge and the development of a fine territory southwest of the city. Had it not been for this suit this development would have been accomplished by this date but although delayed its realization will soon come about.

The Redlands plant and ditches were constructed by Engineer A. J. McCune, formerly state engineer and one time resident of Grand Junction. The plant is considered by all to be one of the finest in the West. Howard Platt of Grand Junction is the resident engineer of the plant now. Carl Osborne of Fruita is president of the Redlands company and Carnahan & Griffith are the company's attorneys.

News Special Service.

WASHINGTON, Feb. 7.--While Secretary Ballinger stated that the decision in the Redlands cases had not yet been formally made public yet the unofficial information at the department is that the Redlands contest cases have been won by the original claimants. The full text of the decision will be made public within a short time.

News Special Service.

DENVER, Feb. 7.--~Governor Shafroth today received a dispatch from the Department of the Interior stating that the verdict in the Redland contest case in which he with John S Lacey of Iowa and J. C. Carnahan of Grand Junction had appeared as attorneys for the original claimants had been awarded in the favor of his clients and that a full copy of the decision would be mailed to him in a few days. Governor Shafroth was elated over the victory which he declares to be one of the most signal victories ever won before the Department of the Interior

Denver Republican, Feb 4 and Daily News, Aug 7 and 9 1910

BIG GOVERNMENT PROJECTS IN RECLAMATION ARE OUTLINED

SECRETARY BALLINGER GIVES A STATEMENT SHOWING HOW THIRTY MILLIONS FROM BONDS WILL BE USED FOR IRRIGATION.

WASHINGTON, D. C., Feb. 4—(Special.)--The enormity of the work undertaken by the reclamation service can be appreciated from reading a letter written by Secretary Ballinger to Chairman Carter of the senate committee on irrigation, setting forth the amount of money needed to complete projects that are now under way, and the amount that will be necessary to develop these projects to the limit of their utmost possibilities. It appears from this letter that the Yakima project, in eastern Washington, is destined in time to be the largest and the most expensive of all projects built by the government, for in addition to amounts heretofore expended, $1,424,000 is necessary to complete those portions of the projects already mapped out, while an additional $14,000,000 will be required to fully utilize the water that is available, or can be made available by storage in the Yakima drainage basin.

In his letter to Senator Carter, Secretary Ballinger gives a statement showing how it is proposed to utilize the $30,000,000 proposed to be raised by a bond issue, or through the sale of certificates of indebtedness, redeemable from the reclamation fund, and gives an intimation of what is to come in later years when this $30,000,000 has been expended.

To complete the projects according to approved plans, he says, will take a little more than two years; to complete them in their entirety, carrying out work not yet mapped out, would require six years or more.

The secretary's letter, which forms the main argument in support of pending legislation, follows:

Secretary of Interior
Furnished Data on Projects

"The reclamation service reports the following data with reference to amounts necessary to complete the approved portions of the reclamation projects now under construction, and the extensions of such projects which could, with advantage, be made were funds available for that purpose:

Denver Republican, Feb 4 and Daily News, Aug 7 and 9 1910

State	Project	Approved Portion	Extensions
Arizona	Salt River	$ 395,000	$ 275,000
Arizona-California	Yuma	1,290,000	....
California	Orland	20,000	750,000
Colorado	Grand Valley	2,637,000	....
Colorado	Umcompahgre	2,545,000	....
Idaho	Minidoka	650,000	5,500,000
Idaho	Fayette-Boise	2,337,000	10,000,000
Montana	Milk River	1,857,000	4,000,000
Montana	Huntley	80,000	....
Montana	Sun River	108,000	7,500,000
Montana-N. Dakota	Lower Yellowstone	338,000	....
Nebraska-Wyoming	North Platte	1,100,000	6,000,000
Nevada	Truckee-Carson	182,000	2,000,000
New Mexico-Texas	Rio Grande	8,790,000	....
North Dakota	Williston-Buford	93,000	261,000
Oregon	Umatilla	50,000	2,500,000
Oregon-California	Klamath	1,397,000	3,000,000
South Dakota	Belle Fourche	890,000	....
Utah	Strawberry Valley	1,033,000	....
Washington	Yakima	1,424,000	14,000,000
Wyoming	Shoshone	2,922,000	....
		$30,138,000.00	---------------
Total		$30,138,000	$55,786,000

"In the foregoing table the amounts in the column 'approved portion' are the amounts which should be added to the expenditure made in 1909, plus the allotments for 1910. The amounts in the last column are only approximate estimates.

The estimate for the Minidoka project, above, applies to the south side only, and does not include irrigation from Jackson lake storage. The $14,000,000 estimate for the completed Yakima project includes the Kittitas and Benton units.

"The average annual receipts into the reclamation fund from 1901 to 1908, inclusive, were $7,237,364; the estimate receipts for 1909 are $7,1771,809; the estimated receipts for 1910 are $7,293,000.

"Assuming that the receipts from the sale of public lands, and from repayment of building charges, will average $7,000,000 per year in future, an additional annual appropriation of $7,000,000, under a bond issue, or otherwise, making $14,000,000 per annum available, would permit of the completion of the approved portion of the projects named in the foregoing table in a little over two years.

"The director and chief engineer of the reclamation service advise me that this amount could be economically and wisely expended. They also advise me that, upon continuance of the same annual addition to the fund, the extensions described could be completed in four years more.

"The table headed 'approved portion,' covering an estimated expenditure of $30,138,000, represents the projects or portions of projects approved by my predecessors and myself. The table

Denver Republican, Feb 4 and Daily News, Aug 7 and 9 1910

headed 'extensions,' covering an estimated expenditure of $55,786,000, has not been approved by me, and it is possible that, in working out the matter when the work is actually reached, the amount could be decreased by the elimination of some of the proposed extensions.

Favors Early Completion of Projects Now Under Way

"As recommended in my annual report to the president, I am heartily in favor of an early completion of existing projects, and of prudent extensions, where such extensions appear necessary to reclaim arid public lands, and to utilize water supplies created in connection with approved projects. The early completion of these projects will, I hope, result not only in meeting the wishes of settlers and prospective settlers upon the lands, but secure an early return to the treasury of the moneys expended in the construction of the reclamation work. It is evident that the sooner the projects are completed and the water available, the larger will be the return to the fund and the greater the contribution to the nation's food supply from the products of these heretofore barren lands.

"The present receipts in the fund are not sufficient to enable this department to conduct the construction of these works to the best advantage, as it is hampered by the insufficiency of moneys to carry forward the construction of many of the projects already approved. I am of the opinion that the suggestion that congress authorize the issuance of certificates of indebtedness or bonds against the reclamation fund, 'to an aggregate of not exceeding $30,000,000, or so much thereof as may be needed,' would, if enacted into law, enable me to complete within a little over two years the approved portions of existing projects, and make the more necessary extensions. The other extensions could be left for future consideration."

THE IDEAL FRUIT VALLEY OF THE UNITED STATES

C. J. BLANCHARD OF RECLAMATION SERVICE VISITS CITY WITH NEWARK NEWSPAPER MAN.

ARE SHOWN THROUGH VALLEY

C. J. Blanchard, statistician of the reclamation service, and H. B. Walker, a Washington representative of the Newark (N. J.) News, spent Sunday and Monday in Grand Junction and the Grand Valley inspecting the High Line canal and the valley.

"I can only say that I know of no reason why work should not soon commence on the Grand Valley project, stated Mr. Blanchard last evening. "Of course the army engineers will soon report on this project. I don't believe the people of the Grand Valley realize how favorably the Grand Valley impresses the visitor and how much superior the Grand Valley is to many of the valleys where the government has projects. This cannot but be a tremendous advantage to the Grand Valley project as it shows the great need of the canal. I believe that in the Grand Valley a wonderful development is forecasted.

Denver Republican, Feb 4 and Daily News, Aug 7 and 9 1910

You who are here don't realize it like we do who are traveling from one district to another and see them all. You accomplish more and do more than in any other valley I know of. I know from talking with your people you don't realize how much you have really gone ahead in the last few years." Mr. Walker accompanied Mr. Blanchard here to secure material for a series of special articles for his paper on the Grand Junction charter. He has been making a study of charter government throughout the United States and declares the Grand Junction charter much ahead of all others.

Griffith to Denver.

Formal petitions will be filed in Denver Monday by Attorney Ben Griffith to reopen the matter of the filings of the Water Storage & Supply company and the Inter Mountain Water company upon waters of the tributaries of the Grand river. Garfield and Mesa counties will jointly bear the expense of the legal proceedings. Attorneys Logan and Miller and Attorney Griffith have prepared the papers which will start the big suit which means so much to the water users of this district.

Daily News, Feb 24 and 27, March 3 1910

SENSATIONAL CHARGES ARE MADE BY PINCHOT AT HEARING

SAYS BALLINGER IS 'WILFULLY DECEIVING THE PRESIDENT'

Witness Characterized Glavis as a Faithful Public Servant and Endorsed His Testimony

BALLINGER'S ATTORNEY OFFERS SOME OBJECTIONS

By Associated Press.

WASHINGTON Feb. 26,—With Gifford Pinchot on the witness stand, the Ballinger-Pinchot inquiry entered on its second phase today. Before being sworn, Pinchot dramatically declared that when his story had been told the country would demand a verdict "in harmony with the moral conviction that the secretary of the interior had been unfaithful to the public, whose property he endangered, and the president, whom he has deceived." Pinchot accused Ballinger of having made an explanation of his conduct to the president that was "essentially false." He charged him with being "a dangerous enemy to conservation" and having made a statement "which will be shown by undisputed documentary evidence" to be absolutely false in three essential particulars. He charged him with "willfully deceiving the president" and being disloyal to the president. Pinchot had not progressed very far, however, when there was an objection by Ballinger's attorney as to the witness repeating conversation with Taft. It was contended that the relation of these conversations put the president's attitude where they have remained silent or appear before the committee as a witness which is declared undesirable.

The matter was put over for consideration by the committee and it is expected a decision will be announced Tuesday. Pinchot followed up the opening statement, which was a vigorous attack on Ballinger, by declaring that he believed in Glavis and was convinced that. all Glavis said was true. He characterized Glavis as a "faithful public servant" and declared that the facts which he presented "proved Ballinger unfaithful to his trust as a guardian of public property of enormous value." The conservation movement begun under Roosevelt progressing splendidly up to that time Taft and Ballinger took office, declared Pinchot. In less than a month thereafter, Ballinger had practically broken the back-bone of the central idea of the conservation movement by restoring previously withdrawn water power sites to public domain and laying them open to private appropriation and monopolistic control, Pinchot declared that when restorations were made Ballinger gave no hint that he would rewithdraw power sites and did not rewithdraw them until after Pinchot had gone to the president and made a protest. One of the most dramatic incidents of the day came when Pinchot declared there was no such decision by the comptroller of the treasury as cited by Taft in a letter to Ballinger, dismissing Glavis' charges and authorizing the removal of Glavis. The president had contended that Ballinger had acted under a decision of the comptroller which permitted no appeal, when he abrogated a cooperative agreement with the Agricultural department wherein the forest service was given control of the forest reserves on Indian lands. The implication that Taft had either been misled or was utterly mistaken brought out a rapid fire of questions from Senator Rook Sutherland and other members of the committee. They read into the record various decisions of the comptroller which contended, had bearing on the matter, but Pinchot would not withdraw from his position.

SENATE PASSES $30,000 IRRIGATION PROJECT BILL

By Associated Press,

WASHINGTON, March 2.—After a brief discussion the senate today passed the bill authorizing the issuance of thirty million dollars worth of certificates of indebtedness for the completion of irrigation projects already started. This amount is supposed to be sufficient to finish work. Certificates are to be redeemed by the income from the sale of lands thus improved. The chief speech in opposition was made by Clay of Georgia, who said he was opposed to the principle of the bill and predicted that an annual appropriation of sixty million dollars would eventually be necessary for reclamation. Bailey and Hepburn said the public lands should be turned over to the several states.

Daily News, Feb 24 and 27, March 3 1910

BALLINGER IS TO RESIGN IS CLAIM

News Special Service.

WASHINGTON, Feb. 23.--It is generally talked among the friends of Secretary of the Interior Ballinger that he will resign as soon as the investigation committee has turned in its report of its findings. It is stated that Ballinger has said that he will resign and resume the practice of law, and that he would have resigned sooner but he feared that if he did it might leave a shadow on this administration, and look as if he was compelled to resign or be thrown out.

Daily News, Aug 7, Oct 23, Nov 3 and 6 1910

RECLAMATION MEN HERE

IN GRAND VALLEY GATHERING DATA ON HIGH LINE PROJECT.

C. J. Blanchard, government statistician, and W. J. Lubken, government photographer, both of the reclamation service, arrived in this city yesterday evening from Montrose, where they have spent several days going over the Uncompahgre canal. The two officials are making a tour of the government projects, and will remain here today and Monday, taking photographs and collecting statistics which will be used by the government in advertising the Grand Valley project.

They will be joined here today by Herman B. Walker, representing the Newark, N. J., News, who will be with them for the next two months. Mr. Walker was in Grand Junction two years ago with the party of newspaper men who came over from Denver from the convention of the National Press club. He is making a tour of the west in the interest of his paper.

NEWELL ASSURES EARLY CONSTRUCTION

F. H. Newell, director of the United States reclamation service while in Denver yesterday gave out the assurity that the work on the High Line will be pushed soon. This project is one that is looked forward to, to greatly develop the lands of this section and which will bring hundreds of homeseekers to the valley.

Mr. Newell stated that the government now has $60,000,000 invested throughout the country and next year it expects to invest about $7,000,000 more in projects of this kind throughout the United States.

While in Denver Mr. Newell was conferring with the officials of the Denver reclamation service. Mr. Newell left for Chicago where the transportation offices of the service are located and from there he goes to Washington.

Daily News, Aug 7, Oct 23, Nov 3 and 6 1910

MEETING OF SLOPE WATER USERS BE HELD SATURDAY

Formulate Plans for Fight Against Diversion of Western Slope Water by Eastern Slope Capitalists.

For the purpose of backing up the fight started in the Grand county court by the water users' associations and ditch companies of the Western Slope to prevent Eastern Slope capitalists from securing a title to the waters of the Grand river, a meeting of water association representatives from all over the Western Slope has been called to meet in the local chamber of commerce offices Saturday afternoon at 2:30 o'clock .

A plan of action will be outlined and the money for fully carrying out the fight will be raised. The Western Slope men have filed a petition with the Grand county court asking that the eastern men be not allowed title to the water. A title has already been asked and they filed a demurrer to the petition.

ARMY ENGINEERS REPORT EXPECTED OUT IN FEW DAYS

D. W. Aupperle Receives Encouraging Advices Regarding Action of Board of Army Engineers.

Secretary Aupperle of the Grand Valley Water Users' association received advice late yesterday afternoon that the President Taft expects a report from the board of army engineers in a few days on the Grand Valley project. There is no intimation of course as to the kind of report the engineers will make but Washington have intimated in the past that a favorable report is expected, in fact all who are at all familiar with this project agree that only a favorable report can be rendered because it is well known that this is a thoroughly feasible project.

Every possible effort has been put forth to get an early report from the engineers with the hope that work may be resumed by or before the first of the year.

Daily News, Aug 7, Oct 23, Nov 3 and 6 1910

ACTION IS TAKEN FOR WATER FIGHT IN GRAND COUNTY

Western Slope Will Give Easterners Run for Their Money in County Court

----

With representatives from Grand Junction, Clifton, Palisade and De Beque present the meeting called for the purpose of outlining a defense to be put up by water users on the Western Slope against the attempted diversion of the waters of the Grand river by eastern slope capitalists showed that a pressing need is upon the people here for immediate action. Upon the recommendation of Ben Griffith, Attorney Howard of Eagle was selected by the united water users of Mesa county to defend the water case which is now pending in the Grand county court and which will be called November 14. The case, in brief is as follows: An old filing made by an eastern slope company for water long ago. A decree giving them water rights would have been granted August 10, 1910, had not a petition been filed by interests on the Western Slope asking that this decree be not granted. The eastern company filed a demurrer to this petition und over this demurrer will be made the fight this week in the Grand county court. Attorney Howard is considered one of the best lawyers in that county. The meeting was held in the chamber of commerce rooms with Secretary Mahoney presiding.

Grand Junction Herald, Feb 1 1908

NOTICE OF IRRIGATION DISTRICT ELECTION

Notice is hereby given that the following resolution was duly adopted by the Board of Directors of The Orchard Mesa Irrigation District of Mesa County, Colorado, at a meeting of said Board regularly held at its office in Grand Junction, Colorado, on the 25th day of January, 1908, and that a special election will be held accordingly in said District at the time and places and for the purposes and in the manner set forth in said resolution:

"Whereas, The Board of Directors of the Orchard Mesa Irrigation District of Mesa County, Colorado, has inquired into and considered the cost of providing a suitable irrigation system for the needs of said District, and now considers it for the best interests of said District that the action of the Board of directors of said District, had on the 6th day of May, 1904, adopting certain plans for the construction of a proposed system of irrigation works for said District, and also the action of the electors of said District at the election held in said District on the first day of June, 1904, authorizing the issuance of the bonds of said District in the amount of $175,000, for the construction of such proposed system, and the action of said board of directors had on the 6th day of June, 1904, ordering the issuance of the bonds so authorized, should be rescinded, and the bonds issued thereunder be cancelled, it appearing to the board that none of said bonds have been sold, and that the following plan should be adopted for furnishing the water necessary for the irrigable lands of said District, to wit: the purchase according to the terms set forth in the contract hereinafter set out of the completed irrigation system of the Grand Mesas Land, Canal, and Power Company, said system being partially completed at this time, and the plans for its completion having been approved by the board. Now, therefore, be it

Resolved, That said action of said Board of Directors had on the 6th day of May, 1904, and on the 6th day of June, 1904, be rescinded, and of no further effect, and that all bonds heretofore issued by the District shall be cancelled without, however, in any manner abandoning any rights heretofore acquired by said District that may be available for use and enjoyment by said District in connection with the plan finally adopted by it for a system of irrigation works for said District, and that the plans last above stated for furnishing the water necessary for the irrigable lands of said District be and the same are hereby adopted as the plans of said District and of said board of directors for said purposes. And be it further

Resolved, That the amount of money necessary to be raised for said purchases of said systems of irrigation works, canal, etc., as aforesaid, and for the purpose of paying the first year's interest on the bonds next hereinafter mentioned, and otherwise carrying out the provisions of an act of the General Assembly of the state of Colorado entitled "An Act in relation to irrigation districts," approved May 3, 1905, upon such inquiry, examination, and consideration, be and the same is hereby estimated and determined to be the sum of Seven Hundred and Fifty Thousand Dollars ($750,000), and that said board deems it necessary and expedient that the bonds of said District be issued in said last named amount for said purposes, under and pursuant to the provisions of said last named act. And be it further

Resolved, That a special election be and the same is hereby called, to be held in said District on the 21st day of February, A. D. 1908, at which election the questions in the form stated below shall be and the same are hereby submitted to the electors of said District who at the time of said election shall be qualified electors within said District and resident freeholders in said District, and who shall have paid a property tax in said District during the year preceding such election.

FIRST QUESTION.

Shall the action of the electors of the Orchard Mesa Irrigation District of Mesa County, Colorado, at the special election held in said District on the first day of June, 1904, authorizing the issuance of the bonds of said District in the amount of $175,000, for the purpose of constructing a system of reservoirs, canals, ditches, and works necessary for furnishing water for the lands in said District and for acquiring the necessary rights and property therefor, and otherwise carrying out the provisions of an act entitled "An Act to provide for the organization and government of irrigation districts, etc." approved April 12, 1901, and acts amendatory thereof be rescinded and said bonds cancelled?

SECOND QUESTION.

Shall a certain contract made and entered into the 25th day of January, A. D. 1908, by and between the Grand Mesas Land, Canal and Power Company and the Orchard Mesa Irrigation District, be ratified, said contract being in words and figures as follows, to-wit:

This agreement, made and entered into this twenty-fifth day of January, A. D. 1908, by and between The Grand Mesas Land, Canal and Power Company, a corporation duly organized and existing under and by virtue of the laws of the State of Colorado, party of the first part (hereinafter designated "the Vendor"), and The Orchard Mesa Irrigation District, of Mesa County, in the state of Colorado, party of the second part (hereinafter designated “The District”). Witnesseth:

That Whereas, The vendor is the owner of a certain partially constructed irrigation system hereinafter described, and is now engaged in the completion of the same, and is also the owner of water rights and rights of way thereto belonging. And Whereas, The District is desirous of acquiring the same.

Now, Therefore, in consideration of the payment to the Vendor of the sum of $686,375 the Vendor agrees to complete, sell and convey or cause to be conveyed to said District all said property rights and franchises described as follows, to-wit:

A certain irrigation system having a carrying capacity sufficient to deliver 1.25 cubic feet of water per second of time at a point near the center of section Two (2), Twp. One (1) South Range Two (2) East of the Ute P. M., the exact point of measurement to be fixed by the engineers as hereinafter provided. The said irrigation system to consist of the following parts, to-wit

(1) A power canal having its headgate at a point on the East Bank of the Grand River at a point whence the northwest corner of Section Thirteen (13), Township Ten (10) South, Range Ninety-eight (98) West of the Sixth P. M. bears South seventy-seven (77) degrees, twelve (12) minutes West, Two Thousand Three Hundred Fifty (2350) feet said ditch running thence in a southerly direction along the left bank of the Grand River to a point from which the

Grand Junction Herald, Feb 1 1908

southwest corner of Section Two (2), Township Eleven (11), South Range Ninety-eight (98) West of the Sixth P. M. bears North Fifty-nine (59) degrees West, Eight Hundred Thirty-seven (837) feet; thence in a southwesterly direction to a point on the left bank of the Grand River, from which the southwest corner of Section Ten (10) of said township and range bears North Thirty-five (35) degrees, Forty-seven (47) minutes East Two Thousand Five Hundred Ninety-five and Five-tenths (2595.5) feet at or near which point the power house is to be located.

The above approximate location is shown on the plat of the Palisades High Line Canal filed on the 16th day of November, A. D. 1907, in the office of the State Engineer of the State of Colorado and afterwards filed in the office of the Clerk and Recorder of Mesa County, Colorado.

(2) Irrigation Canal Number One: Beginning at the end of the Power Canal hereinbefore described, thence in a westerly direction along the left bank of the Grand River to a point at or near the center of the Northeast quarter of Section Twenty-three (23), Township One (1), South, Range One (1) East, Ute Meridian, said point being known as the intake of the Johnson Ditch; thence following generally at a proper grade the line of said Johnson Ditch to a point at or near the east sixteenth (16th) corner of the Northeast quarter of Section Thirty-three (33), Township One (1), South, Range One (1) East, Ute Meridian; thence westerly to a point on the Ute Meridian, thence along the south boundary of The Orchard Mesa Irrigation District to the Gunnison River.

(3) Lateral Number One: Beginning at a point on irrigation canal at or near the East Sixteenth (16th) corner of the Northwest quarter of Section Thirty-three (33), Township One (1) South, Range One (1) East Ute Meridian: thence in a northwesterly direction to the division point of the Mutual Mesa Irrigation Canal at or near the East Sixteenth (16th) corner of the Southeast quarter Section Twenty-nine (29) of said Township and Range; thence generally along the North branch of the Mutual Mesa Irrigation Canal; thence into and along the Lee Ditch to the Grand River.

(4) Irrigation Canal Number Two: Beginning at the end of the discharge pipes of the pumping plant at or near the South Sixteenth (16th) corner of the Southeast quarter of Section Three (3), Township One (1) South, Range Two (2) East Ute Meridian: thence in a southerly direction to the Cactus Ditch, said Cactus Ditch being the southern boundary of The Orchard Mesa Irrigation District; thence along the southern boundary of the said District to a point at or near the center of Section Thirty-one (31), Township One (1), South, Range One (1) East Ute Meridian.

(5) A certain pumping plant to be located at or near the point hereinbefore described. The pumping plant is to have a capacity sufficient to elevate Sixty-nine (69) second feet of water from the power canal hereinbefore described to Irrigation Canal Number Two above described, and to include two units which consist of a horizontal turbine direct connected to a turbine pump and is to be first class in every particular, including foundations and buildings; said irrigation system is to be completed according to certain plans and specifications on file with the respective secretaries of the parties hereto, together with all rights of way, water rights, priorities of water rights, franchises, headgates, flumes and other structures belonging to or neces-

sary to said irrigation system; said property to be free and clear of all liens and encumbrances of whatever character, including taxes. The Vendor agrees to furnish all rights of way hereinafter specifically described required for said system without expense to the District, including the rights of way and the water to be carried therein through two ditches, known respectively as the Frank Lee and The Mutual Mesa Irrigation Ditches wherever required, for the purposes of said District.

In consideration of which the District agrees to pay to the Vendor the sum of $686,375, at the times and in the amounts as follows, to-wit:

(1)	On or before the 15th day of April, 1908................	$ 2,375.00
(2)	Upon the completion and approval of the headgate of the power canal to and including Plateau Creek crossing......	44,650.00
(3)	Upon the completion and approval of the Power Canal from the Plateau Creek crossing to the center of Section Twenty-three (23), Twp. Ten (10) South, Range Ninety-eight (98) West	47,500.00
(4)	Upon the completion and approval of the Power Canal from the center of Sec. Twenty-three (23) aforesaid, to the South line of Sec. Twenty-six (26), same Twp. and Range...............	19,950.00
(5)	Upon the completion and approval of the Power Canal from the South line of Sec, Twenty-six (26) aforesaid to the South line of Sec. Thirty-four (34) same Twp. and Range	28,500.00
(6)	Upon the completion and approval of the Power Canal from the South line of Section Thirty-four (34) aforesaid to the South line of Sec. Two (2), Twp. Eleven (11), South, Range Ninety-eight (98) West...........	82,175.00
(7)	Upon the completion and approval of the Power Canal from the South line of See. Two (2) aforesaid to Power House site..........	12,350.00
(8)	Upon the completion and approval of the Power Pumping Plant.	118.750.00
(9)	Upon the completion and approval of the Irrigation Canal Number One from the Power House site to the center of See. Eight (8), Twp. One (1), South Range Two East, Ute Meridian........	90,250.00

Grand Junction Herald, Feb 1 1908

(10) Upon the completion and approval of Irrigation Canal Number One from the center of See. Eight (8) to the center of Sec. Twenty-seven (27), Twp. One (1) South, Range One (1) East, Ute Mer 21,850.00

(11) Upon the completion and approval of Irrigation Canal Number One (1) and Lateral Number One from the center of Sec. Twenty seven (27) aforesaid to the end of said canal and lateral............ 9,500.00

(12) Upon the completion and approval of Irrigation Canal Number Two from the Power House site to the South line of Section Seventeen (17), Twp. One (1) South, Range Two (East 10,925.00

(13) Upon the completion and approval of Irrigation Canal Number Two from the south line of Section Seventeen aforesaid to .the East line of Sec. Twenty-six (26), Twp. One (1) South Range One (1) East 19,000.00

(14) Upon the completion and approval of said Irrigation Canal Number Two from the East line of Sec. Twenty-six (26) aforesaid to the end of said canal 17,100.00

(15) Upon the completion of said system of canals and Power Plant, the sum of.... 66,500.00

(16) Upon the final delivery and the acceptance of all the property under this contract and not less than Thirty-five (35) days after the completion of all work on said system, the balance of said purchase price, to-wit: 95,000.00

It is understood that the money to be paid by the District to the Vendor as consideration for all of said property is to be realized from the sate of the bonds of said District and in the event that said District is not able to sell its bonds after due advertisement for bids according to law, then the District shall deposit with the Hamilton National Bank of Chicago, Illinois, as Trustee, on or before April 15th, 1908, the bonds of said District in the sum and amount of $722,500, the said bonds to be by said Trustee paid over and delivered to said Vendor in the following amounts and at the following times, to-wit:

(1) On or before the 15th day of April, 1908 .... $ 2,500.00

Grand Junction Herald, Feb 1 1908

(2)	Upon the completion and approval of the headgate of the Power Canal to and including Plateau Creek crossing	47,000.00
(3)	Upon the completion and approval of the Power Canal from the Plateau Creek crossing to the center of Sec. Twenty-three (23) Twp. Ten (10) South, Range Ninety-eight (98) West	50,000.00
(4)	Upon the completion and approval of the Power Canal from the center of Sec. Twenty-three (23) aforesaid to the South line of Sec. Twenty-six (26), same Twp. and Range..	21,000.00
(5)	Upon the completion and approval of the Power Canal from the South line of Sec. Twenty-six (26) aforesaid to the South line of Sec. Thirty-four (34), same Twp. and Range ...	30,000.00
(6)	Upon the completion and approval of the Power Canal from the South line of Sec. Thirty-four (34) aforesaid to the South line of Sec. Two (2), Twp. Eleven (11) South, Range Ninety-eight (98) West	86,500.00
(7)	Upon the completion and approval of the Power Canal from the South line of Sec. Two (2) aforesaid to the Power House site .	13.000.00
(8)	Upon the completion and approval of the Power Pumping Plant....	125,000.00
(9)	Upon the completion and approval of the Irrigation Canal Number One from the Power House site to the center of Sec. Eight (8), Twp. One (1) South Range Two East	95,000.00
(10)	Upon the completion and approval of Irrigation Canal Number One from the center of Sec. Eight (8) to the center of Sec. Twenty-seven (27), Twp. One (1) South, Range One (1) East............	23,000.00

Grand Junction Herald, Feb 1 1908

(11)	Upon the completion and approval of Irrigation Canal Number One (1) and Lateral Number One from the center of Sec. Twenty-seven (27) aforesaid to the end of said canal and lateral	10,000.00
(12)	Upon the completion and approval of Irrigation Canal Number Two from the Power House site to the South line of Section Seventeen (17), Twp. One (1) South, Range Two (2) East	11,500.00
(13)	Upon the completion and approval of Irrigation Canal Number Two from the South line of Section Seventeen aforesaid to the East line of Sec. Twenty-six (26), Twp. One (1) South Range One (1) East	20,000.00
(14)	Upon the completion and approval of said Irrigation Canal Number Two from the East line of Sec. Twenty-six (26) aforesaid to the end of said canal..........	18,000.00
(15)	Upon the completion of said system of canals and Power Plant, the sum of	70,000.00
(16)	Upon the final delivery and the acceptance of all the property under this contract and not less than Thirty-five days after the completion of all work on said system, the balance of said purchase price, to-wit.........	100,000.00

Which said bonds of the District shall be taken and accepted by the Vendor in full payment for the Irrigation System hereinbefore described and to be conveyed by the Vendor to the District.

The Vendor agrees to commence work on said system on or before the fifteenth day of April, A. D. 1908, and further agrees that all of said several items hereinbefore mentioned shall be furnished and paid for by it, and the entire property shall be turned over and delivered to the said District, fully completed, on or before the first day of March, A. D. 1909.

In case of delays by reason of litigation involving the District, or by reason of defects in the voting or issuance of said bonds, or by reason of strikes, or act of God, the Vendor shall be allowed a reasonable further time, both for commencing work and for completing the performance of this contract.

It is understood and agreed, that all work done toward the completion of said irrigation system from the date of this contract shall be carried on under the direct supervision of the engineers of the respective parties hereto; provided, however, that the Vendor shall have immediate charge and direction of the work of the completion of the said system.

It is understood and agreed that all work to be done under this contract in the completion of the system of irrigation works herein mentioned and all material used in the completion of the same shall be in accordance with the plans and specifications hereinbefore mentioned and in case of any disagreement between the supervising engineers of the respective parties hereto in regard to the work done in the completion of said system of irrigation works, then said two engineers shall select a third engineer and the three shall constitute a board with power to decide upon all such disputed questions and the decision of such board or a majority thereof shall be final and binding upon both parties hereto.

In the completion of the said system of irrigation works, pumping plant, etc., material and workmanship are both to be first class. The entire irrigation system is to be completed with necessary waste ways, check ways and spill ways ready for use.

It is agreed that the Vendor and its attorneys shall have the time for, and the privilege of, full examination of the regularity and legality of the bonds heretofore mentioned, and of all steps and proceedings had and taken in the organization of said District, and in the issue of said bonds; and said Vendor, if not satisfied with the regularity and legality of said bonds, may declare this contract at an end at any time within thirty (30) days after the time herein specified for the delivery of said bonds to the trustee as aforesaid.

It is further agreed that at all times after the delivery of said bonds to said trustee as aforesaid, the said Vendor shall have the exclusive right to sell and dispose of said bonds remaining in the hands of said trustee and the said trustee shall at any time deliver to the Vendor, or upon its order, any quantity of said bonds, upon the payment to the trustee by the Vendor of eighty per cent (80 per cent.) of the face value of the bonds so delivered; such proceeds of said bonds to remain in its hands as trustee, in lieu of said bonds, and to be paid out by it at the times hereinbefore specified, with a proportionate reduction in the amounts.

The said bonds are to be regularly voted by said District in accordance with law, the form and terms of said bonds to be in accordance with the statute; said bonds to draw six per cent. (6 per cent.) interest, and to be payable at the office of the: County Treasurer of Mesa County, Colorado, or at the Hamilton National Bank at Chicago, Illinois, or any such other banking institution as may be hereafter agreed upon by the parties hereto.

The Vendor further agrees to expend through the Engineer of the District, before the first day of April, A. D. 1908 the sum of Two Thousand Dollars ($2,0'00), in the repairing of the Mutual Mesa Pumping Plant.

The Vendor further agrees that on or about the first day of March, A. D. 1909, it will execute and deliver, or cause to be executed and delivered, to the said District, a good and sufficient deed or deeds, conveying to said District all the property hereinbefore contracted to be conveyed, and including all rights of way, water rights or franchises which it may have acquired, either by appropriation or in any other manner, to the extent of 125 Cubic feet per second of time, for the purposes of carrying out this contract.

The Vendor further agrees to give to the District a bond in the sum of Fifty Thousand Dollars ($50,000), to be ap-

Grand Junction Herald, Feb 1 1908

proved by the board of directors of said District, for the faithful performance of this agreement; said bond to be executed and delivered to said board of directors upon the ratification of this agreement by the electors of the District, according to law.

It is understood and agreed that if at the time fixed herein for the delivery of the completed system of irrigation works by the Vendor to the District any of the rights of way for said irrigation system shall not be finally settled and determined, then and in that event the District in accepting said system of irrigation works shall withhold sufficient cash or the bonds of said District until such time as said rights of way are fully acquired and settled for or in lieu thereof until the Vendor shall have executed a sufficient special bond to be approved by the district conditioned that, the, Vendor shall acquire and fully pay for all said rights of way. The amount of cash or bonds to be withheld to be determined by arbitration as herein provided for the work to be done under this contract in the completion of the system of irrigation works herein described.

It is further understood and agreed that the rights of way herein provided to be furnished by the Vendor shall include only the rights of way for the Power Canal from the headgate of said canal on the Grand River throughout its course to the boundaries of the Irrigation District and including the site for the Power Plant. It being understood and agreed that the District shall proceed promptly to acquire all the rights of way necessary for said irrigation system within the boundaries of said District.

It is further understood and agreed that the Vendor or its assigns, shall have the right any time within five (5) years from the date of this contract to enlarge the Power Canal described from the intake thereof on the Grand River to a point near the Power Pumping Plant hereinbefore described; and also the other irrigation canal herein described as Irrigation Canal Number Two for the purpose of conveying water for any irrigation project in which the Vendor may be interested during said time; but nothing herein contained shall be construed to permit the Vendor to enlarge the Power Pumping Plant herein provided to be constructed and conveyed by Vendor to the District.

And provided further, That in making any such enlargement the same shall be done without injury or damage to the District nor in any way to interfere with or diminish the water supply of the District and that any such enlargement shall not diminish the water supply of the District, but in the event such enlargement is made and there shall not then be such a supply of water through said canal as is necessary for the needs of the District and the Vendor, then and in that event the rights of the Vendor shall be secondary.

And provided further, That any such enlargement made by the Vendor shall be made during the non-irrigating season.

In the event of such enlargement by the Vendor the management and control of the joint system shall be and remain in charge of the District and the Vendor shall pay to the District annually such proportion of the expense of maintenance and operation of the joint system as the amount of water carried through said joint system by the Vendor shall bear to the total amount of water flowing in said joint system.

It is further agreed That if the Vendor shall fail to furnish said bond for faithful performance, or shall fail, actively and in good faith, to begin work upon: said system within the time herein agreed upon, or shall fail to prosecute said work with reasonable diligence, the board of directors of said District may, at its option, cancel this contract as to all matters then remaining unperformed, and receive back from said trustee any bonds so deposited with it as hereinbefore mentioned, and not earned as hereinbefore provided; and if said Vendor shall fail fully to complete all of said contract as hereinbefore provided, by the time herein limited in that behalf, time being of the essence of this contract as to the matters in this paragraph above referred to, this contract may, at the option of said board of directors, be declared by the District to be null and void and of no further force and effect, except as a basis of settlement between the parties hereto as to the amount payable to the Vendor for the portions of said system then constructed by said Vendor, with proper deductions from the full amount of the contract price thereunder as the same may be ascertained by the District Engineer (subject to provisions as to arbitration herein). In the event of such declaration of the said board, the District shall be entitled to receive from said trustee any of said bonds so placed with it and not earned by the Vendor thereunder, or the value of the same, on the basis of the price obtained therefor as hereinbefore provided in case of disposition thereof by the Vendor; provided, that such declaration shall not relieve either party from any liability theretofore incurred to the other party by reason of the breach of any of the provisions of this agreement.

It is further agreed that nothing hereinbefore contained nor any acceptance or approval by, the District of any part of this contract, or the payment of any of the sums as hereinbefore provided, shall in any manner relieve the said Vendor from its obligation to completely carry out this entire contract according to each and all of its terms, nor shall the right of said District to proceed against the bond of said Vendor, in case said contract is not fully performed, for the damages sustained by said District by reason of such non-performance of this contract, be affected in any way by anything in this contract contained. The foregoing contract is subject to the approval of the electors of said District and in case not so approved shall then be void and of no force and effect.

In witness whereof, The said parties to this contract, pursuant to resolutions of their respective board of directors adopted in that behalf, have caused their presents to be signed in their respective presidents or vice presidents, and attested by their respective secretaries, with their respective corporate seals thereto affixed, the day and year first above written, this contract being executed in duplicate.

THE GRAND MESAS LAND, CANAL AND POWER COMPANY,
By C. C. Magenheimer, President.
[Seal]
Attest:
GEORGE S. HENRY, Secretary.

THE ORCHARD MESA IRRIGATION DISTRICT,
By Fred W. Barber, President.
[Seal]
Attest:
E. E. UDLOCK, Secretary.

THIRD QUESTION.

Shall the bonds of the Orchard Mesa Irrigation District of Mesa County, Colorado, be issued in the amount of $750,000, under and pursuant to the provisions of an act of the General Assem-

Grand Junction Herald, Feb 1 1908

bly of the State of Colorado, entitled "An Act in relation to irrigation districts," approved May 3, 1905, payable in lawful money of the United States in installments as provided by said act, beginning at the expiration of eleven years after the date thereof, and wholly payable in twenty years after the date thereof, and to bear interest at the rate of not to exceed six per centum per annum, payable semi-annually, and the said bonds and the issuance thereof to conform in all respects to the provisions of said act for the purpose of making the foregoing purchases and for the purpose of paying the first year's interest on said bonds, and otherwise carrying out the provisions of said last mentioned act?

That all persons voting on said questions shall vote by ballot, upon which shall be placed the words:

1. Rescission. Yes.
   Rescission. No.
2. Ratification of the contract with the Grand Mesas Land, Canal and Power Co. Yes.
   Ratification of the contract with the Grand Mesas Land, Canal and Power Co. No.
3. Bonds. Yes.
   Bonds. No.

For the purposes of said election, said District has by order of the Board of Directors thereof, been divided into two election precincts, as follows, to-wit: Precinct No. 1 is all that portion of said District lying west of the east township line of Township One South, Range Two East of the Ute Meridian: Precinct No. 2 is all that portion of said district lying east of said township line.

The polling place in Precinct No. 1 for said election will be at the residence of George Smith, located on Lot 2, Grand Junction Orchard Mesa Land Company's Orchard Sub-division, and the polling place for Precinct No. 2 will be at the residence of Thomas Y. Reese, located on north half southeast quarter and northeast quarter southwest quarter and lot 3, Sec. 8, township one south range Two east Ute Meridian.

That the polls at each election precinct shall be opened at 8:00 o'clock A. M., and remain open until and be closed at the hour of 6:00 o'clock P. M. of said day, and the election shall be held between the said hours of 8:00 o'clock A. M. and 6:00 o'clock P. M. of said day.

That notice of said election shall be given, and said election shall be held, and the results thereof determined and declared; in all respects as provided by law, and the secretary of said Board of Directors shall cause such notice to be given over his official Signature and the seal of the said irrigation District, and make due return to said Board of Directors of proper evidence of the due giving of said notice."

That said resolution has been by me as secretary of said board duly entered and remains of record among the proceedings of the said board and the records of my office as secretary of said board.

Witness my hand and the seal of said District, at Grand Junction, this 30th day of January, A. D. 1908.

E. E. UDLOCK, Secretary of the Board of Directors 0f the Orchard Mesa Irrigation District.

Daily News, Harvest Edition 1908

Marvelous Growth of the Grand Valley

By George E. Caldwell

"Grand Valley fruit lands, which twenty years ago were worth only the government price of $1.25 an acre are now selling as high as $4,000 an acre."

The above fact in itself constitutes all that could be written in an endless volume, and it only remains to set down in glowing detail how this is found in the natural constituents of the soil itself, while the magic is that of conquering climatic conditions happily combined with the fructifying favors of exhaustless irrigation waters--a unison which makes the streams of the Grand Valley veritable golden waters and annual floods of finance.

It is conservatively estimated that the first cost of the irrigated reclamation of the valley soil does not exceed 15 per cent of the primary orchard value secured—a percentage of irrigation cost which is steadily diminished as the orchards themselves attain bearing maturity--a fact which eventually results in not only the entire wiping out of irrigation cost, but the transformation of the original irrigating charge into an actual orchard asset and source of annual revenue.

With fruit lands with a perpetual water right starting at a purchase price of $250 an acre, and advancing in market value from year to year to $500, $1,000, $2,000, $2,500 and $4,000 an acre, it can be readily seen that the original and permanent irrigation charge speedily resolves itself into a practically paying investment.

Nature Generous in the Grand Valley.

The Grand Valley--also called "The Gem of the Western Slope"—lies entirely in Mesa county and comprises an area about 40 miles long with a maximum width of ten miles, the country deriving its name from the broad and fertile mesas rising above the current of the truly Grand river, while the beautiful and prosperous county seat owes its christening of Grand Junction to its situation at the confluence of the Grand and Gunnison rivers.

The valley soils are almost wholly comprised of materials derived from sedimentary rock formations along the Grand river itself.

Generous old Mother Nature has beneficently divided these soils

A Bearing Fruit Tree.

Daily News, Harvest Edition 1908

into distinct and separate belts, each of which is especially adapted to its own particular and prolific variety of harvest return.

The Laurel fine sand has a soil depth ranging from one to six feet and is especially and profitably given to market garden culture.

The Laurel sandy loam is from two to six feet deep and produces large returns in the berries and smaller fruits, sugar beets and alfalfa, while heavy truck crops are also secured.

Fruit Belt Rich in Fertile Soil.

The Billings fine sandy loam has a depth of six or more feet, and constitutes the great fruit belt of the valley, growing nearly all the orchards of the superb Palisade peach region. This soil also yields heavy crops of sugar beets and alfalfa. The western portion of this great soil belt has many excellent apple orchards, both stone and seed fruits being of such prolific bearing that it is usually necessary to sacrifice a portion of the product during its earlier growth stages, in order to enable the trees to bear the big burden of fruit at maturity. Alfalfa yields on the Billings fine sandy loam comprise three crops a year, with an aggregate of from four to six tons an acre. Sugar beets run to 15 tons an acre, while potatoes, market garden vegetables, corn and berries all do well.

The Billings clay loam is naturally designed for the growing of apples, pears, wheat, oats and corn, with alfalfa and sugar beets in generous return.

The Billings silt loam is from three to six feet in depth and produces good crops of alfalfa, pears and sugar beets. This is an alluvial soil and its harvest values are being annually added to by an active and intelligent seepage drainage.

An Orchard in Bloom.

Finest Orchards on Mesa Sandy Loam.

The Mesa sandy loam is light and friable, ranges in soil depth from six to twenty feet, and supports some of the finest orchards in the valley. But a small area of this soil is yet in cultivation, but it has demonstrated, in addition to its great orchard values, splendid capacities for alfalfa and market garden croppage, a United States government expert saying that "the Mesa sandy loam is preeminently adapted to fruit and tilled crops."

Mesa clay loam is from four to six feet deep, is rich and productive, and especially adapted to alfalfa, sugar beets and grain. This soil is chiefly found on Orchard Mesa, near Grand Junction and is practically free from alkali.

Fruita sandy loam is six or more feet in depth, is friable and readily cultivated, rarely proving refractory to tillage. This soil principally composes the wide and wealthy fruit areas immediately surrounding the town of Fruita. Some of the finest apple and pear orchards of the valley are upon this soil type, while sugar beets, alfalfa, wheat and oats are all large growths of the fertile Fruita fine sandy loam.

The Fruita loam, a soil separate and distinct from the Fruita fine sandy loam, is at present a good grower of sugar beets and grain, and, with more drainage, will develop into fine fruit lands, while irrigation is the sworn and successful foe of alkali deposits.

There is practically no hard-pan in the soil area of the Grand Valley.

Mt. Garfield. Photo by Dean

Money vs. Mortgage In Orchard Land.

The farmers of the Grand Valley are, as a rule, in a remarkably prosperous condition. Such a thing as a mortgage foreclosure is almost unheard of, while the field, the orchard and the garden vie with each other in harvest production and money making markets.

Peach trees begin commercial bearing at three years old, and attain full production at seven or eight years of age, the fruit netting the grower from $400 to $800 an acre. The great Elberta, frequently attaining a weight of over a pound each, is the pride of the valley peach orchards, while the early and late Crawfords; the Carman and the Arp Beauty are all superb commercial varieties.

The apple orchards chiefly count the great commercial varieties of the Ben Davis, the Gano, the Jonathan, the Winesap and the Minkler.

Cherries, apricots, nectarines, prunes and quinces are also grown with marked success in the Grand Valley, while all the berries flourish in profusion and profit.

Within the last three years, counting the present season, about two million new fruit trees have been set out in the Grand Valley.

Grand Valley fruits have an alike national popularity and market scope, their marketing being facilitated by the traversing of the valley by the Denver and Rio Grande and Colorado Midland railroads, the Midland running over the Rio Grande tracks, while a Denver and Rio Grande line runs from Grand Junction across the mountains to Ouray and the San Juan district, and through Montrose, Delta and other towns, and over other mountains to Gunnison, traversing the Gunnison country and connecting at Salida again with the main Rio Grande rails.

Margery Building. Photo by Dean

Organization Has Enlarged Markets.

The recent organization of active fruit associations and cooperative marketing concerns has resulted in the shipment of Grand Valley fruits to far distant markets, the chief peach markets now being Missouri river cities and towns, while peaches have been shipped to Boston with gratifying results.

Apples are largely sold within the state, with an extended and lucrative Western shipping market, principally to Southern California points.

The now constantly growing sugar beet industry is proving a new and great Grand Valley boon, as it brings into active and prosperous cultivation those

Daily News, Harvest Edition 1908

lands not so well suited for fruit or other general farm culture, and also establishes a new and heavy wage and sale schedule.

The smaller fruits and vegetables meet with a ready and profitable sale to the mining districts and various cities and towns of the state.

Alfalfa, with its production of from four to six tons an acre. and selling price from $8 to $15 a ton, is nearly all consumed in the valley, the fruit growers furnishing a ready alfalfa home market.

Phenomenal Yields of Grain and "Spuds".

The grain and potatoes are of profitable sale, wheat reaching as high as 60 bushels an acre and oats climbing to 125 bushels, while potatoes "dig up" money for yields of 300 and 400 bushels an acre.

Market garden vegetables and corn or potatoes are often planted between the rows of young fruit orchards, with the double good result of a thorough irrigation of the trees themselves and securing of paying crops from the land during the unproductive first years of orchard growth.

Raw valley land is readily and cheaply cleared of its natural growth of sage brush, a riding plow in nearly all locations being capable of at once breaking and clearing the land.

High Line Canal Will Reclaim Vast Acreage.

The late visit to Grand Junction of high officials connected with the Grand Valley United States Government High Line Canal brought with it the welcome assurance of the commencement of actual canal construction by the first of next January.

It is estimated that the canal-will be completed within twelve months from the commencement of construction, will cost an approximation of $2,000,000 and, through the fruit and general farming cultivation of the 70,000 acres covered, will advance the Grand Valley to premier rank in Colorado tillage and insure to Grand Junction a population of 30,000 within the next six years, thus inseparably and imperishably interweaving the names of Garfield, the Grand Valley and Grand Junction in a great sign manual of progress, prosperity and population.

Photo by Dean Dr. H.R. Bull's Res. Dr. G.R. Warner's Res.

Fruit Harvest Worth Two Million Dollars.

Last year's fruit harvest brought $2,000,000 to Grand Junction and the Grand Valley. Last year frost killed practically the fruit of all sections of the nation, but the orchards of the Grand Valley well filled the deficit.

Last year H. O. Harvey marketed $2,750 worth of pears from one and one-quarter acres; M. F. Yeaton banked $4,000 from four acres of pears; Robert Orr averaged a sale of $2,000 per orchard acre; one apple tree produced 47 boxes selling for $60: one apple acre produced $1,700 worth of fruit; a cherry tree has averaged for several years past $30 per year; ten acres of peaches six years old have returned an aggregate receipt of over $10,000; two-thirds of an acre of peaches brought $940; common returns in peach culture count from $400 to $800 per acre.

Poultry Profits in the Grand Valley Great.

Poultry breeding is very profitable in the Grand Valley. The poultry farm of Charles A. Smith, near Grand Junction, counts 1,500 of ordinary and fancy bred chickens, and a large number of homing pigeons from which $700 worth of squabs are marketed each season, while 1,000 dozen eggs are sold every year from the hennery.

Fancy bird eggs for hatching purposes bring $2.25 per setting, about $400 per season being realized from this source, the breeding cocks costing as high as $40 each, with hens worth from $5 to $15. Mr. Smith's total investment is about $2,500.

Irrigation Supply of Valley Exhaustless.

The irrigation water supply of the Grand Valley chiefly comes from the Grand river and includes the application of both the gravity, pressure canal and the water lifting pumping plant driven by water power itself, this irrigation water supply always, doubtless, being greater than can be used by the valley.

The Grand Valley canal has a total length, including the Kiefer extension, of 46 miles, and has an irrigating capacity of 50,000 acres, with a present covering of 40,000 acres and an actual cultivation of 30,000 acres, which is being added to at a rate of 3,000 acres per year, the present water distributing length of the canal system being over 100 miles.

The stock of this canal is all owned and issued by the farmers themselves, the combined cost of irrigation itself and canal maintenance averaging but 65 cents an acre, an annual cheap irrigating canal cost which confidently alike challenges comparison and competition.

The Grand Valley has 150,000 acres of easily irrigable land, of which 85,000 acres are now provided for by canals and already operated or projected pumpage plants, while the coming big United States Government High Line Canal will irrigate the remaining 65,000 acres.

The great pumpage plant of the Red Lands Irrigation & Power company is alike the most powerful and perfect of its character yet known to the West.

This plant will irrigate 5,000 acres of land on the rich Red Lands Mesa across the Grand river from Grand Junction, and will also generate electric power for extended commercial purposes, including the furnishing of electric motors for various coming Grand Junction industrial enterprises and the city's projected street car trolley system. The total cost of this plant and general irrigation system to date has been in excess of $200,000, of which $50,000 has been expended on machinery and comprises two 600 horse-power pumps and two 600 horse-power electric generators. The pumpage has a present lift of 180 feet and can discharge a combined flow of 24.000 gallons of water per minute, which is equivalent to 2,000 irrigating water inches. Only one of the pumps, however, is now in operation, as its water discharge is sufficient for present irrigating need.

Power Plant Could Supply Large City.

The power canal is four miles long 27 feet on the bottom, 44 feet on top, and ten feet deep, while the distributing canal is 15 miles long.

At least 600 electric horse-power can be generated for commercial sale, while the company will establish an ad-

Daily News, Harvest Edition 1908

ditional pumpage irrigation plant farther down the river which will be electric driven and have a water lift of 125 feet. It is estimated that this plant can generate sufficient electricity to fully supply all lighting or industrial power need of a city of 25,000 people, and also operate 50 miles of trolley railroad. The running expenses of the plant itself does not exceed $175 per month.

This year will see 1,500 acres of the rich Red Lands mesa irrigated by this pumpage plant, while the entire 5,000 acres will be irrigation served within the next two years.

The Red Lands Mesa is practically free from frost visitation, the Kiefer Brothers' orchard, for instance, located at the head of the mesa, being in alike blissful and bountiful ignorance of even the word frost.

Irrigation District for Orchard Mesa.

An irrigation district has been organized under state law to cover 12,000 acres on the fertile Orchard Mesa on the south side of the Grand river and opposite Grand Junction.

These lands are to be irrigated by a pumpage power plant similar to that of the Red Lands Mesa, it being expected that 6,000 acres will be receiving irrigation water within the next two years.

Orchard Mesa is one of the choicest of the many choice fruit and farm areas of the Grand Valley. On this mesa a water lift pumpage has been successfully irrigating 1,000 acres of land for many years.

The successful operation of a pumpage water lifting plant renders available to the irrigation the ruggedest and most broken lands lying along the Grand Valley streams or canals. Every barren butte will be crowned with the burdened orchard bough, every haggard hill will hold the halo of heavy horticultural harvest, and every gashed gulch and gulley will be glorious with overhanging fruitage. Even the waste places shall be those of wealth and roughness shall be riches.

The Smudge Pot Slays Jack Frost.

The kindly climate and high land elevations naturally antagonize frost in the Grand Valley, and even the rare frost visitations are now in a fair way to be entirely prevented by the introduction of the "smudge pot" with a burning of oil or coal and a use of from 40 to 80 pots an acre. By the way, the name of "smudge pot" in this connection is a misnomer, as it is evidently the heat from the pots instead of the smudge or smoke which routs the frost fiend. The various Grand Valley chambers of commerce, boards of trade and fruit associations have intelligently and energetically taken up this matter of frost prevention and next spring will see three-quarters of a million coal or oil heated smudge pots in operative burning in the Grand Valley, with the undoubted result of a final "goodbye" to spring orchard frosts and the henceforth annual insurance of full maturing of the naturally magnificent valley fruit harvest.

Grand Valley farms are nearly all of home holding, being actually owned by the farmers themselves, the farms ranging from five to 160 acres. In fact, the only large proprietary areas in the valley, are those of thousands of acres of the Western Land & Sugar company and the 3,000 acres owned by the Reed Investment company in the Loma section.

A generous yearly harvest labor market is furnished by the Grand Valley. the hired workers being generally robust and intelligent young men, many of whom promptly graduate into prosperous home holders themselves.

Grand Junction Beet Sugar Production.

The Western Sugar and Land company is cropping this year 6,000 acres of sugar beets, which will average 14 tons an acre, with a saccharine showing of 15 1-2 per cent per ton. This means a cutting by the sugar factory this year of 75.000 tons of beets and a sugar production of 200,000 bags of sugar of 100 pounds each. The company will distribute in cash this year over $400,000 to the farmers for sugar beets and will pay out more than $300,000 for labor.

Last year the company cut 63,000 tons of beets as against the 75.000 tons to be cut this year. To provide for this additional business, the company is making an improvement in its plant amounting to $100,000, the added expenditure including the purchase of new boilers, new crystallizers and new centrifugal machinery.

General Manager Davie has just completed a tour of all the beet territory subject to cultivation by the Western Sugar and Land company and says that the outlook was never better and could not possibly be improved.

Manager Davie says further that a $500,000 sugar factory will be erected year after next at some point in the Uncompahgre Valley, the location of the factory not yet being finally decided upon.

The Metropolis of the Western Slope.

Grand Junction, the seat of Mesa county, is the center as well as the capital of magnificent Mesa county. It is the metropolis of the Western Slope and is the coming city between Denver, Colorado Springs and Pueblo on the east and Salt Lake City on the west, and is already the practical commercial distributing point of Western Colorado and Eastern Utah.

It has at present a population of 10,000 and is rapidly putting on newer and greater civic garments. It is a center of modern civilization as well as munificent commerce, having a superb system of public schools, commodious and well supported churches and all of the residence and building block improvements of the day and the time, a further evidence of these facts being the present construction of a $60,000 Y. M. C.A. building. James R. Garfield, secretary of the interior, said:

"At Grand Junction I saw a wonderful country. That part of Colorado is as near perfect as nature and the hand of man can make it." Secretary Garfield might have added that while nature had made a wonderful country of the Grand Valley, modern man was making just as wonderful a city called Grand Junction.

Peach Culture at Palisade.

Palisade and peaches are two words largely and luxuriantly interwoven in the great Grand Valley production of this great fruit. The Palisades are sheer and serried cliffs which lift themselves from 1,000 to 1,500 feet above the flow of the Grand river, the fertile peach lands lying along the cliff bases.

Palisade peach culture includes two great cardinal facts: First, the natural soil adaptability to peach growth; and, second, the at once sheltering and further fructifying harvest action of the Palisades themselves.

From the broad and lofty surface of the Palisade cliffs, an alike warm and wise sun reflects its powerful and timely rays upon the Palisade orchard areas, thus uniting with the rich soil for the fi-

Daily News, Harvest Edition 1908

nal production of the perfect Palisade peach. This exceptional and fortunate natural union of soil and sun is yet further improved by the intelligent practice of Palisade peach growers of annually thinning out and opening to the sun's rays the topmost foliage of the peach trees themselves and thus allowing the utmost free action of the sun in the rapid ripening and conquering commercial coloring of the fruit itself.

What is known as the Palisade fruit region comprises 10,000 acres covered by two irrigation water lifts of 35 and 82 1-2 feet, and an aggregate (here we are directed to a page that was not in the package)

GRAND VALLEY HIGH LINE CANAL

Engineer Sands Explains in Detail the Plans of the Government in Regard to the Big Irrigation Project.

On the 17th of June, 1902, the president signed the Reclamation Act. This act provided that all moneys received from the sale of public lands in Arizona, California, Colorado, Idaho, Kansas, Montana, Nebraska, New Mexico, North Dakota, Oklahoma, Oregon, South Dakota, Utah, Washington and Wyoming, excepting land office fees and a few other minor sums, should be set aside and appropriated as a special fund known as the Reclamation Fund, to be used for the examination, survey, construction and maintenance of irrigation works for the storage, diversion and development of water for the reclamation of arid lands in the said states and territories.

The second section gives the secretary of the interior authority to perform the work contemplated by the act and requires a report to congress at the beginning of each regular session.

'The third section authorizes the secretary to withdraw lands to be irrigated by any project and this section also provides that such lands can be entered only as homesteads.

The fourth section authorizes the secretary:

To let contracts.

To give public notice of the irrigable acreage.

To limit the size of entries.

To announce the charges per acre for entered and private lands benefited.

To determine the number of annual instalments, not exceeding ten, in which the charges may be paid.

This section also provides that the entire cost of constructing a project shall be returned to the fund. It also provides that no Mongolian labor shall be employed and that eight hours shall constitute a day's work.

Section five provides that an entryman shall reclaim at least one-half of his land and pay all charges before a patent shall be issued. No water right shall be sold to water private lands in tracts in excess of 160 acres.

The owners of such private lands must reside on the lands or live in the neighborhood, and no water right shall permanently attach until all payments shall have been made.

Section six provides that the secretary shall use the Reclamation Fund for the building and operation of all reservoirs and irrigation works constructed under this act: Provided, that when the payments are made for the major portion of the irrigated lands the maintenance and operation of the irrigation works shall pass to the owners of the lands irrigated thereby, under such regulations as the secretary may prescribe: Provided, that the title to the works shall remain in the government.

Section seven authorizes the secretary to acquire title to rights of way by purchase or condemnation.

Section eight provides that the state irrigation laws shall be conformed with, and that the water rights shall be appurtenant to the land and beneficial use

Daily News, Harvest Edition 1908

shall be the basis, the measure and the limit of the right.

Section nine provides that during each ten year period the major portion of the moneys derived from any state shall be expended in that state.

Section ten authorizes the secretary to make such rules and regulations as may be necessary.

From the above review of the Reclamation Act it is seen that the plan of operation of the Reclamation Service is to expend the money received from.the sale of land in the arid states for the building of irrigation works and assess the charges against the lands benefited; so the fund is a rotating one and will constantly grow larger.

The works undertaken in the arid west are of a magnitude that surpasses the wildest dreams of the promoters of private enterprises. When the works at present begun are completed, over three million acres of the finest land in the world ,will be reclaimed.

During the present year over five hundred thousand acres are being irrigated from government works. The structures built and the engineering difficulties encountered have been without precedent.

At Roosevelt, about eighty miles from Phoenix, Arizona, the great Roosevelt Dam, which will store sufficient water to cover 1,284,000 acres, one foot deep is being built. This dam is 284 feet high and 1,080 feet long at the crest.

To build this dam the government developed 4,400 horse-power from a power canal. They built a road that cost over a quarter of a million dollars to get to the dam and they built a cement mill to furnish the cement at the dam.

Near Yuma, Arizona, on the Colorado river, a dam 4,780 feet long and 19 feet high is being built across the silt which forms the bed of the Colorado river.

At Montrose the Gunnison Tunnel, 30,583 feet long, the longest in the Western Hemisphere, is being built through granite, sandstone, limestone, shale and mud, and will eventually water 146,000 acres of land.

In Bighorn county, Wyoming, a dam is being built to store the waters of the Shoshone river. This dam, from the bottom of the foundation to the crest, will have a height of 310 feet. At its side the Masonic Temple, at Chicago, would look like a dwarf.

The water stored will water 130,000 acres of land near Cody, Wyoming. .

By the end of the present year a total of over $41,000,000 will have been expended by the Reclamation Service.

The project which probably interests most of us who are present more than any other is the Grand Valley project. This project contemplates the irrigation of from 60,000 to 70,000 acres of land lying above that irrigated at the present time. The land to be irrigated lies north and west of Grand Junction, Fruita, Loma and Mack. About 40 per cent of it is private land and the remainder is public land, most of which has been filed upon under the provisions of the Reclamation Act.

In 1902, immediately after the president had signed the Reclamation Act, the government sent a corps of engineers into this valley to investigate this project. A strip of country was investigated and mapped which included every probable canal route.

No line was staked out or decided upon, consequently up to the present time there has never been a government line. However, certain parties have issued maps purporting to show the location of the proposed government line.

In the spring of 1903 a favorable report was made to Washington regarding the feasibility of the project.

The government at that time had the funds and was ready to begin construction after more extended surveys and plans had been prepared. This was prevented because certain engineers and so-called capitalists went to Washington and stated that private capital was ready and anxious to build this project, and as the government had numerous demands on the fund from other projects this one was abandoned for development by private capital.

About a year ago the people of this valley became tired of waiting for private capital and sent a man to Washington to see if the government could again be interested in this project.

The secretary of the interior made two propositions to the people:

First. That he would have surveys begun and plans prepared and would construct the project as soon as funds were available.

Second. That he would restore all lands in the valley to all the usual forms of entry so the project could be built by private capital under the Cary Act, or some similar plan, and that the government would definitely abandon the project.

At a large mass meeting the citizens unanimously accepted the first proposition and the secretary is complying with its provisions at the present time.

An office was established here about the first of March of the present year.

Since that time a party has been continuously in the field and by the end of September the entire irrigable area will be mapped on a scale of one thousand feet to the inch, ten foot contour interval. We consider a first class topographic map absolutely necessary for the economic development of a large irrigation enterprise. Probably a number of those present are familiar with such maps. For those who are not a brief explanation will be given.

The maps we are making show every road, fence, railroad, section line, irrigation ditch, arroyo, and in addition all ten foot contours. A contour is a line joining all points of the same elevation. The easiest way to imagine such a line is as follows: Suppose the ocean should rise until its level was 4, 700 feet higher than at present. If we should sketch its new line we would have the 4,700 contour. Now assume that it rises an additional ten feet and the new shore lines were drawn. This would be the 4,710 contour, and so on. After such a map is completed we are able to determine the location of all canals and laterals. In addition we can estimate the amount of waste or non-irrigable land in any tract. The steepness of the slope of the land is shown by the distance between contours. We can tell if the land be bottom land or mesas.

In case the ocean would rise as we have assumed, the mesas would form peninsulas out into the ocean, and the bottom lands would be inundated and would form bays.

Another feature of our surveys has been an accurate classification of the soil. Every square foot has been classified into one of five classes:

No. 1 is red sandy soil.

No. 2 is fine, black, sandy loam.

No. 3 is adobe.

No. 4 is land too rough to irrigate.

No. 5 is seep land.

Where irrigable lands are located above the probable location of our canal line we have extended our surveys to an elevation of 200 feet above the canal

Daily News, Harvest Edition 1908

line, to include all area that may be watered by pumping.

The main line of our canal has been located from the headworks to Palisade, and plans and estimates are being prepared. Our locations are made with great care, and without leaving the office we can estimate the difference in cost that will result in shifting the line a foot in either direction. Before our line can be considered as finally located the location and plans must be examined by three engineers, who will report to the director. As this board has not yet met, none of our location can be announced as final.

Sufficient study has been made to enable me to announce the general features of the project now, even though they have not yet been approved. Please remember that I use the word probable.

The headgate will be located between the railroad tunnel and the mouth of Plateau creek, about six miles above Palisade. No great advantage and a large increase in cost would result in attempting to go higher. The canon at this point is such that tunnels would require a grade nearly equal to that that would be gained by placing the intake farther up the river.

At the point of diversion we contemplate raising the river nearly 20 feet by means of a dam over 400 feet long. The canal will be taken out on the right hand or west side of the river. It will follow along the right of way of the D. & R. G. railroad for about two and one-half miles. From Cameo to Palisade the line will be largely in tunnel. This first six miles of line will involve 2,500 feet of flume and over 9,000 feet of tunnels. It is proposed to make the flume of timber, lined with galvanized iron, and built on concrete stringers. The tunnels will all be lined with concrete.

There have been a number of wild goose schemes advocated by those not familiar with the conditions. One of these has been a plan to run up Coal Canon and tunnel through to a point west of Mt. Garfield. This would increase the cost nearly $2,000,000, and would not result in any gain in grade.

From Palisade the canal will run in a general northwesterly direction and end at a point near mile post No. 8 of the Uintah railroad. It is estimated that the project will cost from $2,500,090 to $3,000,000, and the cost per acre will probably be about $50. One thing is certain, and that is that such work costs much more than is generally believed, and an earnest effort is going to be made to have our estimates large enough to cover the cost before construction is begun.

As there are a large number of individual land holders to deal with, and as the maintenance and operation of the work will ultimately be turned over to the water users, it is necessary for them to be organized, and the Grand Valley Water Users' association represents the water users in their dealings with the government.

This association has power to levy assessments for any purpose and the government will aid in the collection of these assessments by refusing to give a man water until his assessments are paid.

At the present time the Reclamation Service has a large number of projects nearing completion and which will soon be on a paying basis. For this reason it is not considered good business policy to start new projects.

However, the people of this valley are anxious to get the work started at right and also to secure a contract with the department. There is also a well grounded fear that certain private parties will attempt to break the strong sentiment which exists at present in favor of government construction if construction operations are delayed.

The Water Users' association has, therefore, made a proposition that they will raise half of the funds necessary to get construction well under way during the year 1909. To raise these funds the association proposes to issue certificates bearing 8 per cent interest. The government guarantees to accept these certificates at face value in payment of water charges, and the association guarantees the interest, which will be collected from all the water users by assessments. If this plan is approved in Washington, the government will assist the association in collecting these assessments, as stated above.

For those who are not able to advance cash a plan is being developed to enable a man to work out his water right. This plan has been used on several other projects and has met with marked success. The certificates issued are negotiable.

In all works done under the Reclamation Act there are two distinct branches of the government service to deal with. The Reclamation Service makes the surveys, builds the works, determines the cost and decides the size of the farm units. The land office receives the entry, sees that the residence and other requirements are met, and collects the charges.

When public lands are to be watered on any project, these lands may be segregated by the secretary of the interior by two methods. What is known as the “first form” withdraws the lands from all form of entry, and the lands become a government reserve while so withdrawn. This is done when the land is needed for construction purposes. Entries made prior to the order withdrawing the land from entry are not affected and the entryman should continue in the usual manner. His entry may be contested and the person initiating the contest has a thirty day preference right to enter the land when it is again restored to entry.

Lands segregated by the "second form" can be entered, but only under the provisions of the Homestead Law and Reclamation Act. This is done so that desert, timber and stone and other forms of entry which give a man more than 160 acres cannot be made on lands to be irrigated from a government irrigation system.

In the matter of securing rights of way for canals the government has advantages and rights not enjoyed by individuals or companies. The act of August 30, 1890, provides: "That in all patents for lands hereafter taken up under any of the land laws of the United States or on entries or claims validated by this act west of the one hundredth meridian, it shall be expressed that there is reserved from the lands in said patents described, a right of way thereon for ditches or canals constructed by the authority of the United States.”

The entries or claims validated by that act are such as may be asserted under entries made subsequent to October 2, 1888, as held by the department in the circular of October 5, 1893.

In such cases the Reclamation Service is willing to pay a reasonable amount for improvements. This is not toby based on speculative values. There is no doubt that hardships will be worked in certain instances. However, the abstracts and patents have been specified and there is nothing unfair about the matter. Suppose the government had for years withheld all these lands from entry on the ground that they would later

Daily News, Harvest Edition 1908

be needed for canal rights of way. A general complaint would have arisen that the progress of the communities affected was retarded. The people would have said: "Let us develop these lands and get the benefit of their use. When you are ready to build your canals we will surrender the necessary rights of way."

Without the people requesting such action, that is exactly the thing that congress did, and we are now ready to assert our rights to the right of way. We try to be fair in all such matters, but, on the other hand, do not propose to be unduly annoyed or held up. Occasionally we meet a man who attempts to prevent us from making the necessary surveys, and pulls up our stakes. An Act of Congress approved June 10, 1896, provides: "That hereafter it shall be unlawful for any person to destroy, deface, change, or remove to another place any section corner, quarter-section corner, or meander post, or any government line of survey, or to cut down any witness tree, or any tree blazed to mark the line of a government survey, or to deface, change, or remove any monument or bench mark of any government survey. That any person who shall offend against any of the provisions of this paragraph shall be deemed guilty of a misdemeanor, and upon conviction thereof in any court shall be fined, not exceeding two hundred and fifty dollars, or be imprisoned not more than one hundred days. All the fines accruing under this paragraph shall be paid into the treasury and the informer in each case of conviction shall be paid the sum of twenty-five dollars." Up to the present time we have always been able to turn the offender from his evil ways without recourse to law. However, a few cases of malicious and wilful disregard of this law have been observed, and the time may come when it will be necessary to make an example of some offender.

Our water rights are secured under the state laws the same as those of any other user. Of course it is impossible to conduct an enterprise as large as this one without conflicting with both legitimate and private interests. The result is abuse of the officers of the service that is unwarranted. The members of the service are handicapped by being required "to treat their enemies with- respect and give them an advantage to which they are not entitled."

Indications are that this project will be consummated without undue delay. It will increase the wealth of this community by thirty million dollars in the next fifteen years. It will give farmers water under the most favorable terms. It will supply labor to hundreds of men. It will increase the markets of not only our local merchants but of the entire state. Our equipment, cement, steel and lumber will be largely Colorado products The railroads into the valley will have their local business more than doubled. It is estimated that the population of the valley will be increased from fifty to one hundred thousand.

The lands to be benefited are close to the hills and include a large percent of sandy mesas. These are the best fruit lands known. This project is looked upon by the Reclamation Service as one of the most promising. The land is the most productive, the climate finer, the water supply the most certain. After the "High Line" is built we will stop our printing presses for a few moments and change our advertising circulars, as the "Little Empire of the Western Slope" will be inadequate to describe the magnitude, of the industry, prosperity and resources of the Valley of the Grand.

MEMBER OF THE ASSOCIATED PRESS

THE DAILY NEWS

WEATHER Issued at Denver SHOWERS TODAY

Our Motto: The News Without Fear or Favor--Principle Above Private Interest or Pecuniary Gain

ESTABLISHED 1882. GRAND JUNCTION, COLORADO, SUNDAY MO[illegible]NG, OCTOBER 25, 1908.

DEMOCRATS DESPERATE; APPLY FOR INJUNCTION

At Loss For Real Issue, Concentrate Their Puerile Efforts Against Voting Machines

HAVE LOST ALL HOPE OF WINNING

The Democratic county central committee yesterday filed complaint and notice of application for an injunction in this district court asking that the voting machines be discarded at the coming election and in a lengthy petition giving several reasons (?) why they should not be used. The central committee, consisting of H. L. McClintock, county chairman; H. W. Kluge and J. W. Bucklin, are responsible for the application. Attorneys Henry Rhone, S. N. Wheeler and an imported lawyer, E. C. Stimson from Denver, are attorneys for the plaintiffs. John Hynes, Wm. Dittman and John Secor, county commissioners and Reuben E. Starr are named as the defendants. The hearing has been set for 10 o'clock tomorrow morning.

BY ROY M. M'CLINTOCK.

In political circles it is generally regarded that the application of the Democrats for an injunction restraining the county commissioners from using the voting machines in the coming election is an indication that the remainder of the county ticket is to be sacrificed in a last desperate effort to elect Shrader, Shores and Weiser.

The objection of the Democrats from the very first to the voting machines has been that the use of the machines [illegible]

[illegible] lots, outside of print paper, the use of which would probably not be allowed. It would be somewhat of a question whether, at this late date, the paper could be secured and the ballots printed in time for the election. And what would be the effect of an election held without ballots?

Further, the opposition of the Democrats to the voting machine proposition is very new. Two years ago, when the question was before the people of Mesa county for referendum vote, none of the Democrats who are now so vociferously opposing the use of the ma- [illegible]

ORCHARD HEATER CONTEST NEXT WEDNESDAY NIGHT

[illegible]

SCENE ON THE VANDERBILT CUP COURSE ON LONG ISLAND.

THE PENDING IRRIGATION PROJECTS

[illegible]

THRONGS TENDER TAFT AN OVATION IN INDIANA

Third Visit to Hoosier State Fully Equals the Two Preceeding Trips

DEMOCRATIC RULE MEANS A PANIC

[illegible]

VANDERBILT CUP WINNER MAKES A BETTER RECORD

[illegible]

WANT TO ORGANIZE WESTERN SLOPE [illegible]

[illegible]

O'CONNER ARRESTED FOR CONTEMPT OF DISTRICT COURT

Granted Writ of Habeas Corpus and Released on $500 Bond

HEARING MONDAY

[illegible]

LA JUNTA BOY COMMITS SUICIDE

[illegible]

SATURDAY'S FOOTBALL RESULTS FROM OVER THE COUNTRY

[illegible]

PAGE FACSIMILE: *The Daily News, Oct. 25, 1908, page 1*

PAGE TWO SUNDAY MORNING, OCTOBER 25, 1908. THE DAILY NEWS

THE DAILY NEWS

Office of Publication, No. 102 North Sixth Street, Grand Junction, Colorado.

PUBLISHED BY

The Grand Junction Printing Company

KARL A. BICKEL, ... Publisher
ROY A. M'CLINTOCK .... Business Manager

SUBSCRIPTION RATES

Mail or Carrier, one year in advance $6.00
Mail or Carrier, one month in advance .... .50

ADVERTISING—All objectionable medical, financial and matrimonial advertisements are refused. THE DAIL NEWS IS A HOME NEWSPAPER and can be read by every member of every household it enters.

Entered at the postoffice in Grand Junction, Colorado, as second class matter.

NOTICE TO SUBSCRIBERS.

Under the new postoffice regulations newspapers are placed practically on a cash basis. Paid up subscriptions are necessary to entitle publishers to the pound rate. Isn't it just as easy to pay promptly as to let your account run up and pay in a lump? Help us to obey Uncle Sam by prompt payment.

All display advertising copy must be in the office by three o'clock in the afternoon to be accepted for the next morning.

TYPOGRAPHICAL UNION LABEL GRAND JUNCTION

Republican County Ticket.

Representative	R. H. BANCROFT
County Clerk	RUEBEN E. STARR
Sheriff	MILTON P. SMITH
Treasurer	ALEX SEEGMILLER
County Judge	JUDGE W. S. SULLIVAN
Assessor	N. N. SMITH
Superintendent of Schools	C. G. SARGENT
Coroner	DR. H. S. DAY
County Commissioner No. 1	JOHN HYNES
County Commissioner No. 3	THOMAS DWIRE
County Surveyor	JAMES H. FISK

WHITEWATER CATTLE TAKES PRIZES IN KANSAS CITY

News Special Service.

WHITEWATER, Colo. Oct. 24.—The information is received today that the carload of yearling Herefords selected from the Casement ranch and shipped by Tom Smith, foreman, to the Kansas City show, took two first prizes there. The first prize awarded a the bunch of 20 head exhibited which were selected fro mthe carload, more than paid the freight in this shipment and the sweepstakes awraded as the best Hereford yearling feeders shown was considerably above that.

Mr. Dan Casement was at the show in person and was greatly delighted at the showing made by his stock ranch in the Unaweep canyon.

A fund is being raised by subscription in Whitewater for the purpose of building the Whitewater end of the Clifton-Whitewater road. The Highlands people who see the importance of this outlet both ways are ready to set apart the right of way on the line and build a graded thoroughfare over the top and to the foot of the slope on either side.

The Clifton Chamber of Commerce has taken the matter up and has chosen a point on the Grand river, a quarter of a mile west of the north and south line as the place where a bridge can be most economically and advantageously installed. Every hay raiser in the Whitewater and Kannah Creek county should get on the subscription list at Meserve's office to help push this needed outlet to the hay market in the fruit sections, because it means a short haul and a top price for his product.

Travel on the railroad and wagon roads is once more resumed. Damages from washouts were confined to a few small bridges, which have been repaired or replaced. The D. & R. G. has built a fine bent bridge at the outlet of the Snyder Wash, so that henceforth the town of Whitewater will not suffer from the August cloudbursts.

The cattle men are making heroic efforts to get out the cattle that are stalled in the snow or in pastures on the top of Grand Mesa and the Uncompahgre Plateau. Several cowmen are laid up with snow blindness and many have had hair-breadth escapes from perishing with cold.

If the weather continues unfavorable in the high country, there will be a heavy loss of cattle and horses.

The painters are putting the finishing touches on the fine new Pride of Kannah Creek school house.

Ed Walker has purchased theN. G. Rounds' house on the corner of First and Gaylord streets.

We have received our car of "Front Rank" furnaces and can now attend to your furnace work promptly. J. H. Lane.

Spring chicken like mother cooks at Cady's restaurant.

A Healthy Family.

"Our whole family has enjoyed good health since we began using Dr. King's New Life Pills, three years ago," says L. A. Bartlet, of Rural Route 1, Guilfor, Maine. They cleanse and tone the system in a gentle way that does you good. 25c at the C. D. Smith Drug Co.

Just arrived, Tobyhanna cigar, 5c, the real equal of any 10 cent cigar manufactured. Sturtevant-Quist Drug company.

Van Horn for drugs, fresh and pure.

PAGE FACSIMILE: *The Daily News, Oct. 25, 1908, page 2*

TAKE THE DAILY NEWS

THE "PEER GYNT' MUSIC

The name "Peer Gynt" has for years meant to the public nothing beside an exquisite suite of music, composed by Edward Grieg, the Norwegian. It is one of the most beautiful expressions of musical genius.

The few of broader culture and information knew the meaning of the four heads to the divisions of the suite, "Morning," "Ase's Death," "Anitra's Dance" and "The Hall of the Mountain King." They knew Grieg had found his inspiration in a lofty poetic drama by Henrik Ibsen, "Peer Gynt" by name, phases of which he translated into his tone poem and they knew also of another, second suite, "Peer Gynt" by Grieg, with three divisions: "Ingrid's Plaint," "Solveig's Song" and "Peer's Homecoming."

Richard Mansfield by his notable achievement in producing "Peer Gynt" on the dramatic stage, where it had been a stranger except for endowed performances in a few Royal Theatres of Europe, attracted renewed attention to the "Peer Gynt" music and since Louis James has appeared in the great play, every symphony and concert orchestra in America has played the Grieg music, in many instances the rarely heard second suite.

Ibsen did not believe when he wrote his play that it would ever be acted. The scenic difficulties were too overwhelming. But Grieg seized on the humor and pathos and fantasy of the great poem for his first suite. Then the Royal Theatre of Christiania decided to attempt a production of the drama. The scenes difficult of pictorial realization were eliminated bodily. Louis James presents in the Richard Mansfield version a third more than Ibsen edited for acting. At Ibsen's behest, Grieg wrote more "Peer Gynt" music and the second suite was the result.

All of Greig's "Peer Gynt" music and much other Norwegian characteristic music is made conspicious embroidery of this elaborate production, which is presented here in its original Mansfeldian entirety.

The themes of all the music in "Peer Gynt" have been gathered into an overture by Clarence Lucas, who arranged the musical setting of the play, and as the last strains of the overture die away, the choir of strings take up the delicious strains of "Morning," the tone picture of the rising mists, the refreshed nature, the carolling birds and the brilliant sun. During this the curtain rises on a picture of a Norwegian mountain with the mill, the waterfall, the brook at hand, and the peaks piled up beyond. Peer and his mother come down the mountain as the music fades away and he begins the scene of incomparable imaginative lying and banter.

The scene soon changes to the Hegstad farm during the festival attending the wedding of "Ingrid" and "Mads Moen". An old Norwegian country dance theme is used as the accompaniment of the characteristic dance of lads and lassies on the green. "Peer" drinks to excess, and(provoked by "Solveig's" refusal of him, in sheer bravado, steals his former sweetheart, "Ingraid," and escapes in perilous flight over the mountains. Before act two "Ingraid's Plaint" is played, and directly after illustrated in the scene high up among the now-capped peaks between "Peer" and "Ingraid."

The liquor he has drunk soon shows its final results and "Peer" sinks in a swoon. The w[illegible] and amber s[illegible] change instantly to a supernatural green and the balance of the scene and all the next is thus indicated as the illusion of "Peer's" drugged mind. Every Norwegian knows the folk story of the Trolls or goblins who dwell up in the Ronde mountains. It is the daughter of the "Mountain King" who appears to "Peer" and summons her steed, a huge wild boar, on which "Peer" and she ride over the mountains to the Hall of the "Mountain King." As the curtain falls, the orchestra takes up the slow pizzicato movement of Greig's fourth number to his first suite. It represents the dance of the Trolls or weird goblins of Norwegian folklore. The heavy steps soon become nimble and finally the tempo lashes itself in a mad, swirling, irresistible presto. When the curtain rises again, it is on the assembly of the Troll King's goblin court, a scene which has been compared to the celebrated Walpurgis revelries in the "Broken Scene" of "Faust." Greig's "Hall of the Mountain King" is again used in this scene to accompany the dance of the imps and elves.

When "Peer" awakens from his Troll experience, his adventures take him home to his mother, "Ase," whom he finds dying, and this scene between mother and boy is one of the most exquisite pieces of poetry in all drama, and as the auditor listens he readily understands the inspiration which drew from Grieg one of the most beautiful tone pictures in music, "Ase's Death," which is played before and at the end of the scene.

"Peer" now "fares beyond seas" and after thi[illegible] years, he is encountered in act fou[r] an illustration of the adage "no fool [li]ke an old fool." "Anitra" and a coterie of attendants strip him of his wealth and send him off over the desert celebrating their shady achievement in a ballet and song to the light, careless, tripping of Greig's "Anitra's Dance."

A vision of the faithful "Solveig" awaiting "Peer," is revealed in this act and the poem is sung by "Solveig to Greig's familiar "Solveig's Song." The next musical illustration by Greig is in hte fifth act, "Peer's Home-coming" as the aged [h]ome-wanderer returns to find peace, [j]oy and unending love in the arms of [the] pure woman who has opened he[r h]eart to him as a boy and retained h[er i]deal of him through all her life lo[ng].

The oth[er m]usic incidental to the action and i[llus]trative of the sentiment of the suc[cess]ive scenes is drawn from a variety [of N]orwegian composers and from the [folk] songs of the Scandinavians.

The Daily News Phone is Junction 2.

PAGE FACSIMILE: *The Daily News, Oct. 25, 1908, page 3*

PAGE FOUR SUNDA MORNING, OCTOBER 25, 1908. THE DAILY NEWS

TOOL CHEST IS STOLEN

A robbery of a chest of tools was ported yesterday by a carpenter working on the Reed block, but thec chest was found in the alley in the rear, having been opened and the contents taken. The owner prizes them very highly and the police are at work on the case but nothing which would lead to an arrest was discovered yesterday.

ROY SMITH BUYS JEWELRY STORE

It was announced yesterday that Roy Smith who has been employed in the S. W. Moody jewelry store for the past two years has purchased the Ridley jewelery store of Montrose. He and his wife will leave soon for their new home.

TIGERS DEFEAT TELLER INDIANS

The Indians and the Tigers again clashed yesterday afternoon and again the Tigers won by a score of 25 to 0. The Indians in the second half played better ball than the score indicates but the husky supporters of the orange and black easily defeated them. The game was noticeably full of wrangling and the crowd were in the way of the players many times making fast and open play almost impossible. The Indians were faster than in their last game and worked the short forward pass to an advantage. The locals attempted the forward pass and off side kick many times but fumbled it too often to make any material gain on it. Barber for the locals played a very fast game and his defensive work proved to be a stumbling block in the way of the Indians' fast plays. Budeller stopped many of the rushes and showed up in great shape.

The game was not as fast as it might have been, but the condition of the field, which was very muddy, made the fast plays, which the locals figure so prominently in, hard to get off and for this reason the score was held to 25-0. The Indians have had the advantage of some excellent coaching shown by the way that they get off the forward pass and short line bucks but are still slow on their feet.

MOOREHOUSE CASE BEFORE THE COURT

The case of George Moorehouse charged with forgery and raising a check was yesterday bound over until the December term of the district court. From the evidence it appears that the young man who was acting as R. A. Mathew's foreman in Palisade, received the checks from R. A. Mathews to be paid to the workers on his place in that city. One of the workers by the name of C. A. Walters was given a check for $94 and Moorehouse told him it included Walter's wages, which amounted to $24 and some money for himself.

Walters endorsed the check and Moorehouse received his $70. Walters, suspicioning something wrong, telephoned in to Mathews and asked if all was right. Mathews, believing that it was a forgery scheme put the local sheriff's office on the case and Moorhouse was arrested.

(TOO LATE TO CLASSIFY)

WANTED—To rent, eight to ten room house, either wholly or partly furnished; close in. Phone Junction 484.

MODERN COTTAGE $2,500—Five rooms pantry, hall, closets, bath, good cellar, electric lights, gas, two lots, lawn, fruit trees, small barn, cash $700, balance monthly. A snap. Must sell at once. Phone Red 2362.

HARRY GRAY AT GLENWOOD

Word has been received from Harry Gray that he is safe at Glenwood and will probably arrive in this city today. He made the trip through the gorge and experienced some very hard knocks but arrived without the machine being put out of commission. Cashier Francis of the Globe Express company received the communication this afternoon.

POSTPONED.

Hopkins' lecture tonight (Sunday) at opera house, 8 p. m.

W. E. Dean and wife, who have been on their honeymoon for the last few weeks arrived in the city yesterday afternoon. They will go to housekeeping in the near future.

Frank and Will Ferrier will leave today or tomorrow to bring back their father who is snowed in on the other side of Whitewater. The old man's condition is not serious but the provisions are getting low and his sons are getting uneasy concerning his welfare.

The Elks' have arranged a very fine program for their social session which will take place this coming Monday and all those favored with invitations expect a royal time.

The Daily News Phone is Junction 2.

ROOSEVELT THANKS THE JAPANESE

By Associated Press.

WASHINGTON, Oct. 24—Ambassador O'Brien at Tokio today presented to the emperor of Japan a message from President Roosevelt conveying the thanks of the American people for the treatment accorded the officers and men of the Atlantic fleet on its voyage to that county and saying: "The people of the United States have been deeply touched by this fresh and striking proof of friendship and regard."

CAMPAIGN EXPENDITURES.

Democratic Party Cannot Claim Unsullied Virtue as to Source of Its Revenues.

(From Gov. Hughes' Youngstown Speech.)

"Mr. Bryan also has much to say with regard to corrupt practices and campaign expenditures. But he omits to give due credit to the Republican party for what it has accomplished with regard to these important reforms, an accomplishment the more noteworthy in the light of Mr. Bryan's reiterated criticisms of contributions to Republican campaign funds. In the State of New York a Republican Legislature in 1906 passed a statute prohibiting corporations from making any political contributions directly or indirectly, and providing that officers, directors or stockholders participating in or consenting to the violation of the law should be guilty of a criminal offense. And I know of no more drastic statute in this country with regard to the publicity of campaign contributions and for the prevention of corrupt practices than that passed in New York under Republican auspices. These were not promises of an opposition party seeking power, but enactments by a party in power securing genuine reforms. Proper recognition must, of course, be given to the patriotic Democrats who supported these reforms, but they were enacted by a Republican administration. Congress has also legislated against political contributions by corporations. Purity of elections and free opportunity for the uncorrupted expression of the popular will lie at the foundation of every reform and cannot be too carefully safeguarded. And there should be federal legislation securing proper publicity of and accounting for campaign contributions in connection with federal elections. But it must fairly be recognized that the skirts of neither party have been clean. If we search Mr. Bryan's following we shall find not a few who have sinned, and also those who, if we may judge from their local activities, are still unrepentant. The Democratic party cannot claim unsullied virtue either with regard to the source of its revenues or its readiness to receive them."

All the members of Miss Chester's vocal class will meet at the Odd Fellows' hall Monday at 2:30 o'clock.

E. D. Nelson who was in the hunting party which nearly perished on Pinon Mesa is reported as being worse and will probably be brought to the hospital in this city.

THE SOCIALIST CANDIDATE FOR PRESIDENT ON THE STUMP.

Eugene V. Debs, the Socialist nominee for president, began his campaign before any of the other candidate and has already covered much of the country. His private railroad train, known as the "red special," is scheduled to complete a campaigning tour of 3[illegible]00 miles before election day.

THE NEXT PRESIDENT

It has been conceded by both parties, that we will have a president to succeed President Roosevelt. Whoever the next president of the United States may be, I will continue to fit glasses for the relief of eye-strain which causes headaches, nervousness, chorea, vertigo and stomach trouble. Cross-eyes can, in a great number of cases, be straightened by the use of properly fitted glasses. Consultation free.

CHAS. GASHO, Optometrist. Rooms 1 and 2, Margery Bldg.

Malleable and Universal Malleable ranges. Crown Furniture Co.

J. ARENTSEN.

Expert sewing machine repair man at Lynch's, 236 Main; phone Red 621.

Beef, pork, mutton, lamb, veal, spring chicken, fish, oysters, ranch eggs—the best of everything the local and Denver markets afford, at Cady's.

DO YOU WANT BRICK?

We will sell you white pressed brick a great deal cheaper than you have been paying for other pressed brick, and they are harder and make a beautiful residence or business block, if layed up in black mortar. See Moyer's residence or the Colorado Fruit Growers' association block. Write for sample and prices on our brick and drain tile.

THE DELTA BRICK AND TILE CO.

Where Bullets Flew.

David Parker of Fayette, N. Y., a veteran of the Civil War, who lost a foot at Gettysburg, says: "What the good Electric Bitters have done is worth more than five hundred dollars to me. I spent much money doctoring for a bad case of stomach trouble, to little purpose. I then tried Electric Bitters, and they cured me. I now take them as a tonic and they keep me strong and well." 50c at the C. D. Smith Drug Co.

PAGE FACSIMILE: *The Daily News, Oct. 25, 1908, page 4*

KINGS OF SERVIA AND BULGARIA AND THE LATTER'S STAFF.

Nowadays Ferdinand of Bulgaria is the most conspicuous ruler in Europe, and old King Peter of Servia is scarcely less so on account of his determination to fight the Austrians. Both these Balkan sovereigns are men of progressive ideas.

CLIMATE OF THE GRAND VALLEY

By R. M. Hardinge

The part of the valley of the Grand river treated of in the following discussion is that portion of the Western Slope that lies approximately between Palisade on the east and Fruita on the west, a distance of about 20 miles. The valley is V-shaped, opening to the west and having a miximum width of something less than 20 miles. The land is comparatively level; its altitude is about 4,600 feet above sea level. The valley is enclosed on the north by the Book Cliffs, almost perpendicularbluffs of clay, sand and rock about 1,000 feet high. On the south rise the more gradual levels of the Uncompaghre Mesa, attaining altitudes of something near 1,500 feet above the valey within a distance of 15 miles. At the head of the V stands the majestic Grand Mesa, stretching its rim rock against the eastern horizon at an altitude of 5,400 feet above the valley.

The prevailing drift of the atmosphere at this altitude is toward the east, hence the location of the valley causes it to be enveloped in a comparatively stagnant atmosphere of considerable depth. This checking of the surface winds normal to this latitude and altitude is an important factor in rendering the climatic conditions different from that which obtains over the Western Slope in general. The force of the surface winds is reduced; the winter's cold is modified, and the heat of summer emphasized. Another important factor tending to influence the temperature conditions is the line of bluffs bordering the valley on the north. Facing the south these cliffs receive the full effects of the sunshine, and reflect and radiate the heat into the valey. "Vast warming pans," a term aptly applied.

These conditions, taken together with the low relative humidity of the semi-arid west, an unusually slight change of mean daily temperature, and the small precipitation, combine to make the climate peculiarly attractive to the lover of out-door life. Cold, cloudy and blustery weather is seldom experienced, even for a day, and although midsummer temperatures often reach a high degree the heat quickly passes with the sun. With hardly a single exception in any one year the nights are cool and invigorating. On account of the dry, rarefied air the hottest part of a midsummer day is almost never oppressive. There are few days in the year that do not permit of out-door work being carried on under comfortable conditions.

The average climatic conditions of the valley may be gathered from the records of the weather bureau station at Grand Junction. These data cover the period from 1892 to the present time. The first seven years of record were mad by Dr. Samuel M. Bradbury, who early realized the value that trustworthy data would have in a locality so favorably situated as this. These records show that the annual mean temperature is a little more than 52 degrees. The highest temperature recorded was 104, on the 28th day of June, 1900, but that this temperature is extreme for the valley is shown by the fact that the temperature has not gone above 99 degrees during the past six years. The lowest temperature, 18 below zero, was recorded on the 20th of January, 1898, but the temperature has not dropped below zero since February of 1905. The average percentage of sunshine is 70, which compares very favorably with Denver, Colorado Springs, Santa Fe, El Paso, and others famed for their unclouded weather.

The total yearly precipitation, including rain and snow, averages a little less than eight and a quarter inches. The winter months average 0.48 of an inch; spring, 0.79, summer 0.61. The month of August, averaging 1.06 inch of rain, is the wettest month, or gives about 21 per cent of the total amount. Light thunderstorms are frequent during the summer. The destructive electrical storms frequently experienced east of the Rocky mountains are almost unknown in the Grand valley, occurs, on the average, three or four times a year, but the records of the weather bureau station show that its injurious effects have been slight. The average annual snowfall, unmelter, is 19 inches, being heaviest usually in February. The greatest depth lying on the ground was 9 inches on the level, occurring in January and again in February, 1905.

The average hourly wind velocity is a little more than five miles an hour, being greatest during the spring, and during the early afternoon hours throughout the year. The highest wind during the past ten years was 4 miles an hour from the southwest on April 30, 1902. The prevailing direction is very generally from the southeast to south, from midnight to noon, and northwest from noon to midnight. "The wind follows the sun" is an expression often heard.

The latest date on which a damaging frost occurred in spring was May 14, in 1907, and May 28, 1908. There has never been an entire failure of the fruit crop from this cause, but the output for the two years just mentioned was greatly reduced. In some localities of the valley, however, during these two years, there was little or nothing on the trees, but the loss was largely recovered from root crops and melons planted after the damage was apparent. From the records the earliest date of killing frost in autumn is September 14. The average date on which the last killing frost of spring occurs is April 20, and the first of autumn is October 17. From these dates it can be seen that frosts damaging to the agricultural interests of the valley seldom occur.

Any discussion of the damaging effects of frosts to the fruit crop of the valley should mention the efforts made by several prominent orchardists last spring to save their fruit by burning smudge pots of oil or coal in the orchard during the threatening period. Their efforts were highly successful. The success was fully demonstrated by the fact that the trees thus protected bore abundantly, while all other trees in the immediate locality, and subjected to the same general atmospheric conditions, bore practically no fruit. These demonstrations were conducted under what appear to be extreme conditions, although the same conditions may occur any year, and show without a doubt that the fruit crop of the valley may be protected from frosts by this means under any conditions that are likely to arise during the frost season.

HUNTING PARTY IS NOW SAFE

Walking thirty-five miles to Palisade to tell the anxiously waiting friends and relatives of the hunting party, which had been gone for over a week, Richie Budelier yesterday staggered into that town nearly dead from the cold and exhaustion. The remainder of the party which consists fo well known young men in the valley, remained in camp to look after the horses, and sent Budelier to tell the friends and relatives that the party was safe and had plenty of provisions and would return as soon as the weather would permit. Budelier said that he never had experienced such cold and suffering and many times would have liked to stop and rested but knowing that it meant freezing if he did, he kept on and arrived in Palisade yesterday morning.

The party is situated at the head of Buzzard creek and from the reports of Budelier are snowed in but have a good supply of food and will wait until the snow melts a little until they return. They have been gone over two weeks and only expected to stay a week.

PRACTICAL DEMONSTRATION

Actual Operation of Postal Savings Banks in Philippines.

Two Years' Experiment Indicates the Plan Is Successful.

The only postal savings bank system in operation under the American flag is that in the Philippines, where, after two years of experiment, the institution has been declared a great success. In the report of the Philippine commission, just published, the postal savings bank is discussed briefly, in part as follows:

"The postal savings bank was first opened in Manila on Oct. 1, 1906. The law provides for three classes, with limitations as to the amounts that can be received at any one time in each class. Following the opening of the Manila office, other postal savings bank offices were opened in various cities, of different classes, until at the close of the year 235 had been opened, of which 9 were first, 86 were second and 140 were third class.

Filipinos Adopting System.

"At first the Filipinos did not to any great extent avail themselves of the opportunity of depositing a portion of their receipts in the postal savings bank, anda great bulk of the depositors were Americans and of other nationalities. As the advantages of this system have been disseminated through means of schools, government bureaus, private agencies and others, the proportion of Filipinos is continually increasing, and gives very good promise that the Filipinos will learn to trust their money to the care of the government, and welcome the opportunity.

"The accounts opened up to the close of the fiscal year numbered 2,676, of which 347 were closed, leaving 2,329 still open. The deposits reached the total sum of 786,361.03 pesos—an average of 108.21 pesos. Withdrawals reached the total amount of 280,499.21 pesos, leaving a net deposit of 509,463.81 pesos at the end of the fiscal year—or the first nine months of the operation of the postal savings bank—but by no means the first nine mouths of operation of the average number of offices, the smaller ones having been opened from time to time during the year, so that many had not been opened more than a mouth or two at the time these figures were taken.

Encouraging Small Accounts.

"The device of issuing postal savings bank stamps has met with some response. The stamps are simply a method by which small amounts can be deposited without the expense of issuing a book, figuring interest, and opening an account.

"On the whole, the reception accorded ... has been gratifying, and gives good promise. The investment of the funds is in the hands of a postal savings bank investment board, which carries a certain amount of the funds on deposit, and has limited power of investment. The law provides that 25 per cent of the money held on deposit may be loaned on first-class real estate, under limitations carefully drawn to protect the depositors, and 10 per cent on agricultural lands under still more rigid limitations."

AS ROOSEVELT SEES BRYAN.

"You say that you have advocated more radical measures against private monopolies than either I or my party associates have been willing to undertake. You have, indeed, advocated measures that sound more radical, but they have the prime defect that in practice they would not work. I should not in this letter to you discuss your attitude on this question if you did not yourself bring it up, but as you have brought it up, I answer you that in my judgment the measures you advocate would be wholly ineffective in curing a single evil, and so far as they had any effect at all, would merely throw the entire business of the country into hopeless and utter confusion. I put Mr. Taft's deeds against your words. I ask that you be judged both by the words you wish remembered, and by the words that seemingly you and your party now desire to have forgotten. . . . I hold it entirely natural for any great law-defying corporation to wish to see you placed in the Presidency rather than Mr. Taft. Your plans to put a stop to the abuses of these corporations are wholly ...merical. — Theodore Roosevelt."

If it is true that Roosevelt is going to take to the stump Bryan will have to take to the woods.—Philadelphia Press.

Mr. Bryan no longer refers to Mr. Roosevelt as his imitator, counterpart and understudy.—St. Louis Globe-Democrat.

O. Z. Glenn representing the Frederic Stearns Manufacturing company of Detroit, was in the city yesterday.

L. and A. Baum, prominent merchants of Galena, Kansas, are visiting their old friend W. P. Stough of this city. The young men are well pleased with the prospects for business in this and may locate here.

FRUITA DEPARTMENT

GEORGE W. BELSEY, Jr., Fruita Correspondent.

The branch office of The Daily News in Fruita is located at the office of Bruner & Sharpe, next door to the postoffice, phone Blue 95. All news items or advertisements left there will receive prompt attention.

Methodist Church.

Ten a. m. Sabbath school; good teachers and suitable classes for all who come.

Eleven a. m., Divine worship, sermon by the pastor, subject, "Falling Leaves," a lesson of the season.

Six-thirty p. m. Epworth league, leader, Laura Hiatt, subject, "God in the Christian's Heart."

Seven-thirty p. m., People's Popular service. This is a service especially planned to be interesting. Song service followed by sermon. Subject is: "The Devil," suggested by a much-advertised play book. Young people especially invited to this service.

Congregational Church.

The meeting arranged for last Sunday in the interests of "Men's Day," were postponed on account of the storm and will be held today.

"Men Wanted in Business Life," by Rev. A. L. Coudy.

"Men Wanted in Politics," by Rev. J. A. Walker.

"Men Wanted By and for the Boys," by, Rev. R. Rutt.

"Men Wanted in the Church," by the pastor.

Baptist Church.

All our regular services will be according to schedule today.

The theme in the morning will be "Advancement," in the evening "The Simplicity of the Gospel."

The public is invited.

Christian Church.

Services today in the Masonic hall. Bible school for old and young to study God's word at 9:45. Be present. Morning service at 11:00 a. m. Subject, "Service." Matters of great importance are to come before the congregation at the morning service. Evening service will be announced Sunday morning. We always extend the hand of welcome.

Christian Science Church.

Christian Science church Sunday, 11 a. m. at the residence of C. H. Barnes, Subject, "Probation After Death."

Doctor Porter and Deacon Cranston are back from a trip to the hills after deer. They are still patronizing the butcher shops.

Miss Lillian Weckel is down from Grand Junction over Sunday with her mother.

Miss Edith Moore was a Grand Junction visitor yesterday.

Miss Mae Nugent entertained yesterday afternoon in honor of Miss Margaret Cox who is to be married on Monday to Mr. Chalsey Hobson. The affair was in the nature of a "towel shower" and an enjoyable time was reported by those present. A dainty luncheon added greatly to the pleasure of the afternoon.

The infant child of Mr. and Mrs. G. Howell died on Friday evening. The child was buried yesterday afternoon in Elmwood cemetery, Rev. Dickey of the Methodist church officiating.

Dr. C. B. Masser recently purchased a new automobile and the machine has just arrived. As soon as the mud clears up the doctor will call on his patients in this "devil wagon."

PAGE FACSIMILE: *The Daily News, Oct. 25, 1908, page 5*

PAGE SIX — SUNDAY MORNING, OCTOBER 25, 1908. — THE DAILY NEWS

CLASSIFIED WANT ADS

FOR SALE—REAL ESTATE.

FOR SALE—Modern 7-room house on Chipeta avenue; inquire at 920 Ouray.

TWENTY ACRES—Fine ridge land, full water right, three-fourths mile from Fruita, at $125 per acre. J. H. Schiesswohl, 123 South Sixth street.

BARGAINS in high line lands in small and large tracts. $40 per acre and up. J. H. Schiesswohl, 123 South Sixth.

FOR SALE—Five room house at 248 Chipeta avenue. Bargain for ten days at $1050. Easy terms. 502 Grand ave. or phone Black 789.

HOUSE at 729 Colorado for sale; 10 rooms, modern. Reason for selling, poor health. A bargain if sold quickly. Mrs. M. E. De la Mar. Phone Black 1674.

FOR SALE—Twenty-acre tract on Orchard Mesa; fine lands; best location, on main traveled road, $3,500. See owner, "Z."

FOR SALE—MISCELLANEOUS.

FOR SALE—Side board, good as new, at Cady's cafe.

FOR SALE—Lady's ticket to Lincoln, $10; boy's ticket to Lincoln, $5. Phone Walnut 64.

FOR SALE—Team of horses weighing about 1100 pounds each; also single buggy and harness and good family cow. 502 Grand avenue, or phone Black 789.

FOR SALE—Or will trade for city property, three pieces ranch land; inquire L. C. Hedges, 'phone Red 1362.

FOR SALE—Good horse and buggy, cheap. Phone Black 2171.

FOR SALE—The best meal in the city at Brunnick's cafe, 109 N. Fourth.

FOR SALE—Lady's ticket to Kansas City. Phone Red 3291.

FOR SALE—Alfalfa. Prompt delivery. Phone Red 1623.

FOR SALE—Homegrown Carman, Champion, Salaway and Philips cling peach trees. Also D'Anjou, P. Barry, Walker Wakefield and Bartlett pear trees. Russell Bros., phone Brown 132, Palisade.

FOR RENT.

FOR RENT—9-room house on North First street, $15 per month. Schiesswohl, 23 South Six.

FOR RENT—Furnished room for two gentlemen preferred. 548 Grand ave.

ROOMS—Furnished, 102 W. Rood.

FOR RENT—Six room house, 335 N. Spruce. Phone Red 84.

FOR RENT—Strictly modern office rooms, first door south of courthouse. J. H. Schiesswohl, 123 South Sixth street.

GUNS TO RENT—Lynch, 236 Main.

MISCELLANEOUS.

WANTED— To buy iron safe. 340 Main.

WANTED—Scissors to grind, umbrellas, trunks and valises to mend. B. Lynch, 236 Main St. Phone Red 621.

WANTED—Two or 3 furnished rooms for light housekeeping, close in on north side. H 10, News office.

WANTED—Sewing by the day. Mrs. Outten, 354 Rood.

MIDWIFE—Graduate, 24 years' experience; answer calls day or night. Mrs. H. Feeney, 725 South street.

WANTED—Several copies of The News of October 15. Call up The News.

WANTED—Clean, cotton rags at this office. Hurry up!

WANTED—In Clifton district, 20 acre fruit ranch, from 5 to 10 acres in bearing trees; write particulars to "K. A. G." Daily News.

WANTED—To rent 10 or 12 room house south of Main street and west of Fifth, modern; 'phone 2196 Red.

TO TRADE—Automobile, White steamer for real estate. Call on Rex Graham, Mesa County National bank.

HELP WANTED—MALE.

WANTED—A first class blacksmith, steady employment and good salary for the right man. Address F. N. Jo Hantgen, Box 103, Meeker, Colo.

LOST AND FOUND.

FOUND—At 109 North Fourth, Brunnick's Cafe and grill. Everything first class; serve anything you order. Regular dinner, 25 cents.

FOUND—B. Lynch, locksmith and gunsmith. Saw filing and trunk and valise repairing. 236 Main St. Phone Red 621.

LOST—Small blue beaded pocket book; return to News office and receive reward.

LOST—Black fur glove, right hand; finder will favor owner by leaving same at News office.

LOST—Saturday afternoon on Main st., a $5 bill by a hardworking woman. Mrs. E. Oakes, 641 W. Colorado.

REPUBLICAN MAJORITY IS FULLY CONCEDED

DENVER, Colo., Oct. 24.—It is now fully conceded in the councils of the Democratic party that unless something extraordinary is done the entire Republican ticket will be elected with the usual majorities. The general discontent and disputes among the Democratic candidates have let the cat out of the bag. It is the plan of the Democratic leaders to trade everybody on the ticket in the interest of Shafroth for governor and Hughes for United States senator.

The extraordinary and lavish expenditure of campaign funds has been directed almost wholly to the legislative ticket in the interest of Hughes and in other directions to the further interests of Shafroth.

Every voter should know of this scheme to knife the Democratic state ticket and to steal Republican votes for the legislative candidate and for Shafroth.

There is absolutely no reason for any Republican to make a trade of this kind. Reports from all over the state are uniformly encouraging and these reports from committeemen, chairmen and others coming into headquarters all indicating a solid Republican vote. There are absolutely no defectations except from the Democratic ranks to the Republican ticket. The fact that the Democratic party is sufficiently discouraged to enter into a scheme to knife its own ticket foretells an overwhelming defeat of the Democrats on November 3.

— Nothing Like — a TENDER STEAK

—for luncheon, breakfast or dinner—don't get tired of it as you do of other meats. But, there's steak and steak. The thick cuts, juicy, easily broiled, the "fit to eat kind," aren't picked up everywhere. To be had regularly at Kramer's. And there's the skill in cutting and trimming! And the attention!

Pay cash and get your money's worth at—

C. H. Kramer

337 Main St. Phone Jct. 198

Western University

of Industries, Science and Arts

Grand Junction, Colorado.

Autumn Quarter open October 1st, 1908.

Elementary School

School of Commerce

Western School of Music

Preparatory Engineering School

School of Domestic Science and Art

College Preparatory

Teacher Training

Civil Service and College Course

For information write

O. A. Young, President

The Postoffice Book Store

Phone Red 1971.

Headquarters for all kinds of School Supplies, and Magazines

CIGARS AT WHOLESALE: San Felice, Little Tom, Child's Cabinet, little Chancellor, Morey's Cabinet and New Tariff. All 5-cent Cigars.

Chancellors and Tom Moore, 10-cent Cigars.

M. F. FREDERICKS, Proprietor

Where Are You Going

to plant your orchard? Let me sell you a ten-acre tract in the Bonita valley. It will cost you from $1,250 to $1,500. Good sandy soil, first-class water right and easy terms. I can sell you a good two year old orchard in the Grand valley for [illegible] per acre.

Horace T. DeLong

555 Main Street

The Board of Trade

"A Gentlemen's Resort"

The Finest Wines, Liquors and Cigars to be had on the Market

Ed. McKinney, - Proprietor.

485 Main Street 'Phone Junction 317

THE UNION TRUST AND BANKING CO.

CAPITAL AND SURPLUS, $52,000.

Officers and Directors.

WM. E. DUDLEY, Pres. C. D. SMITH, Vice-Pres.

D. T. STONE, Cashier. ROBERT A. ORR

FRANK A. LYONS THOMAS B. SCOTT

Conducts a Safe Banking and Trust Business.

20 Acres Raw Land $2,000

15 Acres in Rose Vale sub-division 4,500

5 Acres Rose Vale, per acre 250

8-room modern residence, two stories, cement basement, barn, No. 820 North Seventh street. An elegant home.. 5,250

KIEFER BROS.

Room 6 Union Trust & Banking Co. Bldg.

PHONE OFFICE, RED 602. RESIDENCE, RED 2544.

COLORADO IRRIGATED LAND

The Best Investment That You Can Find

Money Invested in Good Irrigated Land is Absolutely Safe--You Can Always Sell and Realize Quick Profits or Have a Farm of Your Own With An Independent Income for Life.

10,000 ACRES

To be Sold at Once in Tracts to Suit.

LOW PRICES—EASY TERMS.

NO BETTER INVESTMENT WAS EVER OFFERED TO YOU. IRRIGATED LAND IS BECOMING SCARCE AND INCREASING IN VALUE FASTER THAN ANY OTHER REAL ESTATE INVESTMENT.

Examine this map, showing location of our land and the Otero canal.

LOOK WHERE YOU MAY

There is no other tract of land in Colorado so favorably located, which combines so many advantages, where the settler or investor gets so much for his money. This land is cheaper now than any other irrigated land in this state, but it is going to advance—and advance quick. There is a reason for this.

Last week a special train carrying the La Junta Board of Trade and prominent business men of the valley left La [illegible] to inspect the [illegible] constructed [illegible] reservoir. They were greatly pleased and predicted that this reservoir will double the value of our land over the prices we are now selling at. This will guarantee investors quick profits on their investment, and those who buy now will be able to take advantage of the raise in prices when they want to sell.

We have published a letter from one of the greatest irrigation engineers in Colorado, certifying to the adequacy of this water-right, and another letter from the president of the La Junta Board of Trade, saying that our land is as good as any in this state.

Cantaloupes Pay

From $75 to $250 Per Acre.

Sugar Beets

From $35 to $75 Per Acre Profits.

Alfalfa

From 18 to 25 Per Cent.

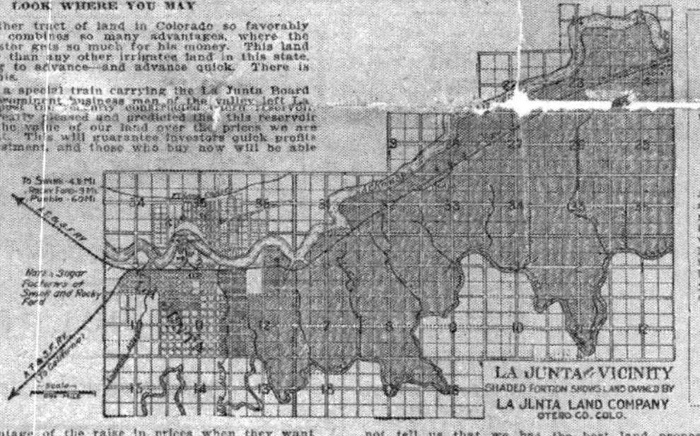

Write For Prospectus

And then come in person to verify our statements. This will stand the most searching investigation. No offer like ours has been made in Colorado, and it will be a long time before you will have another opportunity so good.

BUY LAND NOW

Every person working for a salary ought to begin the acquisition of at least a small tract of irrigated land.

There is a sense of security about the ownership of land. It will make you independent of the whim or caprice of employers.

You can put in the bank from the proceeds of ten acres more than you now earn and have the cost of your living besides.

You will not have to work so hard as you do in the store or office, and the occupation is one that will make you live longer.

That part of our land immediately adjacent to the town of La Junta is divided into 2 1-2, 5 and 10-acre tracts, which can be had on easy terms. This is all practically suburban property, and in time will be regarded as part of the city, and will become very valuable. Children can walk to fine schools in a few minutes.

A young man or woman cannot possibly make a better investment of his savings than in the purchase of these tracts.

In the Otero Irrigation District the land holders are the proprietors of the water-right, of the canal and of the reservoir. Their rights are protected by wisely-drawn statutes of the Colorado Legislature. They elect their own directors, levy their own maintenance assessments, and the county treasurer, under bond, is the treasurer of the district.

No other lands can participate in the benefits of this great irrigation system without the consent of the growers in this district.

In the 10,000 acres we are offering for sale, you cannot get far away from the railway or from the main canal, because the land extends in a long narrow strip between the canal and the Santa Fe for nine miles, with three townsites and three stations within that distance. Furthermore, there are three beet sugar factories within Otero county.

OUR DESCRIPTIVE BOOK

Gives detailed information about crops. It tells what you can do on this land. It contains letters of indorsement from some of the most responsible citizens of La Junta and the state. We don't ask you to take our word for anything. We ask you to investigate every detail personally, and if you do not consider this the opportunity of your life to buy land for a farm or an investment, you will be the first man that made a thorough investigation and did not tell us that we had the best land proposition that could be offered to farmers and investors.

Make up your mind at once to come to La Junta and let us show you our land. We are prepared to take you out and show you the land at any time. Our men are at your service and will give you every attention. Our great irrigation system has just been completed, and we are going to turn this land into one of the most prosperous sections in the Arkansas Valley.

REMEMBER

The purchase of this land is not like buying in a new, unsettled or undeveloped community. It is right in the heart of a rich, populous valley, and our tract extends into the very limits of one of the finest and most promising cities of Colorado—La Junta.

A great transcontinental railway system has its shops at La Junta, and pays its employes $30,000 a month; 2,300 miles of railway are operated from this point. La Junta is an excellent market for all kinds of farm products. The city has a public school system second to none in Colorado, and boasts a library of 12,000 volumes, as well selected as any in this country.

You can buy this land and feel that you are going to live in a community that is keeping step with progress, and you can be sure that your future is provided for by the ownership of land whose fertility never diminishes, being irrigated by the silt-laden waters of the Arkansas river.

FREE BOOK CUT THIS OUT MAIL TODAY.

The La Junta Land Company, 268 Santa Fe Ave., La Junta, Colorado.:

Gentlemen—Kindly send me at once your free book about the 10,000 acres of land you are offering to settlers at La Junta, Colo., in the Arkansas Valley and full details of your plans for settlement.

Name ....

Address ....

Town and State ....

Just Sign Your Name—Mail Coupon in an envelope at once.

The Prices Will Advance Shortly--Investigate Now While You Can Buy at Ground Floor Prices.

The La Junta Land Co., 268 Santa Fe Avenue, LA JUNTA, COLORADO

WE HAVE A FARM FOR YOU

PAGE FACSIMILE: *The Daily News, Oct. 25, 1908, page 6*

THE DAILY NEWS — SUNDAY MORNING, OCTOBER [illegible] 1908. — PAGE SEVEN

A FAMOUS SINGER WHO MAY BECOME A PRINCESS.

Mary Garden, reigning prima donna last season at the Manhattan Oper House, New York, who will begin a second American engagement in November, does not deny that she is going to marry Prince Mavrocordate, a wealth Levantine settled in Paris, where he is building a palace near the Trocader Miss Garden says she is not quite decided, but admits that the prince will com to America to visit her in January.

COMING AGAIN!
TO GRAND JUNCTION
at the St. Regis Hotel
MONDAY & TUESDAY
Oct. 26th and 27th
DR. K. C. SAPERO
Oculist and Aurist

EYE, EAR, NOSE
THROAT
SPECIALIST
Of Denver

is too well known to need any introduction. His references are his patients your neighbors and friends.

Hundreds of Patients in This Vicinity. No Incurable Cases Taken.

Cross-eyes straightened and cataracts removed. Granulated eyelids, sore eyes and catarrh successfully treated. Glasses scientifically adjusted by the latest and most approved, up-to-date methods Difficult cases solicited. Bring your school children.

OVER 800 PATIENTS

treated in Colorado. A large stock of glasses and artificial eyes on hand. A large proportion of headaches, hysteria, insomnia, chorea, St. Vitus dance, nervous prostration, hallucination, stomach troubles and other physical maladies, when of nervous nature, are caused by eye strain. The removal of the cause effects a permanent cure.

PHYSICIANS especially are invited to call and investigate his methods of correcting errors or refraction.

Dr. Sapero has been practicing in Colorado for many years. He is a graduate of the leading medical colleges of Europe and this country, and is licensed by the State Board of Medical Examiners of Colorado and other states.

REFERENCES furnished from some of the leading citizens of Grand Junction and vicinity, now under treatment.

Those unfortunate and unable to pay will receive medical and surgical treatment free.

REMEMBER Dr. Sapero does not experiment. He cures where others fail. Over 17 years practical experience.

THOUSANDS of testimonials from grateful patients, whom he has cured, furnished upon demand.

Dr. Sapero has been visiting in Grand Junction regularly for Many Years. All Work Guaranteed.

If you desire to consult the doctor please make sppointments early.

CONSULTATION AND EXAMINATION FOR GLASSES FREE.

a pleasant visit with friends in the city.

John Goetz was in the city yesterday on business.

W. H. Abrams of Eagle was in the city yesterday.

Roland Parsons was in the city from Montrose yesterday.

Attorney Rozzelo was in De Beque yesterday on business.

Mrs. F. H. Welles left yesterday on the Midland for Denver.

Rev. G. F. LeCiere has returned from an extended trip in Texas.

W. N. Hall of Leadville was in the city yesterday on business.

Allen Humphries of Montrose was a visitor in the city yesterday.

N. D. Cunningham of Ontario is one of yesterday's new arrivals in the city.

C. W. Sanker and wife of Iowa City Iowa, are visiting friends in the city.

J. A. Laurent of Kannah Creek was in the city yesterday for a short time

Mr. and Mrs. V. B. Lee returned yesterday to their home in De Beque after

W. C. Baughman of Pulaski, Iowa, is one of yesterday's new arrivals in the city.

William F. Walker and wife of Eldora, Neb., arrived in the city yesterday for a short visit with friends.

Mrs. Tickler and daughter of Painsville, Ohio, arrived in the city yesterday for a visit and will probably locate here.

Thomas W. Duckles the representative of the Svea Insurance company of Sweden was in the city yesterday on business for his employers.

Rev. D. Luther Edwards returned yesterday from a trip to Boulder where he attended the Presbyterian synod which was held there last week.

E. E. Udlock yesterday started work upon the removal of his present home to the lot back on Ouray and will build a fine residence on his present location.

Charged with the ill treatment of a horse, but the prosecution not having sufficient evidence to convict, Clarence Hoedenpile was yesterday acquitted of the charge in Justice of the Peace Sweeney's court.

NATURES' WARNING

GRAND JUNCTION PEOPLE MUST RECOGNIZE AND HEED IT.

Kidney ills come quietly—mysteriously,

But nature always warns you.

Notice the kidney secretions.

See if the color is unhealthy—

If there are settlings and sediment.

Passages frequent, scanty, painful.

It's time then to use Doan's Kidney Pills,

To ward off Bright's disease or diabetes.

Doan's have done great work in Grand Junction.

Mrs. J. J. Julian, of 141 North Third street, Grand Junction, Colorado, says: "I gladly testify to the great merits of Doan's Kidney Pills. For six years I suffered from kidney complaint. Especially when on my feet I was annoyed with too frequent action of the kidney secretions, they being of an unnatural color. Now and then I had a distressing feeling across my loins, and I would arise in the morning feeling lame and sore. Hearing so much in praise of Doan's Kidney Pills, I procured a box and was so much benefited by their use that I got a second box and was soon freed of all the distressing symptoms. Since using Doan's Kidney Pills, I have been in better health and stronger in every way than I have been for twenty years."

For sale by all dealers. Price 50 cents. Foster-Milburn Co., Buffalo, New York, sole agents for the United States.

Remember the name—Doan's—and take no other.

$ EACH WEEK

BUCK'S STOVES & RANGES

COMFORT INSURANCE

Advertising and Results

If you have articles [illegible] merit to sell it pays to let the people know. If you can't do as you adve[illegible]ise you'd better saw wood. We have only been in Grand Junction six m[illegible]nths. We have advertised and are satisfied with resnlts. Why? We ha[illegible]e the goods and want to please our trade.

Just now perhaps you are [illegible]anting a Range or Heating Stove. The Bucks Ranges and Heaters are O. K. We want to hav[illegible] you call and see them. We have other Ranges, priced as follows: 6-lid Merrit with warming ov[illegible]en for $23.00. 4-lid Reliance, $25.00. 6-lid Climax $33.00. 4 sizes and styles Superior Ranges f[illegible]om $45.00 to $60.00.

Ladies can find the good kind of cooking and kitchen fixings here. Chafing Dishes, 5 O'clock Teas, Carving Sets, Etc.

A word about your Builders' Hardware. We are without a doubt headquarters in Quality Style and Price. We are glac to show you.

The Ross Hardware & Supply Co.

521 Main Street. QUALITY PRICE SERVICE Phone Mesa 424

GREAT COMEDY DRA[illegible] COMING TO PARK THEATER [illegible]EXT MONDAY

Scene From the Comedy-[illegible]rama "Texas."

"Texas" a comedy drama in four acts is a picture drawn in glowing colors and gleaming sunshine of life on the boundless plains bordering on picturesque Mexico. J. Mauldin Feigl, the author, who was born in Texas and passed many years there, gives a truthful representation of the stirring life on the frontier. The story of the play teems with virile breath of the prairies; it fascinates with its powerful plot of love, intringue and adventure and compels many a hearty laugh with its bright comedy softened anon with a gentle touch of pathos.

The beautiful story is so naturally unfolded that it instantly appeals to the h[illegible]t of every man and woman in th[illegible]dience.

"T[illegible]s" is presented by a splendid arra[illegible] talent that has won the unstint[illegible] praise of the press and public. Ever[illegible] will be presented in a careful [illegible] painstaking manner, no expens[illegible] aving been spared in furnishing [illegible]borate scenery and every acces[illegible] from the hide-covered furniture [illegible] the cowboy accoutrements, whi[illegible] are authentic and were secured at [illegible] scene of the play. "Texas" has mad[illegible] annual tour of all the large citie[illegible] the United States for the past five [illegible]ars. This is its first presentation [illegible] the West and it will be at the Park [illegible]pera house Monday, Oct. 26.

Palisade Coal

The best coal on the market
$4.50 per ton delivered
$4.00 at yard.
Yard on Pitkin between Eleventh and Twelfth streets.

Walter Stokes

Phone Junction 284.
Office with Union Tea Co.

Van Horn for drugs, cream and pure.

Haskell

BILLPOSTING COMPANY

Out-Door Advertisers

See them before you tack on [illegible]bees or electric light or telegraph poles or other places.

Try a T-bone steak at Cady's Cafe.

Avoid Rank Speculation

A wise man said, "Real Estate is the basis of all wealth." Look back and call to mind the towns and cities you knew when **"Opportunity"** awaited you and you failed to take advantage of it. **Don't let it occur again. Buy an orchard. Buy a home. Buy city property** of any kind. But above all **Buy Land at this time on**

Beautiful Orchard Mesa

THE RAMEY - UDLOCK INVESTMENT CO.

116-117-119-121 FIFTH STREET. Phone Junction 106.

Low Prices

Cannot be where there is high rent. Have your bicycle work done at 228 Main St., where there is no rent to pay and prices are right.

A. C. FLETCHER

228 Main Street

PAGE FACSIMILE: *The Daily News, Oct. 25, 1908, page 7*

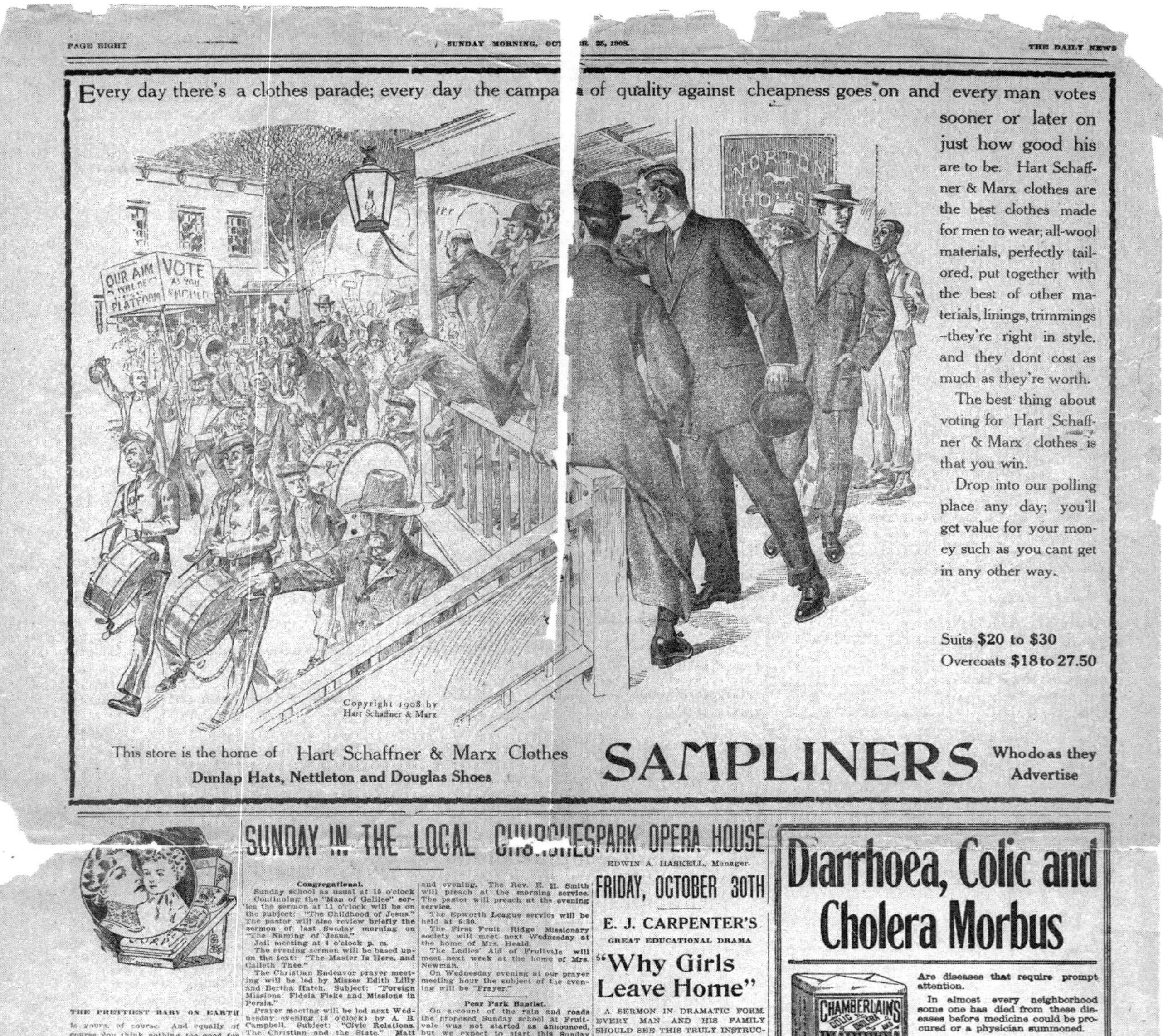

PAGE EIGHT — SUNDAY MORNING, OCT R 25, 1908. — THE DAILY NEWS

SUNDAY IN THE LOCAL CHURCHES

Congregational.

Sunday school as usual at 10 o'clock. Continuing the "Man of Galilee" series the sermon at 11 o'clock will be on the subject: "The Childhood of Jesus." The pastor will also review briefly the sermon of last Sunday morning on "The Naming of Jesus."

Jail meeting at 4 o'clock p. m.

The evening sermon will be based upon the text: "The Master Is Here, and Calleth Thee."

The Christian Endeavor prayer meeting will be led by Misses Edith Lilly and Bertha Hatch. Subject: "Foreign Missions: Fidela Fiske and Missions in Persia."

Prayer meeting will be led next Wednesday evening (8 o'clock) by A. B. Campbell. Subject: "Civic Relations. The Christian and the State." Matt 22:21, Rom. 12:17-18.

Presbyterian.

The pastor left Monday for Boulder to attend the meeting of Synod and will be absent during the week. He will return for today's services and should be greeted with a large audience, both morning and evening.

The Ladies' Aid was postponed until next regular meeting, because of the bad condition of the roads.

The brotherhood are endeavoring to secure the attendance of one hundred men every Sabbath evening in November. Will not each man of the congregation make a pledge to be present at the evning service during November?

The prayer meeting on Wednesday evening will be led by Rev. Johnson and will be a helpful service for all who attend.

Christian.

Bible school Lord's day 10 a. m.
Junior C. E. at 4 p. m.
Senior C. E. at 6:30 p. m.

We expect Bro. Myers of Kansas City with us on Lord's day, Oct. 25, for both services. Definite announcement will be made later.

Word comes from Barboursville, Ky., that he is in a great meeting at that place, and meeting with good success. Let us greet him with a good audience.

Sec. No. 2 of the Woman's Union meets with Mrs. Hutton, 920 Ouray, Wednesday all day. Bring thimbles and scissors. Sec. No. 3 meets with Mrs. Hibbs, 832 North Fourth street Wednesday at 2 p. m. A full attendance is desired.

Prayer meeting Wednesday evening 7:30 p. m.

Orchard Mesa.

"What Is Thy Name?" This question by the "Angel of Jehovah," will be the text of the twenty-minute sermon. Sunday school, as usual at 3 o'clock.

Handy A. M. E.

Preaching at 11 a. m.

Sunday school at 2 p. m. and will elect officers for the year.

Preaching at 7:30 p. m. Subject "More Labors."

Prayer meeting every Wednesday evening.

Methodist.

Services will be held at the Methodist church on Sunday, both morning and evening. The Rev. E. H. Smith will preach at the morning service. The pastor will preach at the evening service.

The Epworth League service will be held at 6:30.

The First Fruit Ridge Missionary society will meet next Wednesday at the home of Mrs. Heald.

The Ladies' Aid of Fruitvale will meet next week at the home of Mrs. Newman.

On Wednesday evening at our prayer meeting hour the subject of the evening will be "Prayer."

Pear Park Baptist.

On account of the rain and roads the proposed Sunday school at Fruitvale was not started as announced, but we expect to start this Sunday without fail. Let all in the community, who are interested in this good work be on hand at 3 p. m. We hope to make this school the very best. Come and help us. At the same hour a like school will be opened at Garfield school house. In both of these districts there are many who do not attend Sunday school anywhere. It is tl meet the wants of these that the work is undertaken.

At the church, 10 a. m., Sunday school, followed by sermon by the pastor, as usual. At the evning service following the B. Y. P. U., Rev. W. T. Davis of Grand Junction will speak on "Temperance," while the pastor is to speak on the same subject t the Bethel M. E. church.

First Baptist.

Sunday school at 9:45 a. m.
Sermon at 11 a. m.
B. Y. P. U. at 6:30 p. m.
Sermon at 7:30 p. m.

A full attendance of the churc and congregation is requested, as w desire to plan for the special me tings to begin on Nov. 8, and to pra and plan for the special work under aken by the young men of the church

St. Matthew's Church.

Corner White and Fourth, services for 18th Sunday after Trinity Celebration Holy Eucharist 7:30. Sunday school 10, morning prayer and litany 11, evening prayer and sermon 7:30. Wednesday feast of S. S. Simon and Jude celebration Holy Eucharist 7:30 a. m. Meeting of St. Matthew's Guild at the rectory 3 p. m. sharp. All women of the congregation invited. Choir rehearsal Thursday 7:15 p. m in the vesting room. Strangers always cordially welcomed to any and all of the services. John W. Ohl, rector.

Earl Graham of Denver, a platform orator of splendid abilities, will speak at the court house tomorrow afternoon at 3 o'clock upon the subject "How to Make Colorado Dry." Mr. Graham comes to the city with the very highest recommendations, and will certainly please those who go to the court house. He is also a violinist of note and will play several selections during the afternoon. Everybody is extended a most cordial invitation to be present to hear Mr. Graham speak upon this subject.

THE DAILY NEWS ADVERTISERS DO AS THEY ADVERTISE.

DID 'HONEST JOHN' DODGE A CHANCE?

News Special Service.

DENVER, Colo., Oct. 24.—"Honest John" Shafroth dodged the only chance he ever had in congress to do something for the beet sugar industry of Colorado. His present efforts to make the people of this state believe that the Democrats will maintain a tariff on Philippine sugar should burn holes in his conscience.

Here are the indisputable facts: Under Cleveland the Democrats in congress took off the two cent bounty on American beet sugar and the protective tariff from foreign sugar placed there by the Republicans. Then the Democrats voted solidly against the Dingley bill, which restored protection to the beet sugar industry.

Shafroth refused to vote for the protection of this Colorado industry by simply answering "present" when his name was called.

Three years ago when the Philippine tariff bill was up in congress the Democrats voted solidly for free sugar from the Philippines. They did more; they introduced a substitute for the Republican measure, which substitute was signed by Congressman John Sharp Williams, Champ Clark, William Bourke Cochran, Oscar W. Underwood and D. L. D. Granger, the Democratic minority members of the committee, which substitute was entitled:

"A bill to secure to the Philippine Islands free trade with the balance of the domain of the United States."

And its first paragraph reads:

"Be it enacted by the senate and house of representatives of the United States of America in congress assembled:

"That from and after the date of the passage of this act all goods, wares, merchandise, or other articles being a growth or product of any of the islands forming a part of the Archipelago called the Philippines, or the island of Guam, when shipped to any port of Hawaii, Porto Rico, or to any other port of the United States, shall enter any said port free of import duty."

In reporting the passage of the Republican bill William J. Bryan said in his newspaper, the "Commoner" under date of July 26, 1906:

"The Democratic position was for absolute free trade with the islands on the theory that if the Philippines were recognized as United States territories no duty should be levied on sugar, tobacco, rice or any other products."

NEW PIANO FIRM.

The Lindholm Piano company, A. S. Lindholm, general manager, has just opened an office in the Schiesswohl building on Sixth street. The new firm has on the road a complete stock of high grade pianos, and will also do tuning. Fred Anderson is associated with the company as salesman. Mr. Lindholm has had many years experience in the piano business in Chicago and Denver and is a first class tuner.

A son was born yesterday morning to Mrs. Mary E. Pickering of Crawford's addition. Mother and son are both doing well.

PAGE FACSIMILE: *The Daily News, Oct. 25, 1908, page 8*

SPECIAL MARKET, RANCH AND RAILROAD NEWS

WHAT IS THE MEANING OF "PEER GYNT?"

By Louis James

"What is the meaning of 'Peer Gynt'? To my mind, it is so simple and direct, that it really requires no explanation. Perhaps, for this reason, so many people, discarding the obvious search for the occult and hidden and read into the character, meanings of their own and interpret it accordingly.

Why, I have heard people give me great dissertations upon the hidden meanings and symbolisms in 'Hamlet,' a play to my way of thinking that is one of the most easily understood that our Shakespeare wrote.

The drama or epic of "Peer Gynt" is a lesson against selfishness, pure and simple. Peer is the egotist who, regardless of others, thinks of himself only; his desires and ambitions. In the end he learns that no kaiserdom, or wealth, or self-gratification can bring lasting happiness; and that peace of mind, eternal love and the salvation of the soul, transcend all earthly possessions.

I do not believe there is any more symbolism in "Peer Gynt" than there is in "Rip Van Winkle" or "Faust"; in fact Ibsen's own words are my authority: "I never painted symbols. I painted men. I never dared to plant a man on the stage until I could plainly see in my mind every one of the buttons on the front and on the back of his coat." What more conclusive evidence can the reader desire?

The characters of the "Green-Clad Woman," "The Ugly Brat," and the unseen "Boyg" are not symbols, but illusions, for we have every authority to prove that Ibsen introduced into this ger of poetic literature a legend of Scandinavian folk-lore that is as well known as the legend of the Catskills, obtained, presumably, when he was commissioned by the Norwegian government to explore the Northland, or Dovre, for the purpose of collecting folksongs and legends for the national archives, and as "Peer Gynt" was not written until five years after that expedition, it is fair to accept this assumption as correct.

I do not believe in presenting upon the stage a play that has any hidden meaning or great psychological problem to solve, the classroom or library is the place for such literature; I have always maintained that the more lucidly andn ationally you exploit a theme or present a character before the dramatic public, the more satisfactory is your own performance and the more popular the presentation.

In playing "Peer Gynt" I try to bring out all the humanity in the character, and I play him, therefore, naturally, and my injunctions to the company have been: "Cut out all this theoretical business and endeavor to convey across the footlights the idea that you are thoroughly in touch with the characters your assume, and that you want the audience to absorb the same idea." Then and only then, can you give a performance that is satisfactory to both player and auditor.

I play Ibsen as I play Shakespeare or any other classic author, and incidentally let me go on record to say that I contend that Henrik Ibsen is the Shakespeare of the twentieth century.

GRAPES BRINGING GOOD PRICES IN CHICAGO MARKET

News Special Service.

CHICAGO, Ill., Oct. 24.—(By Charles Schlosser)—Thirty-three cars of decidious fruit sold at auction here this morning. The assortment was as follows:

11 cars of apples, 3 1-2 cars of prunes, 1 1-2 cars of pears, and the remainder was California grapes, with the exception of a few odd crates of prunes and a few fall and winter pears mixed in with the California grapes.

There was a decided improvement in the prices on grapes today. Although the offering was heavy, Tokays, good stock in good condition brought $1.05 to $1.25, the bulk going out at $1.05 to $1.10, Cornechone 90 cents to $1.20. Taken as a whole, the grapes showed decided improvement.

Italian prunes sold at from 75 cents to 90 cents, mostly at 80 cents and 85 cents, all showing somewhat ripe.

One car of Oregon Clairgeau and D'Anjou pears, the D'Anjous were withdrawn by the receiver and one lot of Clairgeaus sold at $1.70 and the balance withdrawn.

One car of California apples, Bell-Flower, 4-tiers sold at $1.06 and 4 1-2-tiers at 85 cents. The remainder of the apples were mostly fro mIdaho. Jonathans selling at a wide range of price on account of the difference in quality strictly fancy No. 1's bringing from $1.99 to $2.25, while choice grades and small sized fruit sold at prices ranging from $1.25 to $1.65, Baldwins, $1.10, Willow Twigs and Nonesuch $1.05, Bell-Flowers $1.25, Ganos $1.25, Seek-No-Furthers $1.05 to $1.55. This was a pretty heavy offering of apples, considering the quality and the various degrees of quality, the general average was fairly good and, as above stated, the grapes showed decided improvement over last week's results, and it looks as though fair prices would be obtained on everything during the week.

The Daily News 'phone is Junction 2.

3 National Republican Committee Cartoons—1908—B No. 1

PAGE FACSIMILE: *The Daily News, Oct. 25, 1908, page 9*

THE DAILY NEWS

SOCIETY

Conducted by
MERLE M. M'CLINTOCK
Phone Junction 425.

Twenty-six years is not as great an age now as it was in the days of our grandmothers. Still as The News reaches this week its twenty-sixth milestone, it realizes that the time has come when each birthday counts. There is nothing more fitting than upon this anniversary some tribute be paid to Mrs. Edwin Price, wife of the founder of the paper, and one of the pioneers of the town.

Mrs. Price came to Grand Junction November 23, 1882, on the first train on the D. & R. G., the old narrow guage, running through Montrose and Delta. She waited in Delta for several days until the railroad was finished. When she could finally start, she came down on a day coach at the rear of a 40-car freight train. The train left Delta at 8 in the morning and reached here at 5 in the afternoon. The bridge over the Grand was only temporary and the heavy train rocked it to its foundations.

When Mrs. Price reached the town there was no station. At the foot of Third street there was an empty box car which answered for a station. There were no trees north of the river. There were no frame houses. There were only a few white women. Mr. and Mrs. Price lived for a week at the home of the Mobley's, a log cabin at the foot of Main street. Then they moved into a log cabin of their own, and began their social life here.

The influence of the pioneers of the town is hard to estimate. No one doubted that it took heroism to undertake the task of developing the valley. The men who came early and worked long for the success of the valley were brave, but they had their dreams of the future to help them. The women who stayed in the homes and battled with dust, worse by far, than it can ever be again, with the big trees to break the wind, and the fierce heat of the summer, are the ones who felt the discomforts most keenly and who ought to be remembered in any history of the past.

Mrs. Price has always been prominent in the religious and social life of the town and has always been eager to work for its best good.

—o—

Miss Agnew left yesterday on No. 103 for her home in Chesaning, Mich. Just before leaving Miss Agnew received a letter telling her that her father was failing rapidly, so her trip be an anxious one. A number of her friends, teachers and pupils, were at the train to see her off, and to present her with flowers, magazines and candy, to make her journey pleasanter.

—o—

Miss Belle Fleming of Chicago is in the city. She is spending a few days at the home of her cousin, and will then leave for a trip to California.

Kenneth Hinkel of St. Louis is visiting his friend Charles Day of this city.

There is great interest being manifested in the social session of the Elks to be held at the Elks' club tomorrow night.

HOPKINS ADDRESS WILL BE TONIGHT

On account of the lateness of trains from the west, Frederick E. Hopkins, who was to speak under the auspices of the Knights of Columbus last evening, was unable to appear and the affair was postponed until tonight and he will speak at the Park Opera house at 8 p. m. Mr. Hopkins is a brilliant orator and a large crowd is expected to attend the meeting.

Mr. S. E. Rische, the well known basso of Denver, is in the city at present and he has kindly consented to render two sacred songs during the evening.

Words Failed Him.

He was a cowboy, and some comparative stranger had stolen his horse. His friends rallied around him and, anxious to give him every chance, trundled up a barrel for him to stand on while he gave out his views on the matter. He was known as a gifted swearer, and a large audience had assembled in the hope of hearing something special. He got up on the barrel and looked around him. Then he drew a deep breath and, with a sigh, climbed down again. "Boys," he said sadly, "it's no use. I can't do justice to it."

John o' Groat's.

John o' Groat's, the most northerly point of Great Britain, was named after John o' Groat of Groot, a Hollander, who is said to have settled there about 1489. The house which John o' Groat built on Duncan's Head bay was a peculiar one. It was of octagon shape, having one room, with eight windows and eight doors, to admit eight members of the family, to prevent their quarrels for precedence at table. By this contrivance each one came in at his own door and sat down at the octagonal table, at which, of course, there was no "head" or "foot."

The first settlement in Oregon was made at Fort Clatsop, near the mouth of the Columbia river, on the 23d day of March, 1811. The men comprising the group were prospective fur traders sent to the Pacific coast by John Jacob Astor. They had sailed from New York on board the ship Tonquin

NEW BLACKSMITH.

Frank Carkuhuff, blacksmith and wagon repairer. New shop. All kinds of repairing and wagon work executed in most workmanlike manner. Call and give me a trial. 206 South 5th st.

Had a Close Call.

Mrs. Ada L. Croom, the widely known proprietor of the Croom Hotel, Vaughn, Miss., says: "For several months I suffered with a severe cough, and consumption seemed to have its grip on me, when a friend recommended Dr. King's New Discovery. I began taking it and three bottles affected a complete cure." The fame of this life saving cough and cold remedy and lung and throat healer is world wide. Sold at the C. D. Smith Drug Co. 50c and $1.00. Trial bottle free.

Van Horn for drugs, fresh and pure.

The Daily News Phone is Junction 2.

More Than Enough is Too Much.

To maintain health, a mature man or woman needs just enough food to repair the waste and supply energy and body heat. The habitual consumption of more food than is necessary for these purposes is the prime cause of stomach troubles, rheumatism and disorders of the kidneys. If troubled with indigestion, revise your diet, let reason and not appetite control and take a few doses of Chamberlain's Stomach and Liver Tablets and you with soon be all right again. For sale by all druggists.

We have been in the heating business for nineteen years. It is not an experiment when we install your furnace. All work guaranteed. J. H. Lane.

AN AMERICAN GIRL AND AN IRISH EARL.

When the Ogden Millses returned from abroad and went to their Newport summer residence they had as their guest a young Irish earl who is reputed to be the fiance of Miss Beatrice Mills. This titled gentleman responds to the name of Bernard Arthur William Patrick Hastings Forbes and is the eighth Earl of Granard and owns an estate of 21,300 acres, named Castle Forbes, in Longford, Ireland.

PAGE FACSIMILE: *The Daily News, Oct. 25, 1908, page 10*

HARVEST EDITION

THE DAILY NEWS

Grand Junction Section

ESTABLISHED 1882. GRAND JUNCTION, COLORADO PRICE TEN CENTS

Marvelous Growth of the Grand Valley

A Bearing Fruit Tree. Photo by Dean.

An Orchard in Bloom. Photo by Dean.

By George R. Caldwell.

"Grand Valley fruit lands, which twenty years ago were worth only the government price of $1.25 an acre are now selling as high as $4,000 an acre."

The above fact in itself constitutes all that could be written in an endless volume, and it only remains to set down in glowing detail how this [illegible] found in the natural constituents of the soil itself, while the magic is that of conquering climatic conditions happily combined with the fructifying favors of exhaustless irrigation waters—a unison which makes the streams of the Grand Valley veritable golden waters and annual floods of finance.

It is conservatively estimated that the first cost of the irrigated reclamation of the valley soil does not exceed 15 per cent of the primary orchard value secured—a percentage of irrigation cost which is steadily diminished as the orchards themselves attain bearing maturity—a fact which eventually results in not only the entire wiping out of irrigation cost, but the transformation of the original irrigating charge into an actual orchard asset and source of annual revenue.

With fruit lands with a perpetual water right starting at a purchase price of $250 an acre, and advancing in market value from year to year to $500, $1,000, $2,000, $2,500 and $4,000 an acre, it can be readily seen that the original and permanent irrigation charge speedily resolves itself into a practically paying investment.

Nature Generous in the Grand Valley.

The Grand Valley—also called "The Gem of the Western Slope"—lies entirely in Mesa county and comprises an area about 40 miles long with a maximum width of ten miles, the country deriving its name from the broad and fertile mesas rising above the current of the truly Grand river, while the beautiful and prosperous county seat owes its christening of Grand Junction to its situation at the confluence of the Grand and Gunnison rivers.

The valley soils are almost wholly comprised of materials derived from sedimentary rock formations [illegible] along the Grand river itself.

Generous old Mother Nature has beneficently divided these soils into distinct and separate belts, each of which is especially adapted to its own particular and prolific variety of harvest return.

The Laurel fine sand has a soil depth ranging from one to six feet and is especially and profitably given to market garden culture.

The Laurel sandy loam is from two to six feet deep and produces large returns in the berries and smaller fruits, sugar beets and alfalfa, while heavy truck crops are also secured.

Fruit Belt Rich in Fertile Soil.

The Billings fine sandy loam has a depth of six or more feet, and constitutes the great fruit belt of the valley, growing nearly all the orchards of the superb Palisade peach region. This soil also yields heavy crops of sugar beets and alfalfa. The western portion of this great soil belt has many excellent apple orchards, both stone and seed fruits being of such prolific bearing that it is usually necessary to sacrifice a portion of the product during its earlier growth stages, in order to enable the trees to bear the big burden of fruit at maturity. Alfalfa yields on the Billings fine sandy loam comprise three crops a year, with an aggregate of from four to six tons an acre. Sugar beets run to 15 tons an acre, while potatoes, market garden vegetables, corn and berries all do well.

The Billings clay bam is naturally designed for the growing of apples, pears, wheat, oats and corn, with alfalfa and sugar beets in generous return.

The Billings silt loam is from three to six feet in depth and produces good crops of alfalfa, pears and sugar beets. This is an alluvial soil, and its harvest values are being annually added to by an active and intelligent seepage drainage.

Finest Orchards on Mesa Sandy Loam.

The Mesa sandy loam is light and friable; ranges in soil depth from six to twenty feet, and supports some of the finest orchards in the valley. But a small area of this soil is yet in cultivation, but it has demonstrated, in addition to its great orchard values, splendid capacities for alfalfa and market garden croppage, a United States government expert saying that "the Mesa sandy loam is preeminently adapted to fruit and tilled crops."

Mesa clay loam is from four to six feet deep, is rich and productive, and especially adapted to alfalfa, sugar beets and grain. This soil is chiefly found on Orchard Mesa, near Grand Junction, and is practically free from alkali.

Fruita sandy loam is six or more feet in depth, is friable and readily cultivated, rarely proving refractory to tillage. This soil principally composes the wide and wealthy fruit areas immediately surrounding the town of Fruita. Some of the finest apple and pear orchards of the valley are upon this soil type, while sugar beets, alfalfa, wheat and oats are all large growths of the fertile Fruita fine sandy loam.

The Fruita loam, a soil separate and distinct from the Fruita fine sandy loam, is at present a good grower of sugar beets and grain, and, with more drainage, will develop into fine fruit lands.

Mt. Garfield. Photo by Dean.

while irrigation is the sworn and successful foe of alkali deposits.

There is practically no hard-pan in the soil area of the Grand Valley.

Money vs. Mortgage in Orchard Land.

The farmers of the Grand Valley are, as a rule, in a remarkably prosperous condition. Such a thing as a mortgage foreclosure is almost unheard of, while the field, the orchard and the garden vie with each other in harvest production and money making markets.

Peach trees begin commercial bearing at three years old, and attain full production at seven or eight years of age, the fruit netting the grower from $400 to $800 an acre. The great Elberta, frequently attaining a weight of over a pound each, is the pride of the valley peach orchards, while the early and late Crawfords; the Carman and the Arp Beauty are all superb commercial varieties.

The apple orchards chiefly count the great commercial varieties of the Ben Davis, the Gano, the Jonathan, the Winesap and the Minkler.

Cherries, apricots, nectarines, prunes and quinces are also grown with marked success in the Grand Valley, while all the berries flourish in profusion and profit.

Within the last three years, counting the present season, about two million new fruit trees have been set out in the Grand Valley.

Grand Valley fruits have an alike national popularity and market scope, their marketing being facilitated by the traversing of the valley by the Denver and Rio Grande and Colorado Midland railroads, the Midland running over the Rio Grande tracks, while a Denver and Rio Grande line runs from Grand Junction across the mountains to Ouray and the San Juan district, and through Montrose, Delta and other towns, and over other mountains to Gunnison, traversing the Gunnison country and connecting at Salida again with the main Rio Grande rails.

Organization Has Enlarged Markets.

The recent organization of active fruit associations and cooperative marketing concerns has resulted in [illegible] to far distant markets, the chief peach markets now being Missouri river cities and towns, while peaches have been shipped to Boston with gratifying results.

Apples are largely sold within the state, with an extended and lucrative Western shipping market, principally to Southern California points.

The now constantly growing sugar beet industry is proving a new and great Grand Valley boon, as it brings into active and prosperous cultivation those lands not so well suited for fruit or other general farm culture, and also establishes a new and heavy wage and sale schedule.

The smaller fruits and vegetables meet with a ready and profitable sale to the mining districts and various cities and towns of the state.

Alfalfa, with its production of from four to six tons an acre, and selling price from $8 to $15 a ton, is nearly all consumed in the valley, the fruit growers furnishing a ready alfalfa home market.

Phenomenal Yields of Grain and "Spuds".

The grain and potatoes are of profitable sale, wheat reaching as high as 60 bushels an acre and oats climbing to 125 bushels, while potatoes "dig up" money for yields of 300 and 400 bushels an acre.

Market garden vegetables and corn or potatoes are often planted between the rows of young fruit orchards, with the double good result of a thorough irrigation of the trees themselves and securing of paying crops from the land during the unproductive first years of orchard growth.

Raw valley land is readily and cheaply cleared of its natural growth of sage brush, a riding plow in nearly all locations being capable of at once breaking and clearing the land.

High Line Canal Will Reclaim Vast Acreage.

The late visit to Grand Junction of high officials connected with the Grand Valley United States Government High Line Canal brought with it the welcome assurance of the com[illegible]tion by the first of next January.

It is estimated that the canal will be completed within twelve months from the commencement of construction, will cost an approximation of $2,000,000, and, through the fruit and general farming cultivation of the 70,000 acres covered, will advance the Grand Valley to premier rank in Colorado tillage and insure to Grand Junction a population of 30,000 within the next six years, thus inseparably and imperishably interweaving the names of Garfield, the Grand Valley and Grand Junction in a great sign manual of progress, prosperity and population.

Fruit Harvest Worth Two Million Dollars.

Last year's fruit harvest brought $2,000,000 to Grand Junction and the Grand Valley. Last year frost killed practically the fruit of all sections of the nation, but the orchards of the Grand Valley well filled the deficit.

Last year H. O. Harvey marketed $2,750 worth of pears from one and one-quarter acres; M. F. Yeaton banked $4,000 from four acres of pears; Robert Orr averaged a sale of $2,000 per orchard acre; one apple tree produced 47 boxes selling for $60; one apple acre produced $1,700 worth of fruit; a cherry tree has averaged for several years past $30

(Continued on Page Eight.)

Photo by Dean. Dr. H. R. Bull's Residence. Dr. G. R. Warner's Residence.

Margery Building. Photo by Dean.

PAGE FACSIMILE: *The Daily News, Harvest Edition, page 1*

The Wallace Orchard

A NEW FRUIT ADDITION TO THE FAMOUS GRAND JUNCTION DISTRICT.

(R. R. Red Book.)

Somehow the impression has become prevalent that all of the available fruit lands in and about famous Grand Junction, Colo., are now in bearing orchards—rated at prices per acre that sometimes startle, but prove to be a safe investment.

That land is scarce is so far from the actual existing conditions that the investor, homeseeker and horticulturist will do well to investigate this world-famous fruit district.

There is elbow room for 20,000 more people in and around Grand Junction, and not alone elbow room, but abundant opportunity to become well-to-do horticulturists.

One of the most notable enterprises now in the preliminary stage of development in this district is that of The Wallace Orchard company. This company owns a large tract of land on what is known as Ruby Mesa, about five miles from Grand Junction. The original intent of the company was to raise young fruit trees for the market, but their land proved to have such excellent fruit-growing qualities that 13,060 pear trees, in four most favored varieties, were planted on 120 acres in the spring of 1907.

To irrigate this orchard until the government high line ditch is constructed (expected in 1910) a pumping plant was installed, taking water from the present old high line ditch.

A careful record was made of the actual cost of supplying water by this temporary expedient and for the first year, 1907, the figures showed a cost of $2.91 per acre. This will probably be reduced in 1908 to $2.50 per acre.

The loss of trees in this rather expensive planting was less than five per cent, and the trees showed a remarkably healthy growth.

During the season of 1908 a careful record has been kept of the amount of water actually used in the growing [illegible]rd and it is computed by Mr. Sands of the United States Reclamation Service that the amount of water used exclusive of rainfall was 82-100 of an acre foot or about one-third the amount usually supposed to be necessary. An inspection of the orchard shows it to be in perfect condition and to have made a growth during the season of 1908 exceeded by no orchard in the valley of the same age and equaled by very few, regardless of the amount of water used.

The secret of these results are an ideal soil and perfect and constant cultivation.

The orchard was planted as a strictly commercial proposition and no pains have been spared in striving for the best possible results. The orchard is under the direct personal charge of Dr. S. M. Bradbury, than whom there is no better man in that line of work in Colorado today and every employe under him is an expert in his line.

It is a foregone conclusion that this commendable method of placing fruit lands on the market will bring abundant financial returns to the men who lend capital to the project, and are willing to wait a few years for matured development.

The officers of the company are men of noted integrity and extensive business interests in the Grand River Valley.

They are as follows:

Dr. S. M. Bradbury, president and general manager; Mr. W. P. Ela, vice president; Mr. C. H. Arthur, treasurer; Mr. W. S. Wallace, secretary. Their office is in Grand Junction.

CHULA VISTA.

The phenomenal success of The Wallace Orchard company in planting their orchard on Ruby Mesa has induced the company to organize the adjoining 160 acres into a district to be known as "Chula Vista."

The land in Chula Vista is exactly similar to that of the Wallace company orchard, being a rich red sandy loam, comparatively level, and of the quality famous for producing the highest pric[illegible] peaches, [illegible] apples and other fruits.

Chula Vista will be und[er] the new government ditch when [c]ompleted, but in order to advance the district into bearing, a pumping pla[n]t will be installed at once and in the [s]pring of 1909 the entire 160 acres w[il]l be set in fruit trees.

It is not the intention of [t]he company to sell land in Chula [V]ista until the trees come into bearing, and then it will be sold in ten-acre tracts.

This land has all the qualities that years of experience have [gi]ven essential to the production of [hig]h-grade and most profitable fruits. To the north towers a range of protecting cliffs, which guard against th[e] colder winds.

Chula Vista rises about [illegible] feet above the surrounding country [in]suring excellent drainage, the v[alu]e of which is becoming recognize[d] as a positive requirement for co[ntin]uous life of orchards.

Mesa lands have another excellent quality, in that they admit of free circulation of air, which prevents early spring frosts.

Chula Vista will start off with the conveniences of city life; rural free delivery, telephones, daily visits of the grocery wagon, and the incidental advantages that go with a closely settled rural district adjacent to an up-to-date going city like Grand Junction.

The orchards will be planted and cared for by expert horticulturists, and the most profitable varieties of fruits cultivated.

The object being to make each ten acres a commercial orchard and the land to be placed on the market at five years old. Dr. Bradbury, the president of the Wallace orchard will have charge of the new Chula Vista orchard and under his able management the success of the new enterprise is assured.

The officers of the Chula Vista Orchard company are: A. R. Wells, president; Thomas H. Rader, vice president; W. S. Wallace, secretary, and M. F. Fredericks, treasurer, with offices in Grand Junction, Colorado.

THE UINTAH RESERVATION

By E. F. Colbor[n]

Until August, [illegible] a very [illegible]able portion of Utah—the U[intah] Indian reservation—occupying practically the entire northeastern portion of the state—was prohibited ground. The rich areas of grazing and farming land, the metal-bearing ledges and the vast deposits of various forms of hydro-carbon contained in the reservation, have for years occasioned persistent appeals to congress for the adjustment of the Indian rights and the throwing open of this valuable area to settlement.

The much desired end has at last been reached and the steady inflow of settlers, prospectors and miners has already begun.

Vernal, the county seat of Uintah county, is the principal town and the location of the United States land office, having jurisdiction of the reservation lands. Vernal is in the beautiful Ashley valley, and is surrounded with fine farms and orchards. Its streets are paved and it has gas, banks, churches, fine schools and a number of important business institutions. Other towns springing into vigor are Myton, Moffat, Leland, Theodore, Stockmore and Roosevelt.

Prospecting for minerals has only just begun, and the metal contents of the reservation are yet only to be guessed at. Prospectors will head for the North Fork country, where the best float and the strongest mineral indications have been found. A large number of copper-bearing properties have been located, and much is said in a whisper of a gold mine of fabulous richness found many years ago, and hidden to await the time when title to it could be lawfully obtained.

Probably there is not elsewhere in the world such remarkable deposits of ozocerite, elaterite and gilsonite, and such springs and veins of asphalt as the reservation contains. Of these hydro-carbons, there are enough to supply mankind for generations. The hydro-carbon area covers at least one thousand square miles, and the values it contains are incalculable. Two of these substances, gilsonite and elaterite are distinctly Utah curios, and another—ozocerite—is said to be found in but one other place in the world.

By treaty terms, the Uintah Indians, who by the way although living, are "good Indians", have been allotted three hundred thousand acres of fine farming land which they are rapidly learning to till. The unallotted land, comprising approximately two million acres, is being steadily taken up under the general land laws.

A large portion of this land is suitable for farming, and much of the remainder is of fine grazing character.

All told, the population of the reservation is about 8,000, and the estimate is that there is room for one hundred thousand.

The addition of this rich region to Utah is an important step in her progress and will rapidly increase her population and wealth.

The scenery of the region is most interesting and will well repay a visit. To see the best of it, and to [se]e it most comfortably, one should [tak]e the Uintah railroad, which con[nect]s with the Denver and Rio Grande [at M]ack, Colorado, twenty miles west [of G]rand Junction, and extends north[illegible] to Dragon, [whe]nce automobile and stage connections may be made for Vernal, Duchesne and other points.

This little railroad is a big thing in its way. It winds about through miniature canons of the Colorado, crosses the picturesque Book Cliffs—the like of which for singularity or form there is not—climbs grades at times as great as 7 1-2 per cent; passes by topographical features bearing such suggestive names as "Thimble Rock," "Hell's Hole," "Excavation Canon," "Coyote Basin," "Dead Man's Bench," and gives the traveler such a whirl of ragged ruggedness, natural amphitheaters, obelisks, temples and pinnacles, as he could not experience along any other fifty-four miles of railroad on earth.

Baldwin locomotives, coaches, observation cars and gasoline track autos, afford the traveler a choice of accomm[od]ations. The [illegible] is regular, charges are moderate and every employe is a gentleman. What more than this can any one wish?

The time is coming when the trip from Mack to, and through the reservation, will be an experience sought and enjoyed by a host of trans-continental tourists.

Where Bullets Flew.

David Parker of Fayette, N. Y., a veteran of the Civil War, who lost a foot at Gettysburg, says: "What the good Electric Bitters have done is worth more than five hundred dollars to me. I spent much money doctoring for a bad case of stomach trouble, to little purpose. I then tried Electric Bitters, and they cured me. I now take them as a tonic and they keep me strong and well." 50c at the C. D. Smith Drug Co.

C[o]ly's Cafe makes a specialty of [illegible] for [illegible] take [illegible] arrangements.

PAGE FACSIMILE: *The Daily News, Harvest Edition, page 2*

THE DAILY NEWS THE HARVEST EDITION PAGE THREE

My prayer: Don't let me rest complacent, day or night, but put coals under my feet and thorns in my pillow while there is a slave at the wheel, a babe in a dungeon, or a prostitute on the street!—Eugene V. Debs.

SOCIALISM

The other day I saw the wretched and sexless victims of the sweat-shop. Oh, the life they live; the death they die!—Eugene V. Debs.

"Workers of the World Unite; You Have Nothing to Lose but Your Chains; You Have the World to Gain."--Karl Marx.

Is Socialism Possible, Human Nature Being What It is?

Socialism is possible because human nature **is** what it is. **What is it?** A very complex thing which is changed by evolution and environment. The human nature of a man at 20 is often very different from the same man at 40. The character of our country today is very different from the character of our country 100 years ago.

We are told that Socialism will only be possible after we become angels, and that then there would be no need for it.

The question is: Do we want Socialism? If we want it we shall get it as near to our ideal as we can.

Do we want a better state of things? Some of us do, and we think Socialism would be a much better state than the present.

A thief said: "If there were no watches in the world I should be an honest man." He "couldn't help" stealing watches.

So today, in order to live, people must be hard and mean and grasping, and think a good deal of self.

Well, take away the **conditions** which make them hard and mean and grasping, and those bad qualities would have a chance of disappearing.

Socialism would remove those conditions.

Why do you send your child to school? Why do you keep him from evil companions. Why do you give him good books? Why do you teach him morals?

There are thousands of charitable institutions—for what? For rescuing and reforming the unfortunate.

Christians pray every day, "Thy kingdom come!" Do they believe it possible? Then why is Socialism impossible?

Socialism would provide decent and healthy conditions for all. **Every** child would be fed and clothed and educated under Socialism. Every child would be surrounded with good influences.

Today we teach the children to be moral, and then we turn them out into a horrible, immoral system, which shouts at them all the time "Be right and you will go under."

What can we expect of such contradictory methods?

To say that Socialism is only possible **after** we have all became perfect is as foolish as telling a sick man that health is only possible after he becomes well.

To say that Socialism would not be needed when we were all perfect is as foolish as saying to the sick man, "It is ridiculous to think of trying to get health, because when you are well, health will not be needed."

Do we **want** to change this system of strife and competition and devil take the hindermost for a system of cooperation and mutual help?

If we do, then that is a proof that human nature is changing. First comes the desire, then the attempt to satisfy that desire. We may fail, we may fall short of our ideal, but we must **try** to achieve it.

Are you going to help?

A People's Right

From the Declaration of Independence.

"We hold these truths to be self-evident: that all men are created equal, and are endowed with certain inalienable rights; that among these rights are life, liberty and the pursuit of happiness; that to SECURE THESE RIGHTS governments are instituted among men, deriving their JUST powers from the CONSENT OF THE GOVERNED, and that when any form of government BECOMES DESTRUCTIVE OF THESE ENDS it is the right of the people to alter or ABOLISH it, and to SUBSTITUTE SUCH FORM as shall in THEIR OPINION best secure those ends."

Labor, Make the Demand

By Robert Hunter.

Some time ago I said that I would like to hear some American workingman say what he wanted.

A number answered me.

One of them says: "Labor wants free speech, justice, industrial liberty and an opportunity to exercise body, mind, hands and soul, to earn and eat its own bread by the sweat of its own brow, to provide for its own home, household and family, to enjoy all the fruits and products of its own labor."

This is all Labor asks. It demands no more. It is all Socialists ask.

I should like to read that demand from every pulpit of every church in the land.

I should like to read that demand to the Senate and to the House of Representatives.

I should like to see Labor make that demand of every politician in the land.

Labor wants an opportunity to exercise body, mind, hands and soul, to earn its own bread.

That is denied unemployed millions at the present time.

Labor wants by the sweat of its own brow to provide for its own home, household and family.

That is denied to millions at ALL times.

It is denied the fathers of two million child slaves in this country.

It is denied the fathers, husbands and brothers of five million women toilers in this land.

Labor wants to enjoy all the fruits and products of its own labor.

That is denied every man who labors for wages.

But while this is all Labor demands, millions refuse even to whisper that demand.

Leaders of Labor now begging in the halls of the legislatures never make that demand.

Socialists suffer persecution, misrepresentation and martyrdom because they have formed a movement that will not die until it sees the fulfillment of that demand.

To exercise body, mind, hands and soul, to earn its own bread, to provide for its own home and family, to enjoy all the fruits and products of its own labor—who could ask less?

Yet Christian ministers scoff at that demand; politicians ignore that demand; capitalists deny that demand.

All right. But notice!

Two thousand years ago Jesus made that demand—NOT THE WORKERS.

Sixty years ago Karl Marx made that demand—NOT THE WORKERS.

Today a WORKING MAN and million of his comrades in all lands make that demand.

Do you know what that means? I will tell you.

Labor awakens!

As soon as Labor, all Labor, makes that demand, the hour if its fulfillment will be at hand.

Would Socialism destroy incentive? There is no doubt that it would destroy numerous incentives, such as, for instance, the accepting of bribes from railroad companies in the form of "free transportation"; the incentive to give and take boodle, the incentive to shun sweat, but absorb the surplus of other's labor; the incentive to run a county or state government in the interests of politicians than in the interests of all the people. Yes, Socialism would indeed be very hard on incentives—of some kind. And—this is no joke.

Warning Words of Lincoln

"I see in the near future a crisis that nerves me, and causes me to tremble for the safety of my country. As a result of war, corporations have been enthroned, and an era of corruption in high places will follow, and the money power of the country will endeavor to prolong its reign by working upon the prejudices of the people, until all the wealth is aggregated in a few hands and the republic is destroyed............I BID THE LABORING PEOPLE BEWARE OF SURRENDERING THE POWER WHICH THEY POSSESS, and which, IF SURRENDERED, will surely be used to shut the door of advancement for such as they, and fix new disabilities, and burdens upon them until all of liberty shall be lost."

A Clarion Call

Come, comrades, let us band together.
Let us encircle the earth with our fellowship.
For we must build a new world.

We need toilers, men who can bake and brew and weave, who can break ground, sow it with good seed, and harvest the good harvest.

Are you a farmer? Do you know how to raise cattle and grain and till the soil and make the earth blessed? Then we need you.

Are you a driver of horses or of engine? Can you speed a train by day or through the night? Then we need you. Can you make shoes and clothing for mankind? Can you bake bread and cook and feed the world? Can you cut stone and polish marble and build houses? Can you rivet iron girders and make great domes to shelter mankind? Then we need you.

Men of brain, men of brawn, men who feed and clothe and shelter the world. Men of architecture, men of invention, men of mechanics, we need you.

We are going to build a new world. We want all who labor. We can spare none of you. We must have all of you; every producer, every worker with hand and brain.

Idlers, loafers, exploiters, you who are cushions, sip wines and dance—we can spare you.

You who take interest and rent; you profit-seekers and slave-drivers—we can spare you.

But all you who clear forests, who bridge streams, who till soil, who reap grain, who feed the world; all you who build spacious avenues, drive tunnels, clean drains, build palaces—we need you.

You capable of labor, you users of hand and brain—we need you ALL.

This world of today is of your making, but it is not YOUR world.

For you it is a world of slums, of poverty, or insecurity of fretted childhood, of dependent, anxious old age.

For those who sit idle it is a world of palaces of luxury, and of leisure.

AND IT MUST BE MADE ANEW.

Come, comrades, let us band together.
Let us encircle the earth with our fellowship.
For we must build a new world—and we need YOU.

Socialist Candidates for Mesa County

District Attorney	GEORGE BULLOCK
Treasurer	W. M. VAN BUREN
County Clerk	CHARLES LYNN
Assessor	L. F. FOX
Sheriff	ISAAC SAWYER
Coroner	RILEY D. MOORE
Sup't. of Schools	MRS. IDA M'MILLEN
County Commissioner	G. H. MILLER
County Judge	GEO. N. FALCONER
Surveyor	BART LYNCH
Representative	W. W. GOODMAN

THE SOCIALISTS OF MESA COUNTY. Met Sept. 7 and placed a complete ticket in the field. The candidates on this ticket favor a government, city, county, state and national, that would secure to each man his own, to secure to the farmer what he produces from the soil, the mechanic what he invents and makes, and to each worker, in whatever capacity, the full product of his toil. Socialism is a movement by the World's Workers to obtain the wealth they create and so banish poverty and its concomitant evils. Your help is needed in this world-wide movement.

Socialism does not propose to "destroy private wealth," but to transfer SOME forms of private wealth, that are ESSENTIAL TO THE LIFE OF ALL, to the possession of ALL. So far as the right of private property in things produced BY THE PRODUCER is concerned, Socialism would RESTORE private property. It would make it impossible for a few to "invade that sacred right" of the many.

LET THE WORKERS CONTROL THEIR POLITICS.

The Corporations own our industries, our natural resources, our political parties, our government, our courts.

It is time for action. Let the workers control **their** politics. It must be the people organized in their own political party. Conscious of their end and aim. Selecting and controlling their candidates, capable themselves of running their government. To attain this end, there must be no political trading, no compromise, no buying and selling of votes. No mental and moral debauchery. Principles and not personalities must be the slogan of the Party of the People.

SOCIALISM.

"A theory of society that advocates a more precise, orderly and harmonious arrangement of the social relations of mankind than that which has hitherto prevailed."—Webster's Dictionary.

GOVERNMENT.

A government should exist not only to protect one individual from another, the weaker against the stronger, but to **protect ourselves from ourselves;** for the worst enemies of the people are always found in their own midst, in their passions and vanities. And the most dangerous foes of a nation do not advance with drums beating and colors flying, but creep upon it insidiously, with the noiseless feet of a fatal malady.

Man wants but little here below and gets it.

Lincoln's Voice in '65

It is well to read the prophetic and thrilling words of Abraham Lincoln uttered in 1865 in an address to a body of New York laboring men. It was an appreciation of labor that time has not changed.

President Lincoln said:

"It is assumed that labor is available only in connection with capital; that nobody labors unless somebody else owning capital, somehow by the use of it, induces him to labor. Labor is prior to and independent of capital. Capital is only the fruit of labor, and could not have existed if labor had not first existed. Labor is the superior of capital and deserves higher consideration. I bid the laboring people beware of surrendering the power which they possess, and which, if surrendered, will surely be used to shut the door of advancement for such as they, and fix new disabilities and burdens upon them until all of liberty shall be lost.

"In the early days of our race the Almighty said to the first of mankind: 'In the sweat of thy face shalt thou eat bread,' and since then, if we except the light and air of heaven, no good thing has been or can be enjoyed by us without first having cost labor. And inasmuch as most things have been produced by labor, it follows that all such things belong of right to those whose labor has produced them. But it has so happened in all ages of the world that some have labored and others have without labor enjoyed a large portion of the fruits. This is wrong and should not continue.

"To secure to each laborer the whole product of his labor, as nearly as possible, is a worthy object of any government.

"It seems strange that any man should dare to ask a just God's assistance in wringing bread from the sweat of other men's faces."

A LINE OF TALK FOR FARMERS

By Eugene Wood.

It is evident that if there is nothing in Socialism for the farmers of this country, however much there may be in it for the rest of us, we might as well make up our minds that we shall have to wait a dog-goned long while to get it. The farmers are very good in their specialty entitled "Saving the Country," and unless they come to the front with their act the performance cannot make a popular hit.

I beg to submit the following sentiment as applicable to the farmer: "To the laborer the full product of his labor." It is not an unfamiliar sentiment to many. I do not claim that it is original with me. But I do claim that it is a good deal easier to see how the farmer can get the full product of his labor than how the individual in a shoe factory can get the full value of the product of his labor.

If the railroads were owned and operated by the government, so that instead of the object being to make all the money possible for stockholders who don't do one useful tap of work from one year's end to another, the object were to haul to market the farmer's crops and haul back the farmer's wants in exchange for his crops at the cost of the trainmen's wages, fuel, oil, and maintenance of the rolling stock and road bed, there would be a big boost in the farmer's wages right away. And if the stable manure which town people hire men to take away for pity's sake, but which costs so much to get out into the country, where it is needed, were transported at cost, we shouldn't hear so much about the deterioration of the soil. Just government ownership of railroads would help a lot. Does the Republican or the Democratic party advocate this? Nay, nay. But the Socialist party does.

Take postal savings banks. That would bring to the government a lot of money which could be loaned out to farmers at a very low rate indeed. Take government insurance at a rate just enough to cover the cost and wouldn't that help the farmer to get and keep the full value of his labor? Life insurance, accident insurance, fire insurance, old-age insurance, sickness, and even crop-failure insurance. We have to make that good now by increased prices for produce.

It looks to me as if we could go right after the farmer and get him in a minute. He knows who skins him alive. He sees that plain enough. All that is necessary is to point out to him that the same crowd is taking the hide off the town mechanic, and that he must join in with the town mechanic in the demand: "To the laborer the full value of the product of his labor."

A Liner on Religion

By Geo. N. Falconer.

Is Socialism opposed to religion? Yes, say the grouchy ones. Nay, nay, Pauline, they are off their bat. The Socialist platform declares:

"The Socialist party is primarily an economic and political movement. It is not concerned with matters of religious belief."

Religion is an individual matter, and as an individual Socialist I am going to tell you some of the religious beliefs and views held by certain other individual Socialists whom I have talked with and read about. There is nothing like hearing the other side, is there? These Socialists, I find, differ with each other on many things, but they are all dead sure on this proposition: "Human life depends upon food, clothing and shelter." They say we must first be citizens before we are angels. Only with food, clothing and shelter assured, they argue, are "freedom, culture and higher human development possible". Do you get their view-point? These Socialist [illegible] are trying mighty hard to make this world a better place to live in. But they are surely having a devil of a time trying to convert the "non-elect". But in spite of opposition, petty, ignorant opposition, much of it, they are constantly gaining and some day will win out.

Where we shall be and what we shall do in some other world, is no business of theirs, they say.

They have turned their attention to this world. They refuse to worry about the hell of the future, when so many hells exist here. They think that physiology is quite as important for boys and girls as the catechism; that good manners are as valuable as sound doctrine; that fair-dealing in business is more needful than creeds; that homes are as sacred, and a little more so, than cathedrals. They declare it a holier thing to keep your honest word than to keep Lent. Whether we shall know each other in heaven is a problem that can be delayed, but they say our conduct here should be such that we would not be ashamed to know each other there. They demand a kind of taxation which takes money from those who have money a-plenty, not bread from starving children's mouths; they want laws which favor the helpless, not laws which favor the fortunate; they ask for knowledge that shall emancipate us from slavery in every form. Let us have, they say, an unroofed temple of the mind wherein honest thought can receive the benedictions of all men good and true. They want governments with brains enough in them, and humanity enough to keep children out of the factories and keep them in school. Let there be a popular demand, they say, as strong as the demands of death, that newspapers and drinking water alike be clear of sewage. They would have an educational system by which boys and girls would come to despise selfishness and snobbery, as they dread a mistake in the waltz or deplore a wrong color in their bonnets. They would have society ladies take as deep an interest in the sufferings of poor children as they now do in **poodle dogs.** They want established an ethical standard by which campaign lies and political slanders and business frauds be stamped down to the level of common theft and falsehood; a standard by which drunkenness and prostitution are made as disreputable in men as in women. For every "fallen woman", they declare, there is a "fallen man". When these things are attended to it will then be time to consider the problem of the future. One reason, they say, for the very slow advancement of civilization, is that men have ever busied themselves with supernatural concerns and in their anxiety for another world, have neglected this. Man has tried to make peace with the skies instead of peace with his neighbor. And so, Pauline, these Socialist folk are not the gol-darned atheistic, irreligious disbelievers you thought they were. True, their beliefs and views are not exactly yours, but aren't they human, practical, possible of realization? Of course they are. Food, clothing, shelter—once these things are assured, guaranteed, humanity would enter upon a new era. "It would be an era", some wise one has said, "comparable to that wondrous age in ancient Greece, when Plato, Pericles and Socrates lived and thought and worked.

"There would be this tremendous difference, that whereas the great ones in Greece were able to do something more than exist, because the multitude were slaves, all could rise above existence, and only the dumb forces of nature would remain enslaved."

The men and women by the way, who entertain these beliefs and views, are not always able to live them, for you must know, they are like the rest of the crowd, very busy working each other in order to **procure** "food, clothing and shelter". Working others is hell; working **for** others is heaven. See! Some day we shall work each for the other, and then, my! my! The religion of the future, Pauline, will be **humanism**, just plain everyday humanism. To help bring it about, I am going to cast a straight Socialist vote for Eugene V. Debs, the people's candidate. Who are you going to vote for, Taft? Bryan? All right, go ahead. Some day you shall know better. Today you look through a glass darkly.

Come, Lend a Hand

From the Milwaukee Socialist Platform.

Socialism combats not alone the conditions which exploit and oppress the wage-working classes, but every kind of exploitation and oppression, whether directed against a class, a party, a sex, or a race. All its measures benefit not only the wage-working class, but the whole people and while the working people are the banner bearers in this fight in the last analysis, everybody—the merchant, the professional man and the small storekeeper—will profit thereby. Therefore, we invite every honest and well meaning voter, without regard to occupation, race or creed, to join in our undertaking for the emancipation of mankind.

REFERENCE: Equip yourself by reading Socialist literature. The following can be had from your local bookdealer: Chicago Daily Socialist, $3.00 per year; Blatchford's "Merrie England," 10c.; Spargo's "Common Sense of Socialism," 25c.; "Scientific Socialism," by Vail, 25c.; "Evolution Social and Organic," 50c.; the "Christian Socialist," 50c per year. Ask any Socialist for free leaflets, including the "Socialist Party Platform."

PAGE FACSIMILE: *The Daily News, Harvest Edition, page 3*

WESTERN SUGAR AND LAND CO.

Buildings and Yards of the Western Sugar and Land Company, Grand Junction.

The story of the inception and growth of a great industry that has brought thousands of dollars and hundreds of happy and prosperous homes to the Western Slope.

Standing side by side, in point of importance, with the fruit rasing industry of the Grand Valley; year by year growing to great and greater proportions, and giving exceptionally men, the raising of sugar beets and their manufacture into sugar, is, without doubt one of the greatest industrial enterprises of the Grand Valley.

As, from the beginning, every project must have its Nestor, so has been the Western Sugar and Land company, of Grand Junction, the prime encourager and mover in this great industry from its earliest infancy. Far-sighted men, realizing even in the early days that the production of sugar beets would eventually mean much to the Western Slope, in the year 1899 formed a company and erected a factory to manufacture the sugar.

It was small in those days. The capacity at that time being 350 tons daily and while that would be considered a fairly good sized plant in some places even today, this factory now handles 600 tons of beets daily. The first year it sliced 9,000 tons and this year will slice 75,000 tons.

Also, as the majority of other new projects and enterprises experience their vicissitudes, so has this industry has its ups and downs.

It was called the Colorado Sugar Factory then, and, for that matter, it still has a prior right to the name—if it wanted to keep it—for it was the first factory ever erected in the state of Colorado. Like all other Grand Valley people the projectors of this factory have led others along the path of knowledge, and profiting by what they learned from the Western Slope, the Eastern Slope of the state now has several factories, and the beet sugar industry is one of the most flourishing industries of the state.

In the year 1903 the company owning the Colorado factory was re-organized, the name was changed to the Western Sugar and Land company, and the present management assumed control. When the factory was constructed it represented an outlay of $750,000, but since its re-organization many important improvements have been made and all of the latest equipment added, until today as it stands the factory is valued at $1,500,000.

The directors of the company are: R. P. Davie, J. R. McKinnie, Alfred Hand, E. C. Sharer and O. P. Shoup. The officers are: R. P. Davie, president; J. R. McKinnie, vice-president; E. C. Sharer, secretary and treasurer; Fred G. Holmes, manager; George B. Bushby, general superintendent and J. R. McKinnie, Jr., resident superintendent.

The first year of the management of the new company 30,000 tons of beets were handled. In 1905 the amount grew into 40,000 tons; in 1906 to 53,000 tons; in 1907 to 67,000 tons, and this year the company will handle 75,000 tons, which will be manufactured into 160,000 bags of sugar. This is a remarkable growth, but when it is observed that more and more people of the valley are beginning to raise beets it will be readily seen that the growth in the future is destined to be even more remarkable and rapid.

At the factory 200 men are employed and the pay roll monthly amounts to the large sum of $20,000.

Not only does this company manufacture sugar from the beets, but it is now enaged in raising them as well. The company operates a farm at Loma, about 16 miles west of Grand Junction, which contains over 5,000 acres, irrigated by the Kiefer extension canal, owned jointly by the company and the Verner Z. Reed interests. Hundreds of tons of beets are shipped from this farm each year to the factory at Grand Junction. On this farm the company is now planting some fruit trees. Last year, as an experiment, 20 acres were planted, and this year 150 more will be put in. In connection with the Reed people they have plotted a town at Loma, and lots and fruit land will soon be placed on the market. A large hotel will be constructed in the near future and arrangements are already made to build several new cottages on the farm. This land will be sold in tracts of from 10 to 100 acres. The company lands are in charge of K. F. Collins at Loma.

In addition to the factory and farm, this company owns thirty weighing stations over the Western Slope, and from 1 to 5 men are employed at each place.

The field department of the company is in charge of two men, and is Valley is under the supervision of Mr. J. H. Brown, who resides in Grand Junction, and the Gunnison Valley is in charge of Mr. J. W. Douthitt, who lives at Delta.

Several counties of the Western Slope furnish beets to the factory. Each county furnishes in proportion in the following order: Mesa, Delta, Montrose and Garfield. This is the only factory on the Western Slope and secures all beets raised.

Beet pulp and molasses are the principal by-products of the factory, and 3,000 head of cattle are fattened for market each year on this refuse. The cattle firm of Clayton & Murnam have contracted for the entire output, and have at present 1,000 cattle feeding. The pulp does not have to be fed at once, but is placed in silos and used at discretion.

The Western Land and Sugar company has recently purchased the property of the Eastern Sugar company at Glendale, Arizona, and Mr. George Bushby, the general superintendent, has charge of both this factory and the one in that city. Mr. J. R. McKinnie, Jr., resident superintendent in Grand Junction, has charge of the practical end of the factory in Grand Junction, and Mr. Fred G. Holmes is the manager.

It is such men as these that have made the Grand Valley of the Colorado stand out with such prominence among the manufacturing and agricultural sections of this country and every citizen of this section knows and appreciates the value of such enterprises in building up the valley to its present preeminence among the agricultural and fruit growing sections of Colorado. Taking over the factory at a time when it seemed impossible to put it on a paying basis, when the future looked dark indeed, the present directors displayed a faith in the valley's future that has by this time been abundantly justified.

FIRST SUGAR FACTORY IN COLORADO

PAGE FACSIMILE: *The Daily News, Harvest Edition*, page 4

THE DAILY NEWS — THE HARVEST EDITION — PAGE FIVE

WHITEWATER

By W. C. Meserve.

More Whitewater property has changed hands in the last twelve months than in all the previous years of settlement. Also the school room capacity, which is a barometer of progress, has increased 66 per cent in the same period.

The difference between present and future land values is greater today in Whitewater than elsewhere in Mesa county.

The county's leading real estate operators, who formerly ignored Whitewater territory, have during the past year been quietly acquiring holdings in the district. This movement has almost escaped the notice of those who never get into anything on the ground flood.

Agriculturally the county tributary to Whitewater includes 30,000 acres of land showing similar variety in its soil to the Grand Valley. About one-half of this total acreage has been patented from the government for years and 15,000 acres have been taken from the government under the various land acts within the last twelve months. This practically exhausts the government lands that are fit for cultivation. But on Pinon Mesa, and slopes there are still immense tracts of well watered grazing lands and hundreds of choice spots for summer homes still open to the land claimant.

Topographically the Whitewater country is divided into eight districts each possessing peculiarities of soil, elevations or climate.

Back of the town of Whitewater and lying around the three sides of the Horseshoe Bend which the Gunnison river makes at this point, are 8,500 acres of land of which the town itself is the natural center.

The Highlands, in plain sight from Whitewater, Grand Junction, Clifton and Palisade divide the Grand and Gunnison River Valleys and consist of several benches aggregating 2,500 acres of the finest fruit lands known.

Every acre of these mesas has been taken by Grand Junction people and a reservoir association has been organized by H. R. Bull, H. J. Gray, J. Ernest Leaverton, A. H. Meserve, Sherman W. Moody, Edwin Price, C. G. Simpson, J. W. Swire, H. F. Vorbeck and F. Bernard Welsh, to fully irrigate it. Work on one of the reservoirs is in progress at this writing, on the top of Grand Mesa at the head of Whitewater creek where the Highlands people have one immense and several smaller reservoir sites.

On the line of the proposed Clifton-Whitewater boulevard the top of the Highlands is within two miles of the Whitewater depot.

Upper Whitewater creek includes 3,500 acres of rich lands at the foot of Grand Mesa. The 1,200 acre stock ranch of Mr. James Nelson is located here as well as the A. D. A. reservoir now nearing completion. Messrs. James Page and H. B. Ennor of Whitewater and John O'Boyle and James Cosgrove of Grand Junction are the prime movers in this enterprise.

The Reeder Mesa, named in the early days for John D. Reeder, now the Palisade merchant, consists of 3,500 acres of red mesa soil. Here the fruit crop has never failed and the ideal ranches of R. T. Anderson and James A. Laurent illustrate its possibilities.

Purdy Mesa embraces 3,000 acres of soil similar to the Reeder, from which it is separated by the canyon of the North Fork of Kannah creek.

Kannah Creek proper has 4,000 acres largely in cultivation. While the principal crop is alfalfa, everything which can be raised in Mesa county is grown here in perfection. There is less per capita debt in Kannah Creek than in any other section of Mesa county, and an inspection of the large ranches of W. L. Farmer, Charles Bowman and William Ternahan is enough to excite the envy of the little tract men from the thickly settled districts.

Pansford Valley extends seven miles toward Delta. It has 3,000 acres of beautifully smooth fertile soil and most of the land is taken from the government, but is awaiting the development of a water supply.

Unaweep Canyon has for its north and south boundaries, granite walls which in places rise perpendicularly 2,000 feet above the valley which has an elevation of 6,000 to 7,000 feet. Between these rock walls are 5,000 acres of a soil composed of disintegrated granite and sandstone with vegetable mould. Hundreds of veins of copper pierce the granite walls and the development of what promises to be a bonanza mining district some day, advances every year.

The scenery in the valley and from the precipices on either hand is magnificent. No tourist route in the state of Colorado shows anything to compare with it. One travels three miles on an unexcelled natural granitoid road in passing the Casement ranch which is the show place of the Unaweep and situated about midway.

The acreage given for these eight districts cover the best lands that are by common consent admitted to be now available, but as more people settle in the Whitewater country the reclaimed acreage will exceed this estimate.

In an air line Whitewater is seven miles from Grand Junction and ten miles from Palisade, but many people have the impression that the distance is greater on account of the uncultivated areas of Orchard Mesa, the Highlands and the north Whitewater slope. Within the memory of all old timers a similar raw land belt separated both Fruita and Palisade from our county seat and irrigation enterprises are now under way which will soon reclaim these intervening open lands.

With the exception of the bottoms on the Gunnison river which are watered by current wheels or small pumping plants in no single case supplying more than 200 acres; the present irrigation water supply comes from Whitewater, Kannah, East and West creeks and reservoirs located upon their headwaters.

These sources of water while not fully developed are evidently insufficient to irrigate half of this great territory. There are, however, water propositions in variety under consideration and eventually every acre of good land in the whole Whitewater country will have its adequate water supply.

The crying needs of the Whitewater end of Mesa county are more and more progressive settlers and money with which to develop its varied resources.

A Pioneer Business

MESA FLOUR MILL COMPANY.

While we contemplate with a great deal of satisfaction the remarkable progress made by the Queen City of the Western Slope in the past few decades in all lines of commercial and industrial endeavor, therefore, it is not to be wondered at that we view with a great deal of optimism the wonderful accomplishments that the future will bring forth for this great industrial center.

Grand Junction is yet but an infant compared with many other longer established cities. She has not yet cast off her swaddling clothes, and it is with pride that we chronicle her past achievements and look with no pessimistic eye towards her natural future growth and development.

Among those great business enterprises that have ably assisted in making Grand Junction the city it is today is the Mesa Flour Mill company. This great concern succeeded the Grand Junction Milling & Elevator company, and is incorporated under the state laws of Colorado, with ample capital to carry on its vast business.

The founder of this enterprise was David Roberts. The first buildings were erected in 1887. The Grand Junction Milling & Elevator company was afterwards incorporated with David Roberts as president, and his two sons—J. D. and Paul D. Roberts—as vice-president and treasurer and secretary, respectively. The entire plant, with the exception of the engine room, was destroyed by fire in 1898. The following year it was rebuilt and the latest type of Nordyke and Marmon machinery installed. In 1904 it was incorporated under the name of the Mesa Flour Mill company, this change calling into service a trio of men whose past record, individually, was one that embodied all the traits and characteristics that go with the ideal man. The distribution of offices was measured with the fitness of capacity, and with this grade of excellence, J. O'Boyle was appointed president; James Hynes, vice president, and L. M. O'Boyle, secretary and treasurer. This coterie of officials represent all that is good in commercial life. Being practical Christians they have instilled in the business of their concern the spirit and letter of the laws governing a good life.

The capacity of this mill is 125 barrels of flour per day, and this output is readily disposed of in western Colorado and eastern Utah. Only the choicest grades of wheat are used in the productions of this mill. Located on the Denver & Rio Grande main lines and on the Colorado Midland railway, these mills are well situated for receiving the grain and forwarding the flour. Each brand of flour that they produce bears an imprint that guarantees its value. Honesty, being the principal precept of the life of the individual officer, it is giving real existence in the products of the mill. Not asking for credit, they give none in return. With them, excellence and perfection in their line is the beginning and end of business ambition. That this aim is realized is evident in the demands made for their brand of goods in the face of all eastern and western competition.

Home of J. H. Samples.

PAGE FACSIMILE: *The Daily News, Harvest Edition, page 5*

GRAND VALLEY HIGH LINE CANAL

Engineer Sands Explains in Detail the Plans of the Government in Regard to the Big Irrigation Project.

On the 17th of June, 1902, the president signed the Reclamation Act. This act provided that all moneys received from the sale of public lands in Arizona, California, Colorado, Idaho, Kansas, Montana, Nebraska, New Mexico, North Dakota, Oklahoma, Oregon, South Dakota, Utah, Washington and Wyoming, excepting land office fees and a few other minor sums, should be set aside and appropriated as a special fund known as the Reclamation Fund, to be used for the examination, survey, construction and maintenance of irrigation works for the storage, diversion and development of water for the reclamation of arid lands in the said states and territories.

The second section gives the secretary of the interior authority to perform the work contemplated by the act and requires a report to congress at the beginning of each regular session.

The third section authorizes the secretary to withdraw lands to be irrigated by any project and this section also provides that such lands can be entered only as homesteads.

The fourth section authorizes the secretary:

To let contracts.

To give public notice of the irrigable acreage.

To limit the size of entries.

To announce the charges per acre for entered and private lands benefited.

To determine the number of annual instalments, not exceeding ten, in which the charges may be paid.

This section also provides that the entire cost of constructing a project shall be returned to the fund. It also provides that no Mongolian labor shall be employed and that eight hours shall constitute a day's work.

Section five provides that an entryman shall reclaim at least one-half of his land and pay all charges before a patent shall be issued. No water right shall be sold to water private lands in tracts in excess of 160 acres. The owners of such private lands must reside on the lands or live in the neighborhood, and no water right shall permanently attach until all payments shall have been made.

Section six provides that the secretary shall use the Reclamation Fund for the building and operation of all reservoirs and irrigation works constructed under this act: Provided, that when the payments are made for the major portion of the irrigated lands the maintenance and operation of the irrigation works shall pass to the owners of the lands irrigated thereby, under such regulations as the secretary may prescribe: Provided, that the title to the works shall remain in the government.

Section seven authorizes the secretary to acquire title to rights of way by purchase or condemnation.

Section eight provides that the state irrigation laws shall be conformed with, and that the water rights shall be appurtenant to the land and beneficial use shall be the basis, the measure and the limit of the right.

Section nine provides that during each ten year period the major portion of the moneys derived from any state shall be expended in that state.

Section ten authorizes the secretary to make such rules and regulations as may be necessary.

From the above review of the Reclamation Act it is seen that the plan of operation of the Reclamation Service is to expend the money received from the sale of land in the arid states for the building of irrigation works and assess the charges against the lands benefited; so the fund is a rotating one and will constantly grow larger.

The works undertaken in the arid west are of a magnitude that surpasses the wildest dreams of the promoters of private enterprises. When the works at present begun are completed, over three million acres of the finest land in the world will be reclaimed. During the present year over five hundred thousand acres are being irrigated from government works. The structures built and the engineering difficulties encountered have been without precedent.

At Roosevelt, about eighty miles from Phoenix, Arizona, the great Roosevelt Dam, which will store sufficient water to cover 1,284,000 acres, one foot deep is being built. This dam is 284 feet high and 1,080 feet long at the crest.

To build this dam the government developed 4,400 horse-power from a power canal. They built a road that cost over a quarter of a million dollars to get to the dam and they built a cement mill to furnish the cement at the dam.

Near Yuma, Arizona, on the Colorado river, a dam 4,780 feet long and 19 feet high is being built across the silt which forms the bed of the Colorado river.

At Montrose the Gunnison Tunnel, 30,583 feet long, the longest in the Western Hemisphere, is being built through granite, sandstone, limestone, shale and mud, and will eventually water 146,000 acres of land.

In Bighorn county, Wyoming, a dam is being built to store the waters of the Shoshone river. This dam, from the bottom of the foundation to the crest, will have a height of 310 feet. At its side the Masonic Temple, at Chicago, would look like a dwarf. The water stored will water 130,000 acres of land near Cody, Wyoming.

By the end of the present year a total of over $41,000,000 will have been expended by the Reclamation Service.

The project which probably interests most of us who are present more than any other is the Grand Valley project. This project contemplates the irrigation of from 60,000 to 70,000 acres of land lying above that irrigated at the present time. The land to be irrigated lies north and west of Grand Junction, Fruita, Loma and Mack. About 40 per cent of it is private land and the remainder is public land, most of which has been filed upon under the provisions of the Reclamation Act.

In 1902, immediately after the president had signed the Reclamation Act,

PAGE FACSIMILE: *The Daily News, Harvest Edition, page 6*

the government sent a corps of engineers into this valley to investigate this project. A strip of country was investigated and mapped which included every probable canal route. No line was staked out or decided upon, consequently up to the present time there has never been a government line. However, certain parties have issued maps purporting to show the location of the proposed government line.

In the spring of 1903 a favorable report was made to Washington regarding the feasibility of the project. The government at that time had the funds and was ready to begin construction after more extended surveys and plans had been prepared. This was prevented because certain engineers and so-called capitalists went to Washington and stated that private capital was ready and anxious to build this project, and as the government had numerous demands on the fund from other projects this one was abandoned for development by private capital.

About a year ago the people of this valley became tired of waiting for private capital and sent a man to Washington to see if the government could again be interested in this project. The secretary of the interior made two propositions to the people:

First. That he would have surveys begun and plans prepared and would construct the project as soon as funds were available.

Second. That he would restore all lands in the valley to all the usual forms of entry so the project could be built by private capital under the Cary Act, or some similar plan, and that the government would definitely abandon the project.

At a large mass meeting the citizens unanimously accepted the first proposition and the secretary is complying with its provisions at the present time.

An office was established here about the first of March of the present year. Since that time a party has been continuously in the field and by the end of September the entire irrigable area will be mapped on a scale of one thousand feet to the inch, ten foot contour interval. We consider a first class topographic map absolutely necessary for the economic development of a large irrigation enterprise. Probably a number of those present are familiar with such maps. For those who are not a brief explanation will be given.

The maps we are making show every road, fence, railroad, section line, irrigation ditch, arroyo, and in addition all ten foot contours. A contour is a line joining all points of the same elevation. The easiest way to imagine such a line is as follows: Suppose the ocean should rise until its level was 4,700 feet higher than at present. If we should sketch its new line we would have the 4,700 contour. Now assume that it rises an additonal ten feet and the new shore lines were drawn. This would be the 4,710 contour, and so on. After such a map is completed we are able to determine the location of all canals and laterals. In addition we can estimate the amount of waste or non-irrigable land in any tract. The steepness of the slope of the land is shown by the distance between contours. We can tell if the land be bottom land or mesas.

In case the ocean would rise as we have assumed, the mesas would form peninsulas out into the ocean, and the bottom lands would be inundated and would form bays.

Another feature of our surveys has been an accurate classification of the soil. Every square foot has been classified into one of five classes:

No. 1 is red sandy soil.
No. 2 is fine, black, sandy loam.
No. 3 is adobe.
No. 4 is land too rough to irrigate.
No. 5 is seep land.

Where irrigable lands are located above the probable location of our canal line we have extended our surveys to an elevation of 200 feet above the canal line, to include all area that may be watered by pumping.

The main line of our canal has been located from the headworks to Palisade, and plans and estimates are being prepared. Our locations are made with great care, and without leaving the office we can estimate the difference in cost that will result in shifting the line a foot in either direction. Before our line can be considered as finally located the location and plans must be examined by three engineers, who will report to the director. As this board has not yet met, none of our location can be announced as final.

Sufficient study has been made to enable me to announce the general features of the project now, even though they have not yet been approved. Please remember that I use the word probable.

The headgate will be located between the railroad tunnel and the mouth of Plateau creek, about six miles above Palisade. No great advantage and a large increase in cost would result in attempting to go higher. The canon at this point is such that tunnels would require a grade nearly equal to that that would be gained by placing the intake farther up the river.

At the point of diversion we contemplate raising the river nearly 20 feet by means of a dam over 400 feet long. The canal will be taken out on the right hand or west side of the river. It will follow along the right of way of the D. & R. G. railroad for about two and one-half miles. From Cameo to Palisade the line will be largely in tunnel. This first six miles of line will involve 2,500 feet of flume and over 9,000 feet of tunnels. It is proposed to make the flume of timber, lined with galvanized iron, and built on concrete stringers. The tunnels will all be lined with concrete.

There have been a number of wild goose schemes advocated by those not familiar with the conditions. One of these has been a plan to run up Coal Canon and tunnel through to a point west of Mt. Garfield. This would increase the cost nearly $2,000,000, and would not result in any gain in grade. From Palisade the canal will run in a general northwesterly direction and end at a point near mile post No. 8 of the Uintah railroad. It is estimated that the project will cost from $2,500,000 to $3,000,000, and the cost per acre will probably be about $50. One thing is certain, and that is that such work costs much more than is generally believed, and an earnest effort is going to be made to have our estimates large enough to cover the cost before construction is begun.

As there are a large number of individual land holders to deal with, and as the maintenance and operation of the work will ultimately be turned over to the water users, it is necessary for them to be organized, and the Grand Valley Water Users' association represents the water users in their dealings with the government.

This association has power to levy assessments for any purpose and the government will aid in the collection of these assessments by refusing to give a man water until his assessments are paid.

At the present time the Reclamation Service has a large number of projects nearing completion and which will soon be on a paying basis. For this reason it is not considered good business policy to start new projects. However, the people of this valley are

PAGE FACSIMILE: *The Daily News, Harvest Edition, page 7*

anxious to get the work started at once in order to perfect their water right and also to secure a contract with the department. There is also a well grounded fear that certain private parties will attempt to break the strong sentiment which exists at present in favor of government construction if construction operations are delayed.

The Water Users' association has, therefore, made a proposition that they will raise half of the funds necessary to get construction well under way during the year 1909. To raise these funds the association proposes to issue certificates bearing 8 per cent interest. The government guarantees to accept these certificates at face value in payment of water charges, and the association guarantees the interest, which will be collected from all the water users by assessments. If this plan is approved in Washington, the government will assist the association in collecting these assessments, as stated above.

For those who are not able to advance cash a plan is being developed to enable a man to work out his water right. This plan has been used on several other projects and has met with marked success. The certificates issued are negotiable.

In all works done under the Reclamation Act there are two distinct branches of the government service to deal with. The Reclamation Service makes the surveys, builds the works, determines the cost and decides the size of the farm units. The land office receives the entry, sees that the residence and other requirements are met, and collects the charges.

When public lands are to be watered on any project, these lands may be segregated by the secretary of the interior by two methods. What is known as the "first form" withdraws the lands from all form of entry, and the lands become a government reserve while so withdrawn. This is done when the land is needed for construction purposes. Entries made prior to the order withdrawing the land from entry are not affected and the entryman should continue in the usual manner. His entry may be contested and the person initiating the contest has a thirty day preference right to enter the land when it is again restored to entry.

Lands segregated by the "second form" can be entered, but only under the provisions of the Homestead Law and Reclamation Act. This is done so that desert, timber and stone and other forms of entry which give a man more than 160 acres cannot be made on lands to be irrigated from a government irrigation system.

In the matter of securing rights of way for canals the government has advantages and rights not enjoyed by individuals or companies. The act of August 30, 1890, provides: "That in all patents for lands hereafter taken up under any of the land laws of the United States or on entries or claims validated by this act west of the one hundredth meridian, it shall be expressed that there is reserved from the lands in said patents described, a right of way thereon for ditches or canals constructed by the authority of the United States."

The entries or claims validated by that act are such as may be asserted under entries made subsequent to October 2, 1888, as held by the department in the circular of October 5, 1893.

In such cases the Reclamation Service is willing to pay a reasonable amount for improvements. This is not to be based on speculative values. There is no doubt that hardships will be worked in certain instances. However, the abstracts and patents have been specified and there is nothing unfair about the matter. Suppose the government had for years withheld all these lands from entry on the ground that they would later be needed for canal rights of way. A general complaint would have arisen that the progress of the communities affected was retarded. The people would have said: "Let us develop these lands and get the benefit of their use. When you are ready to build your canals we will surrender the necessary rights of way."

Without the people requesting such action, that is exactly the thing that congress did, and we are now ready to assert our rights to the right of way. We try to be fair in all such matters, but, on the other hand, do not propose to be unduly annoyed or held up. Occasionally we meet a man who attempts to prevent us from making the necessary surveys, and pulls up our stakes. An Act of Congress approved June 10, 1896, provides: "That hereafter it shall be unlawful for any person to destroy, deface, change, or remove to another place any section corner, quarter-section corner, or meander post, or any government line of survey, or to cut down any witness tree, or any tree blazed to mark the line of a government survey, or to deface, change, or remove any monument or bench mark of any government survey. That any person who shall offend against any of the provisions of this paragraph shall be deemed guilty of a misdemeanor, and upon conviction thereof in any court shall be fined not exceeding two hundred and fifty dollars, or be imprisoned not more than one hundred days. All the fines accruing under this paragraph shall be paid into the treasury and the informer in each case of conviction shall be paid the sum of twenty-five dollars."

Up to the present time we have always been able to turn the offender from his evil ways without recourse to law. However, a few cases of malicious and wilful disregard of this law have been observed, and the time may come when it will be necessary to make an example of some offender.

Our water rights are secured under the state laws the same as those of any other user. Of course it is impossible to conduct an enterprise as large as this one without conflicting with both legitimate and private interests. The result is abuse of the officers of the service that is unwarranted. The members of the service are handicapped by being required "to treat their enemies with respect and give them an advantage to which they are not entitled."

Indications are that this project will be consummated without undue delay. It will increase the wealth of this community by thirty million dollars in the next fifteen years. It will give farmers water under the most favorable terms. It will supply labor to hundreds of men. It will increase the markets of not only our local merchants but of the entire state. Our equipment, cement, steel and lumber will be largely Colorado products. The railroads into the valley will have their local business more than doubled. It is estimated that the population of the valley will be increased from fifty to one hundred thousand.

The lands to be benefited are close to the hills and include a large percent of sandy mesas. These are the best fruit lands known. This project is looked upon by the Reclamation Service as one of the most promising. The land is the most productive, the climate finer, the water supply the most certain. After the "High Line" is built we will stop our printing presses for a few moments and change our advertising circulars, as the "Little Empire of the Western Slope" will be inadequate to describe the magnitude of the industry, prosperity and resources of the Valley of the Grand.

Chief of Police O'Neil

No city of its size in Colorado has a more efficient or competent man at the head of its police department than has Grand Junction. For nearly six consecutive years, C. C. O'Neil has been connected with the police department of this city. He has received appointments from three different city councils. For three years and six months he served as captain of the night police. Then on January 1, 1907, he was made chief of police to succeed Charles S. Wallis, resigned. In April 1907 O'Neil was elected by a unanimous vote of eight aldermen of the city council to a term of two years as chief of police, the term expiring April, 1909. Chief O'Neil also holds the position of truant officer, being given this appointment by the board of school directors.

In his official capacities Chief O'Neil has given splended satisfaction and has made a record of which any officer could well be proud.

Chief O'Neil has been a resident of Mesa county for about fifteen years and of Grand Junction continuously for eight years. He and his family occupy a pretty home in the northeastern portion of the city.

MARVELOUS GROWTH OF THE GRAND VALLEY

(Continued from Page One.)

per year; ten acres of peaches six years old have returned an aggregate receipt of over $10,000; two-thirds of an acre of peaches brought $940; common returns in peach culture count from $400 to $800 per acre.

Poultry Profits in the Grand Valley Great.

Poultry breeding is very profitable in the Grand Valley. The poultry farm of Charles A. Smith, near Grand Junction, counts 1,500 of ordinary and fancy bred chickens, and a large number of homing pigeons from which $700 worth of squabs are marketed each season, while 1,000 dozen eggs are sold every year from the hennery. Fancy bird eggs for hatching purposes bring $2.25 per setting, about $400 per season being realized from this source, the breeding cocks costing as high as $40 each, with hens worth from $5 to $15. Mr. Smith's total investment is about $2,500.

Irrigation Supply of Valley Exhaustless.

The irrigation water supply of the Grand Valley chiefly comes from the Grand river and includes the application of both the gravity, pressure canal and the water lifting pumping plant driven by water power itself, this irrigation water supply always, doubtless, being greater than can be used by the valley.

The Grand Valley canal has a total length, including the Kiefer extension, of 46 miles, and has an irrigating capacity of 50,000 acres, with a present covering of 40,000 acres and an actual cultivation of 30,000 acres, which is being added to at a rate of 3,000 acres per year, the present water distributing length of the canal system being over 100 miles.

The stock of this canal is all owned and issued by the farmers themselves, the combined cost of irrigation itself and canal maintenance averaging but 65 cents an acre, an annual cheap irrigating canal cost which confidently alike challenges comparison and competition.

The Grand Valley has 150,000 acres of easily irrigable land, of which 85,000 acres are now provided for by canals and already operated or projected pumpage plants, while the coming big United States Government High Line Canal will irrigate the remaining 65,000 acres.

The great pumpage plant of the Red Lands Irrigation & Power company is alike the most powerful and perfect of its character yet known to the West.

This plant will irrigate 5,000 acres of land on the rich Red Lands Mesa across the Grand river from Grand Junction, and will also generate electric power for extended commercial purposes, including the furnishing of electric motors for various coming Grand Junction industrial enterprises and the city's projected street car trolley system. The total cost of this plant and general irrigation system to date has been in excess of $200,000, of which $50,000 has been expended on machinery and comprises two 600 horse-power pumps and two 600 horse-power electric generators. The pumpage has a present lift of 180 feet and can discharge a combined flow of 24,000 gallons of water per minute, which is equivalent to 2,000 irrigating water inches. Only one of the pumps, however, is now in operation, as its water discharge is sufficient for present irrigating need.

Power Plant Could Supply Large City.

The power canal is four miles long, 27 feet on the bottom, 44 feet on top, and ten feet deep, while the distributing canal is 15 miles long.

At least 600 electric horse-power can be generated for commercial sale, while the company will establish an additional pumpage irrigation plant farther down the river which will be electric driven and have a water lift of 125 feet. It is estimated that this plant can generate sufficient electricity to fully supply all lighting or industrial power need of a city of 25,000 people, and also operate 50 miles of trolley railroad. The running expenses of the plant itself does not exceed $175 per month.

This year will see 11,500 acres of the rich Red Lands mesa irrigated by this pumpage plant, while the entire 5,000 acres will be irrigation served within the next two years.

The Red Lands Mesa is practically free from frost visitation, the Kiefer Brothers' orchard, for instance, located at the head of the mesa, being in alike blissful and bountiful ignorance of even the word frost.

Irrigation District for Orchard Mesa.

An irrigation district has been organized under state law to cover 12,000 acres on the fertile Orchard Mesa on the south side of the Grand river and opposite Grand Junction.

These lands are to be irrigated by a pumpage power plant similar to that of the Red Lands Mesa, it being expected that 6,000 acres will be receiving irrigation water within the next two years.

Orchard Mesa is one of the choicest of the many choice fruit and farm areas of the Grand Valley. On this mesa a water lift pumpage has been successfully irrigating 1,000 acres of land for many years.

The successful operation of a pumpage water lifting plant renders available to the irrigation the raggedest and most broken lands lying along the Grand Valley streams or canals. Every barren butte will be crowned with the hardened orchard bough, every haggard hill will hold the halo of heavy horticultural harvest, and every gashed gulch and gulley will be glorious with overhanging fruitage. Even the waste places shall be those of wealth and roughness shall be riches.

The Smudge Pot Slays Jack Frost.

The kindly climate and high land elevations naturally antagonise frost in the Grand Valley, and even the rare frost visitations are now in a fair way to be entirely prevented by the introduction of the "smudge pot" with a burning of oil or coal and a use of from 40 to 80 pots an acre. By the way, the name of "smudge pot" in this connection is a misnomer, as it is evidently the heat from the pots instead of the smudge or smoke which routs the frost fiend.

The various Grand Valley chambers of commerce, boards of trade and fruit associations have intelligently and energetically taken up this matter of frost prevention and next spring will see three-quarters of a million coal or oil heated smudge pots in operative burning in the Grand Valley, with the undoubted result of a final "goodbye" to spring orchard frosts and the henceforth annual insurance of full maturing of the naturally magnificent valley fruit harvest.

Grand Valley farms are nearly all of home holding, being actually owned by the farmers themselves, the farms ranging from five to 160 acres. In fact, the only large proprietary areas in the valley are those of thousands of acres of the Western Land & Sugar company and the 3,000 acres owned by the Reed Investment company in the Loma section.

A generous yearly harvest labor market is furnished by the Grand Valley, the hired workers being generally robust and intelligent young men, many of whom promptly graduate into prosperous home holders themselves.

Grand Junction Beet Sugar Production.

The Western Sugar and Land company is cropping this year 6,000 acres of sugar beets, which will average 14 tons an acre, with a saccharine showing of 15 1-2 per cent per ton. This means a cutting by the sugar factory this year of 75,000 tons of beets and a sugar production of 200,000 bags of sugar of 100 pounds each. The company will distribute in cash this year over $400,000 to the farmers for sugar beets and will pay out more than $300,000 for labor.

Last year the company cut 63,000 tons of beets as against the 75,000 tons to be cut this year. To provide for this additional business, the company is making an improvement in its plant amounting to $100,000, the added expenditure including the purchase of new boilers, new crystalizers and new centrifugal machinery.

General Manager Davie has just completed a tour of all the beet territory subject to cultivation by the Western Sugar and Land company and says that the outlook was never better and could not possibly be improved.

Manager Davie says further that a $500,000 sugar factory will be erected year after next at some point in the Uncompahgre Valley, the location of the factory not yet being finally decided upon.

The Metropolis of the Western Slope.

Grand Junction, the seat of Mesa county, is the center as well as the capital of magnificent Mesa county. It is the metropolis of the Western Slope and is the coming city between Denver, Colorado Springs and Pueblo on the east and Salt Lake City on the west, and is already the practical commercial distributing point of Western Colorado and Eastern Utah.

It has at present a population of 10,000 and is rapidly putting on newer and greater civic garments. It is a center of modern civilization as well as munificent commerce, having a superb system of public schools, commodious and well supported churches and all of the residence and building block improvements of the day and the time, a further evidence of these facts being the present construction of a $60,000 Y. M. C. A. building. James A. Garfield, secretary of the interior, said:

"At Grand Junction I saw a wonderful country. That part of Colorado is as near perfect as nature and the hand of man can make it."

Secretary Garfield might have added that while nature had made a wonderful country of the Grand Valley, modern man was making just as wonderful a city called Grand Junction.

Peach Culture at Palisade.

Palisade and peaches are two words largely and luxuriantly interwoven in the great Grand Valley production of this great fruit. The Palisades are sheer and serried cliffs which lift themselves from 1,000 to 1,500 feet above the flow of the Grand river, the fertile peach lands lying along the cliff bases.

Palisade peach culture includes two great cardinal facts: First, the natural soil adaptability to peach growth; and, second, the at once sheltering and further fructifying harvest action of the Palisades themselves.

From the broad and lofty surface of the Palisade cliffs, an alike warm and wise sun reflects its powerful and timely rays upon the Palisade orchard areas, thus uniting with the rich soil for the final production of the perfect Palisade peach. This exceptional and fortunate natural union of soil and sun is yet further improved by the intelligent practice of Palisade peach growers of annually thinning out and opening to the sun's rays the topmost foliage of the peach trees themselves and thus allowing the utmost free action of the sun in the rapid ripening and conquering commercial coloring of the fruit itself.

What is known as the Palisade fruit region comprises 10,000 acres covered by two irrigation water lifts of 35 and 82 1-2 feet, and an aggregate

(Continued on P. 3, DeBeque Section)

Colorado Fruit and Commercial Company's Building.—Photo by Dean.

PAGE FACSIMILE: *The Daily News, Harvest Edition, page 8*

PRODUCTION NOTES

The following details the process used to conserve *The Government High Line Canal Newspaper Article Archive, 1904 - 1920.* The original scrapbook still resides at GVWUA.

Newspapers are notoriously hard to preserve. They weren't designed to last forever. Keeping newspapers out of light, out of humidity, using acid-free pages between sheets, storing them unfolded, and handling lightly are the best techniques for conservation. More permanent methods for conservation are to replicate them as microfilm or as electronic data. Digitization as electronic text was the method available for this project.

The digitization of the archive took about 3 months or 240 man-hours. The number of pages in the scrapbook with pasted-on articles is 138, with 24 loose articles – some placed at the back and others at specific locations between pages. Some of the loose items were complete pages. Those pages, (except for two sections within the Harvest Edition), were not turned into search-able text, but are available as images at the end of the book.

The process started with capturing every article as a jpeg (Joint Photographic Experts Group for the name of the committee who created the standard). This is a format for compressing the image content and is the most common place to start. A flat bed scanner is usually the most convenient tool for this. A flat bed scanner was used for the small number of loose articles in the scrapbook. For the bulk of the articles that had been glued onto pages, a hand-held, wand-type scanner, a TaoTronics Model TT-DS011 in this case, was needed to avoid too much handling and to get a more complete image from the binding edge of the book. In some cases a personal digital camera was needed to maneuver around tight spots.

The next step was to open the jpegs into an OCR (Optical Character Reader) application. There are many choices available for a reasonable cost, in this case Prizmo was used. The OCR software takes the solid image that a jpeg creates, recognizes the characters, and transforms them into editable text.

From the OCR application, the text was copied and pasted into Microsoft Word. The OCR makes a good first attempt at the text, but on 100-year old newsprint, some of the text is indecipherable. Microsoft Word will flag the obvious errors, allowing a clean-up of those areas on a first pass. But it can miss other areas that have become complete gibberish. It takes a human eye at that point, reading from context to correct text and referring back to the original page for further confirmation. Also during this phase in Word, an attempt was made to replicate the formatting and headline styles of the news page. An Internet search on 'newspaper fonts' found close matches for the text styles and were downloaded for no cost.

Images in the clippings were handled individually in Adobe Photoshop and enhanced as much as possible before importing into the Word document.

The individual article pages were saved as PDF (Portable Document Format, introduced by Adobe Systems to ease the sharing of documents between computers and across operating system platforms while maintain formatting and fonts.)

The final book production was collated and designed with Adobe InDesign. The individual PDF articles were imported onto pages in this application. The title page, table of contents, introduction, postscript and page numbering were compiled with InDesign. The entire book file was then saved as a single PDF and is available in print or on disc. (For a search-able pdf on disc, you may make a written request to: Hugo Madden, P.O. Box 1521, Palisade, CO. 81526.)

Printed in Dunstable, United Kingdom

91222495R00226